# Preface

*Gene Johnson*
Co-Author, Artist &
'Eco-centric' Sage

*Reginald G. Taylor*
Co-Author & Ebullient
Anatomical Instructor

*Jim Wehtje*
Graphics & Acuminous
Format Enforcer

By virtue of this learning-friendly auto-tutorial we will be your friends and guides, to assist — and even soothe you — through the rigorous, and sometimes stressful, process of learning the formidable encyclopedic information that is always required in a standard anatomy & physiology course.

Whereas the world is awash with information, **information well-presented**, is a scarce commodity. In this tutorial each page presents a single anatomical, or physiological concept, in the appropriate style, format, and size, **most conducive to efficient learning**.

Although other anatomy study guides and workbooks are available this one clearly and strikingly stands out because of its clarity and learning-friendly style.

Each exercise requires two sets of neurological outputs known as motor responses: one for number matching and one for coloring. After performing these two sets of motor activities, you will be well on your way to having the anatomical information encoded in your cerebral memory bank, and you will perform better on your quizzes and exams.

A text-guide page which accompanies each exercise, not only walks you through the exercise, but also provides additional key information.

Arm yourself with a quality set of 24 colored pencils, approach each exercise as an aesthetic anatomical adventure, and this tutorial will take the sting out of learning anatomy & physiology ... we guarantee it!

# Anatomy & Physiology
## Learning-Friendly Auto-Tutorial
## Second Edition

By

## *Gene Johnson*
## *&*
## *Reginald G. Taylor*

ISBN 1453703403

Drawings by Gene Johnson
Graphics by Jim Wehtje

© Copyright 2010 Gene Johnson
All Rights Reserved

Johnson & Taylor Publishing
Printed in the USA

---

For questions or comments contact us at:
TheTeam@AnatomyAndPhysiologyTutorial.com

To place an order please visit:
www.AnatomyAndPhysiologyTutorial.com
or
www.createspace.com/3469651

# Instructions

Do the number matching exercises first. Place the correct numbers from the drawings in the underlined spaces provided in front of the words in the word banks.

*For each drawing there is an adjacent guide page to lead you logically and sequentially through the exercise.*

When you are finished with the number matching you can then begin the coloring exercise. Acknowledging the wonderful fact that the human body is as much art as it is science, take pride in your work and purchase a quality set of 24 colored pencils. For any given anatomical part, choose a color, then, after coloring in the box by the correct word in the word bank, proceed to color in the chosen anatomical part or region with the same color. (Refer to the exercise shown on the cover of this book.) **Colored pencils, with their softer pastel tones, and more controllable applications, will do the best job.**

Which exercises you do first will depend on the course outline of your teacher, but *Human Anatomy & Physiology* is a remarkably standardized course, which almost always follows the sequence presented in this tutorial. The Table of Contents will allow you to quickly find any subject area.

Do not think the coloring exercises to be a trivial waste of time, for although this auto-tutorial is well worth its price even if you do only the number matching exercises, most students will, for the following three reasons, greatly benefit from performing the coloring exercises:

- The coloring process provides a quiet and unhurried frame of time wherein the anatomical names quietly and subconsciously take firmer root in the cerebral memory bank.

- The coloring process not only aids in the learning of the names but also helps set the shapes and forms of the anatomical parts, or regions, in the mind.

- *By providing an unhurried interlude, to what is often a much too hurried — and sometimes frenzied — learning endeavor, the coloring process provides precious, therapeutic moments of aesthetic calm and relaxation. And it is precisely these moments of calm and relaxation that many students need to properly facilitate the learning process.*

Here, then, is an opportunity to commit yourself to personalizing and creating your own anatomical reference book. A book that your children and grandchildren will be pleased and proud to look at many years later … so bon voyage!

# Table of Contents

### Body Overviews
- 6 Body Systems
- 8 Body Regions
- 10 Body Cavities

### Lives of Cells
- 14 General Cell Structure
- 16 Endoplasmic Reticulum
- 18 Ribosome
- 20 Golgi Apparatus
- 22 Protein Synthesis
- 24 Centriole & Cilia
- 26 DNA Structure
- 28 DNA — A Universal Code
- 30 Mitochondrion
- 32 Plasma Membrane
- 34 Metaphase of Mitosis & Meiosis I

### 36 Tissue

### 38 Integumentary System
- 40 Integument
- 42 Hair

### 44 Skeletal System
- 46 Osteon
- 48 Long Bone

#### Axial Skeleton
- 50 Infant Skull
- 52 Skull — Lateral
- 54 Skull — Inferior
- 56 Cranium Base
- 58 Paranasal Sinuses
- 60 Mandible
- 62 Cervical & Thoracic Vertebrae
- 64 Thoracic Vertebrae Articulations
- 66 Lumbar Vertebra
- 68 Atlas & Axis
- 70 Sacrum & Coccyx
- 72 Sternum
- 74 Rib

#### Appendicular Skeleton
- 76 Scapula
- 78 Clavicle
- 80 Humerus
- 82 Lower Arm
- 84 Wrist & Hand
- 86 Coxa
- 88 Femur
- 90 Lower Leg
- 92 Ankle & Foot

### 94 Articulations (Joints)
- 96 Shoulder
- 98 Knee
- 100 Other Articulations

### 102 Muscular System
- 104 Skeletal
- 106 Sarcomere
- 108 Power Stroke
- 110 Neuromuscular Junction
- 112 Head & Neck — Superficial
- 114 Neck — Deep
- 116 Body
- 118 Shoulder & Neck
- 120 Thorax
- 122 Upper Arm
- 124 Lower Arm — Anterior
- 126 Lower Arm — Posteriolateral
- 128 Leg — Lateral
- 130 Upper Leg — Anterior
- 132 Upper Leg — Posterior
- 134 Lower Leg
- 136 Tongue
- 138 Eye

### 140 Nervous System
- 142 Neuron
- 144 Action Potential
- 146 Synaptic Events
- 148 Brain — Midsagittal
- 150 Brain Stem
- 152 Brain — Frontal
- 154 Ventricles
- 156 Cranial Nerves
- 158 Spinal Nerves
- 160 Spinal Nerve Origin
- 162 Spinal Nerve — Transverse

# Table of Contents

164 Spinal Cord — Transverse
166 Autonomic Nervous System

### Special Senses
168 Gustatory
170 Olfactory
172 Lacrimal Apparatus
174 Eyeball
176 Aqueous Humor Circulation
178 Photoreceptor
180 Outer Ear
182 External & Middle Ear
184 Middle Ear
186 Cochlea
188 Osseous Labyrinth
190 Membranous Labyrinth
192 Macula & Crista

### 194 Endocrine System
196 Pituitary
198 Thyroid
200 Pancreas
202 Adrenals

### 204 Circulatory System
206 Blood Composition
208 Hemopoiesis
210 Capillary Net & Microcirculation
212 Heart — Anterior
214 Heart — Frontal
216 Cardiac Center
218 Abdominopelvic Vessels
220 Aorta — In Situ
222 Aortic Arterial Map
224 Back & Thorax Veins
226 Neck Arteries    Posterior
228 Head & Neck Arteries — Lateral
230 Circle of Willis
232 Head & Neck Veins
234 Arm & Shoulder Arteries
236 Arm & Shoulder Veins
238 Leg Arteries
240 Leg Veins
242 Hepatic Portal System
244 Skeletal Muscle Blood Pump

### 246 Immune/Lymphatic System
248 Lymphatic Vessels
250 Lymph Node

### 252 Respiratory System
254 Thoracic Cavity
256 Heart & Lungs
258 Respiratory System
260 Alveolar Cluster
262 Larynx
264 Respiratory Center

### 266 Digestive System
268 Salivary Glands
270 Tooth
272 Alimentary Canal
274 Stomach
276 Upper Abdominal Organs
278 Liver Lobule
280 Small Intestine
282 Villus
284 Brush Border
286 Large Intestine
288 Defecation Reflex

### 290 Urinary System
292 Abdominal & Retroperitoneal Cavities
294 Urinary System
296 Renal Arterial Circulation
298 Kidney
300 Nephron
302 Renal Corpuscle

### 304 Reproductive System
306 Male — Lateral
308 Male — Anterior
310 Spermatogenesis
312 Sperm
314 Female — Anterior
316 Female — Lateral
318 Mammary Gland
320 Ovarian Cycle

### 322 Famous Faces

# Body Systems

Because of the human penchant for "shoe-horning" things into specific, arbitrary, categories, coupled with the fact that it is difficult to study "everything" at once, most conventional anatomy & physiology texts follow what is called a **"systems approach"** to the study of the human body. Most texts offer eleven or twelve systems (twelve if the immune system is separated from the lymphatic system).

The human body is of course not actually divided into "systems," and certainly not into separate "systems." The inseparable, interconnected, and interactive components of our body are operationally quite like an ecosystem, and one day we may indeed study and treat it as an ecosystem — but until then we defer to the conventional systems approach, in part because we too find it difficult to study everything at once, and in part because we wish to prepare you, as best we can, for your upcoming "systems" quizzes and tests. Hence, we offer, on the opposite page, a "matching" opportunity which will provide you a gentle introduction to the eleven body systems.

Note: In this exercise we do not separate the lymphatic system from the immune system, thus, we have at least kept some things together!

# Body Systems

Match the body parts (all 26!) with their appropriate body system:

**Body Parts**

A. Adrenal Gland
B. Biceps and Triceps
C. Blood Vessels
D. Bones
E. Brain
F. Bronchi
G. Esophagus
H. Hair
I. Heart
J. Kidneys
K. Liver
L. Lungs
M. Lymph Node
N. Nails
O. Nerves
P. Ovaries
Q. Pharynx
R. Pituitary
S. Skin
T. Spinal Cord
U. Stomach
V. Thyroid
W. Trachea
X. Ureter
Y. Uterus
Z. Vagina

**Body Systems**

_____ Cardiovascular

_____ Digestive

_____ Endocrine

_____ Integumentary

_____ Immune

_____ Muscular

_____ Nervous

_____ Reproductive

_____ Respiratory

_____ Skeletal

_____ Urinary

# Body Regions

We continue with our gentle introduction to the study of the human body by inviting you to match the numbers of the common names with their appropriate technical names. Then label the drawing below with the appropriate technical names. Use your textbook as a reference for doing these exercises, and then relax while having a classmate check your answers.

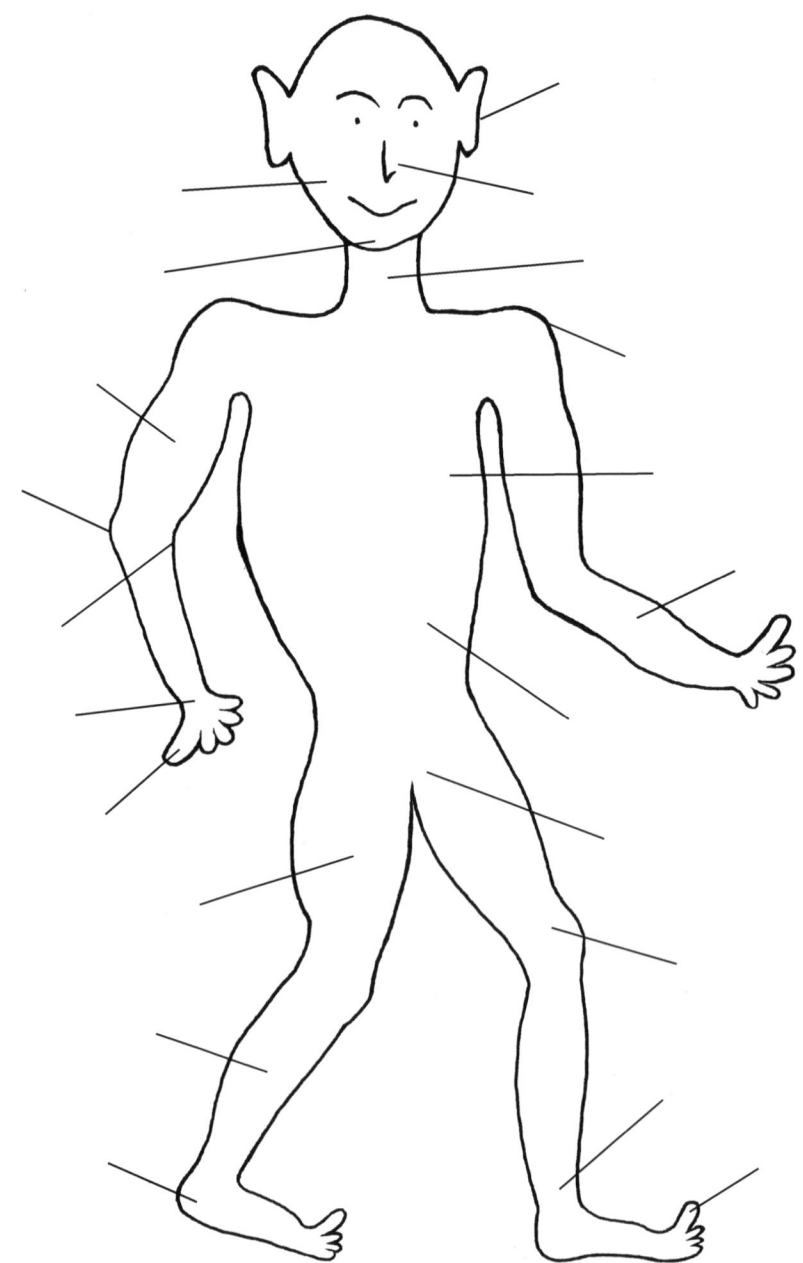

## Common Names

1. Abdomen
2. Ankle
3. Arm
4. Back
5. Back of Elbow
6. Back of Knee
7. Back Base of Skull
8. Between Hips
9. Breast
10. Breastbone
11. Buttock
12. Calf of Leg
13. Cheek
14. Chest
15. Chin
16. Ear
17. Eye
18. Fingers and Toes
19. Foot
20. Forearm
21. Forehead
22. Front of Elbow
23. Front of Knee
24. Front of Lower Leg
25. Genital
26. Great Toe
27. Groin
28. Hand
29. Head
30. Heel
31. Hip
32. Loin
33. Mouth
34. Neck
35. Nose
36. Palm
37. Pelvis
38. Point of Shoulder
39. Shoulder Blade
40. Sole of Foot
41. Spine
42. Thigh
43. Thumb
44. Wrist

## Technical Names

___ Abdominal
___ Acromial
___ Antebrachial
___ Antecubital
___ Brachial
___ Buccal
___ Calcaneal
___ Carpal
___ Cephalic
___ Cervical
___ Coxal
___ Crural
___ Digital
___ Dorsum
___ Femoral
___ Frontal
___ Gluteal
___ Hallux
___ Inquinal
___ Lumbar
___ Mammary
___ Manus
___ Mental
___ Nasal
___ Occipital
___ Olecranal
___ Oral
___ Orbital
___ Otic
___ Palmar
___ Patellar
___ Pedal
___ Pelvic
___ Plantar
___ Pollex
___ Popliteal
___ Pubic
___ Sacral
___ Scapular
___ Sternal
___ Sural
___ Tarsal
___ Thoracic
___ Vertebral

# Body Cavities

The body cavities are shown in this drawing without viscera (organs and tissues).

The **cranial cavity** (1) and the **spinal cavity** (2) comprise the **dorsal body cavity** (3).

The **thoracic cavity** (5) and the **abdominopelvic cavity** (8) comprise the **ventral body cavity** (9).

The abdominopelvic cavity is further subdivided into the **abdominal cavity** (6) and the **pelvic cavity** (7).

Whereas the abdominopelvic cavity is separated from the thoracic cavity by a muscular **diaphragm** (4), the division between the abdominal cavity and the pelvic cavity is only an imaginary line.

# Body Cavities

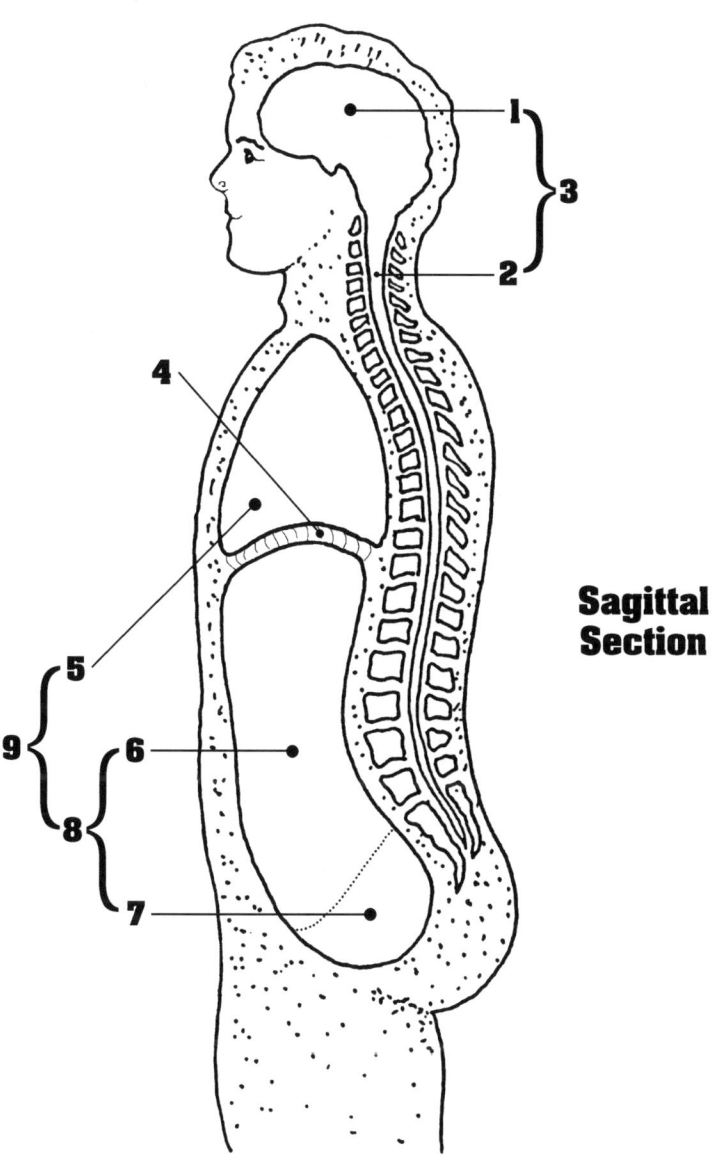

**Sagittal Section**

___ Abdominal Cavity
___ Abdominopelvic Cavity
___ Cranial Cavity
___ Diaphragm
___ Dorsal Cavity
___ Pelvic Cavity
___ Spinal Cavity
___ Thoracic Cavity
___ Ventral Cavity

# Lives of Cells

Shhh! In the beginning, we venture to let you in on a very big "secret." The study of the "human body" as conventionally set forth in most present-day textbooks is only a "half study." The human body, like all multi-cellular "bodies," is a symbiotic cellular symphony, comprised of approximately 10 quadrillion eukaryotic ("human type") cells and approximately 100 quadrillion prokaryotic ("bacteria type") cells.

Even as the "good health" of an ecosystem is dependent on microbial populations which inhabit its every nook and cranny, so also does our "individual" good health depend upon a variety of microbial populations which natively inhabit every nook and cranny of our human "bodies."

Like giant sponges in the ocean, "we" are symbiotic "apartment houses" of diverse life-forms. Without the ministrations of our microbial "inti-mates" we would remain well for only a few hours. **Biologically there is no such thing as a "self-sufficient individual." All multi-cellular life forms are complex symbiotic communities.** It is perhaps appropriate to wake up in the morning and psychologically think of yourself as an "individual," but we do ourselves great harm when we "treat ourselves," biologically and medically, as "autonomous individuals."

Only in the past few years has there been a serious effort to systematically discover, describe, and catalog the symbiotic life forms that intertwine with what we call "our body." But on the basis of what we already know it appears our prokaryotic microbial partners do indeed constitute our "greater half!" — They outnumber our eukaryotic cells by about ten to one!

# Lives of Cells

With additional investigations the list of our indigenous symbiotic partners, together with the essential benefits we derive from them, will grow by logarithmic leaps and bounds. Although in this tutorial we acknowledge the importance of our prokaryotic symbiotic partners, we have, as have almost all conventional anatomy & physiology textbooks, given them short shrift. The following exercises are of the eukaryotic ("human") type.

Intra-cellular structures which have particular functions are called organelles. After completing the next eleven exercises, return to this page and test your cytological knowledge.

**Organelles**

___ Centriole
___ Endoplasmic Reticulum
___ Golgi Apparatus
___ Lysosome
___ Mitochondrion
___ Nucleus
___ Ribosome

**Functions**

A. ATP Synthesis
B. Initiates Formation of Spindle Apparatus
C. Lysing and Digestion
D. mRNA Transcription
E. Protein Packaging
F. Protein Synthesis
G. Rough and Smooth Membrane Factories

# General Cell Structure

The discrete membranous components of the cell are called organelles. Various organelles are supported and stabilized by a **cytoskeleton** (10) infrastructure composed of **microtubules** (9) and **microfilaments** (8).

Information for making proteins is encoded in deoxyribonucleic acid (DNA) molecules contained in the **nucleus** (5). The DNA, together with its packaging and stabilizing proteins, constitute the **chromatin** (16) and the **nucleolus** (14) within the nucleus.

Protein synthesis begins by using DNA as a template for the synthesis of messenger ribonucleic acid (mRNA) molecules. The mRNA then moves out through the **nuclear membrane** (15) via **nuclear membrane pores** (3) and becomes associated with either **free** (2) or **attached** (1) **ribosomes**. Attached ribosomes are embedded in membranous pathways called endoplasmic reticulae. A membranous pathway with embedded ribosomes is called **rough endoplasmic reticulum** (18). A membranous pathway without embedded ribosomes is called **smooth endoplasmic reticulum** (17) (see page 17).

The ribosome translates the RNA message into a protein (see page 19). The energy for synthesis, as well as for all other cellular energy-requiring activities, is provided by **mitochondria** (4) (see page 31).

Many newly-synthesized proteins undergo post-synthesis modification and packaging at the **Golgi apparatus** (11) (see page 21), after which the proteins may be used by the cell itself, or exported to other cells.

Two **centrioles** (13), contained in a **centrosome** (12), initiate the processes of cell division by migrating to the poles and assisting in the synthesis of the microtubules that form the spindle apparatus (see page 35).

The cell is bounded by a **plasma membrane** (6) (see page 33), which, in some cases, produces specialized extensions such as cilia, flagella (see page 25), or **microvilli** (7). Most cells also have two kinds of membranous inclusions: peroxisomes and lysosomes (not shown). Peroxisomes contain oxidase enzymes and lysosomes contain potent digestive enzymes.

*Further details of some of the major cell organelles are presented in the following exercises.*

# General Cell Structure

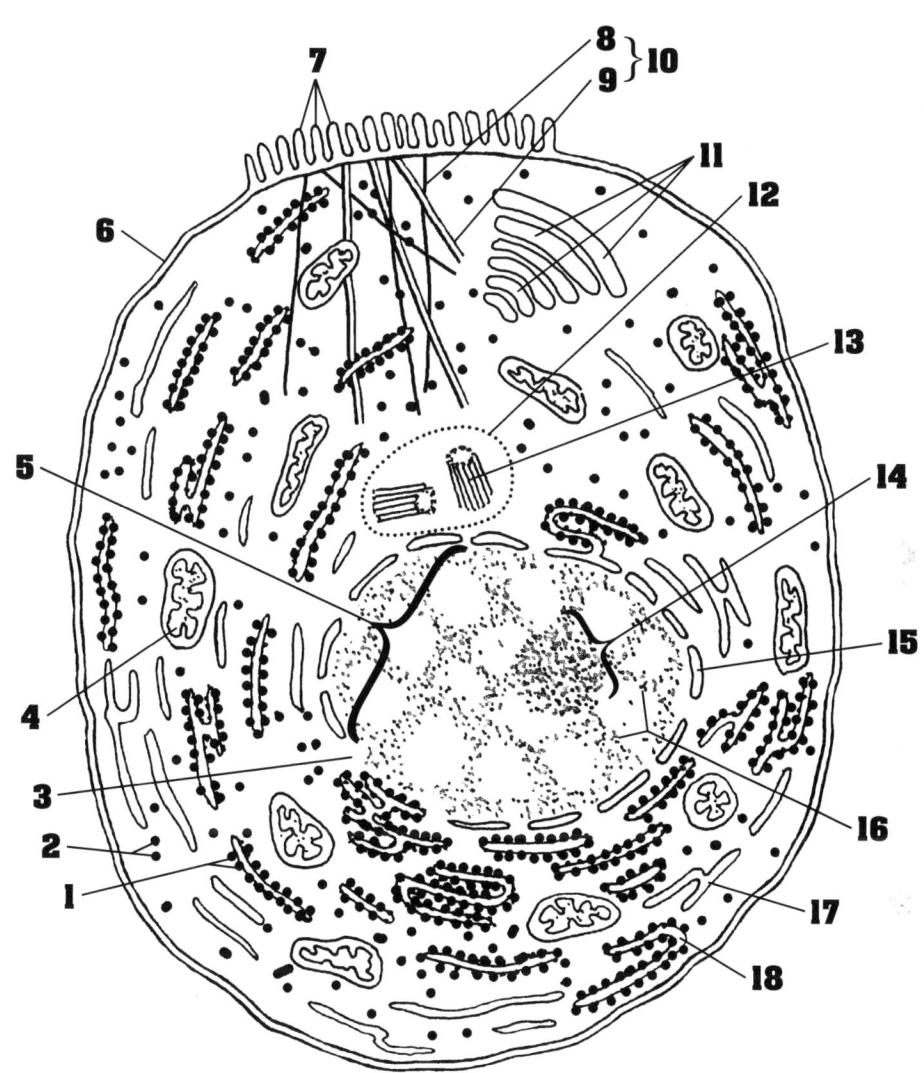

___ Centriole
___ Centrosome
___ Chromatin
___ Cytoskeleton
___ Endoplasmic Reticulum
___   Rough
___   Smooth
___ Golgi Apparatus
___ Microfilament
___ Microtubule

___ Microvilli
___ Mitochondrion
___ Nuclear Membrane
___ Nuclear Pore
___ Nucleolus
___ Nucleus
___ Plasma Membrane
___ Ribosomes
___   Attached
___   Free

# Endosplasmic Reticulum

Endoplasmic reticulum (ER), which is continuous with the **nuclear membrane** (5), consists of an extensive system of interconnected, fluid-filled tubes and cavities called cisternae, which collectively account for about half of typical cells membranes. Two distinct types of ER occur:

## Rough ER

The external surfaces of **rough ER** (2) are studded with **ribosomes** (3). Although not all ribosomes are attached to ER, all proteins secreted from cells are synthesized in ER-attached ribosomes, thus rough ER is abundant in secretory cells. Most ribosomes have the ability, via signal-recognition particles (SRPs) to attach to, or detach from, the ER. Ribosomes not attached to the ER are called **free ribosomes** (1).

ER is also the "membrane factory" of the cell. Integral proteins and phospholipids which are a part of cell membranes are manufactured there. ER cisterns have a **nuclear face** (6) and a **cytosolic face** (7). Most of the enzymes needed for lipid synthesis occur on the cytosolic faces where the needed substrates are more readily available.

## Smooth ER

Although **Smooth ER** (4), which consists of looping networks of tubules, plays no role in protein synthesis, it has other important functions:

- Lipid, lipoprotein, and cholesterol synthesis.
- Synthesis of steroid-based hormones.
- Detoxification of some drugs, pesticides, and other carcinogens, especially in the liver and kidneys.
- Catabolism, in the liver, of stored glycogen to form free glucose.
- Specialized smooth ER, called sarcoplasmic reticulae (see page 111), play an important role in calcium storage and release during muscle contraction.

# Endosplasmic Reticulum

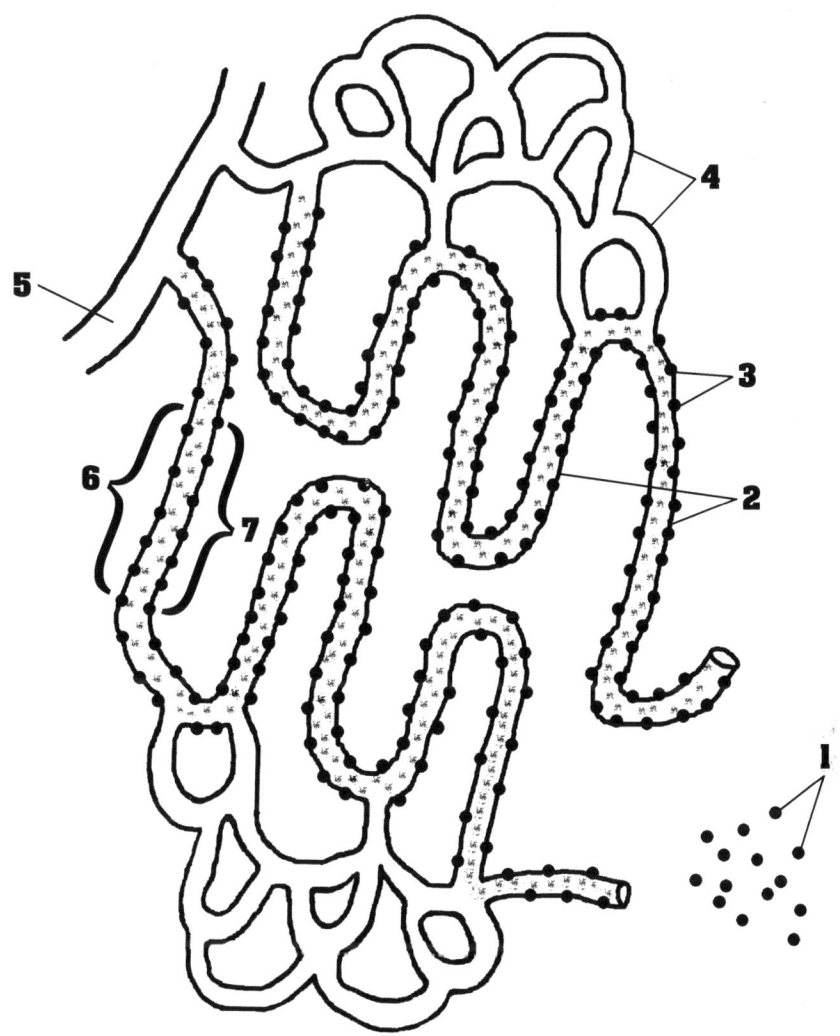

___ Cytosolic Face
___ Free Ribosomes
___ Nuclear Face
___ Nuclear Membrane
___ Ribosome
___ Rough ER
___ Smooth ER

# Ribosome
## In Translation (Protein Synthesis)

An **mRNA (messenger RNA)** (14) transcript is clamped between a **ribosomal large subunit** (4) and a **ribosomal small subunit** (15). As three-pronged **codons** (13) on mRNA arrive at the **A site** (7), the codons are counter-matched by **anticodons** (12) on transfer RNAs (tRNAs). Twenty different tRNAs carry twenty different **amino acids** (9).

Sequentially, from right to left, we see:

- **tRNA loading** (10)
- **tRNA transporting** (11)
- **tRNA at the A site** (8)
- **tRNA at the P site** (2)
- **tRNA unloaded** (1)

"Loaded" tRNAs (carrying amino acids) move sequentially from the A site to the **P site** (5), and while two tRNAs are adjacent to each other in the A and P sites, a **peptide bond** (6) forms between their amino acid "heads." As each new loaded tRNA arrives at the A site, the tRNA, in the P site, loses its amino acid head to the newly forming **peptide chain** (3). The unloaded tRNA is then available to return and pick up another amino acid.

Thus, in the translation process, the sequence of bases on mRNA is translated into a sequence of amino acids in a protein (polypeptide). The energy requirements for the translation process are, of course, supplied by ATP (adenosine triphosphate) from nearby mitochondria. When the newly synthesized protein is released from the ribosome it will likely be transported, by a transport vesicle, to the Golgi apparatus for further modification and packaging.

The synthesized proteins will be used either to form structural tissues or to serve as regulatory enzymes. Because all chemical reactions in living systems are enzyme-regulated, it is no wonder that protein synthesis is the major business of a cell.

*In electron micrographs the ribosomes appear to be as abundant as the stars in the sky. "Twinkle, twinkle, little protein-synthesizing stars, we wonder how many of you there are?"*

# Ribosome
## In Translation (Protein Synthesis)

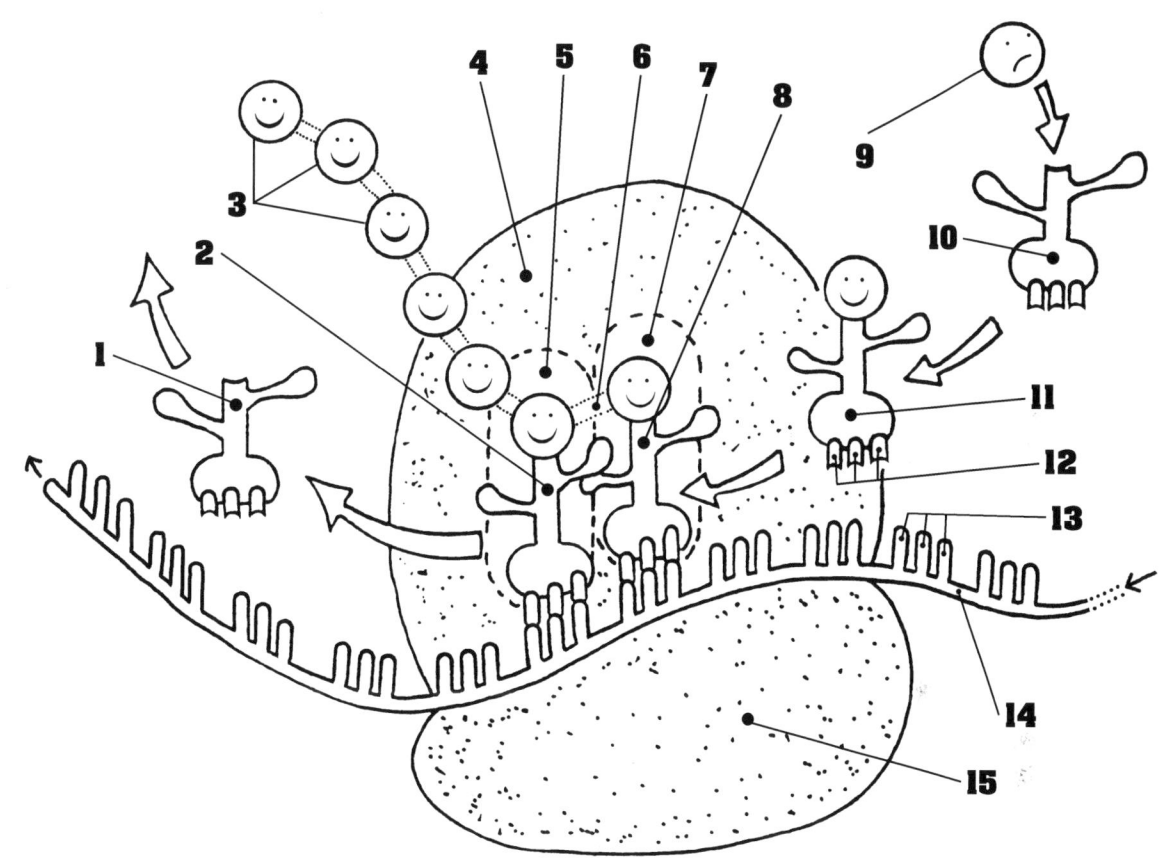

___ Anticodon
___ A Site
___ Codon
___ Free Amino Acid
___ mRNA
___ Peptide Bond
___ Peptide (Protein) Chain
___ P Site
___ Ribosomal Large Subunit
___ Ribosomal Small Subunit
___ tRNA at the A Site
___ tRNA at the P Site
___ tRNA Loading
___ tRNA Unloaded
___ tRNA Transporting

# Golgi Apparatus

As potters form astonishing varieties of vessels from clay, cells form astonishing varieties of organelles from phospholipid bi-layers. Nuclear membranes, plasma membranes, mitochondria, endoplasmic reticulae, and Golgi apparati, all have a phospholipid bi-layer infrastructure.

In the case of the Golgi apparatus (GA) the phospholipid bi-layers are arranged somewhat like a stack of hollow pancakes. Each "pancake" is referred to as a **cisterna** (6).

Via a process called **endocytosis** (1), **transport vesicles** (3) fuse with membranes at the **Cis (receiving) face** (4) and unload **unmodified proteins** (2). As the unmodified proteins pass through the GA they become **modified proteins** (5), and are subsequently packaged and prepared for delivery from the **trans (shipping) face** (12) of the GA.

Whereas proteins designated for intra-cellular use are picked up by transport vesicles, proteins designated for extra-cellular use are picked up by **secretory vesicles** (8) and carried to the **plasma cell membrane** (11) where, by **exocytosis** (9), they are **exported** (10).

The GA also packages a variety of digestive enzymes into spherical membranous organelles called **lysosomes** (7). Under headings such as "demolition" or "digestion," lysosomes, sometimes called "little stomachs," perform a variety of important catabolic cell functions.

The GA also receives, modifies, and delivers some lipid materials.

# Golgi Apparatus

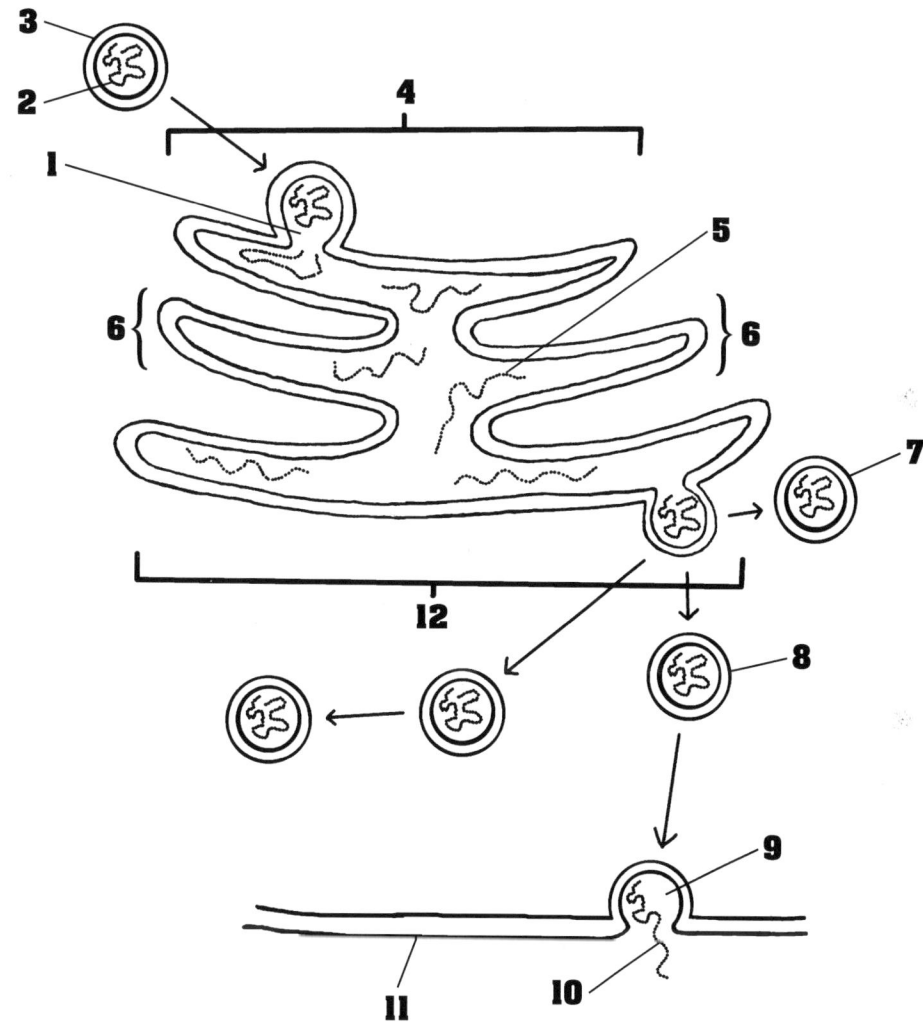

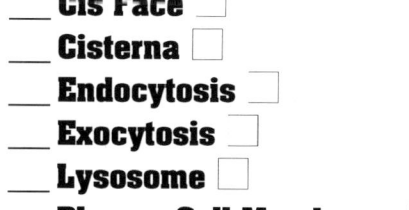

___ Cis Face
___ Cisterna
___ Endocytosis
___ Exocytosis
___ Lysosome
___ Plasma Cell Membrane

Proteins
___ Exported
___ Modified
___ Unmodified
___ Secretory Vesicles
___ Trans Face
___ Transport Vesicles

# Protein Synthesis

Because it is the major business of a cell most cellular organelles are involved with protein synthesis. **DNA** (1) in the nucleus transcribes **mRNA (messenger RNA)** (2), which goes out through **nuclear pores** (10), and passes through the **endoplasmic reticulum** (3) to the **ribosome** (4). Clamped within the ribosome mRNA serves as a code for the synthesis of a particular **protein** (5) (see page 19).

A newly-synthesized protein is picked up by a **transport vesicle** (6) and delivered to the **Golgi apparatus** (7). After processing and packaging within the Golgi apparatus, a modified protein is picked up by either a secretory vesicle or a transport vesicle, depending on whether or not it will be utilized within the cell or exported.

**Mitochondria** (8) secrete **ATP** (9) which supplies energy for all cellular processes, including protein synthesis.

One reason protein synthesis occupies so much of a cell's energy and time is because enzymes are proteins, and all chemical reactions in living cells are enzyme regulated. Additionally, enzymes degrade over time and must be replaced.

*Just think of the challenge of being a DNA molecule and having to continuously send out all those mRNA transcripts for making new proteins.*

*And all this reminds us of the words on a t-shirt of a man we saw sleeping in the science lounge:*

*"On the cell-molecular level I'm really very busy."*

# Protein Synthesis

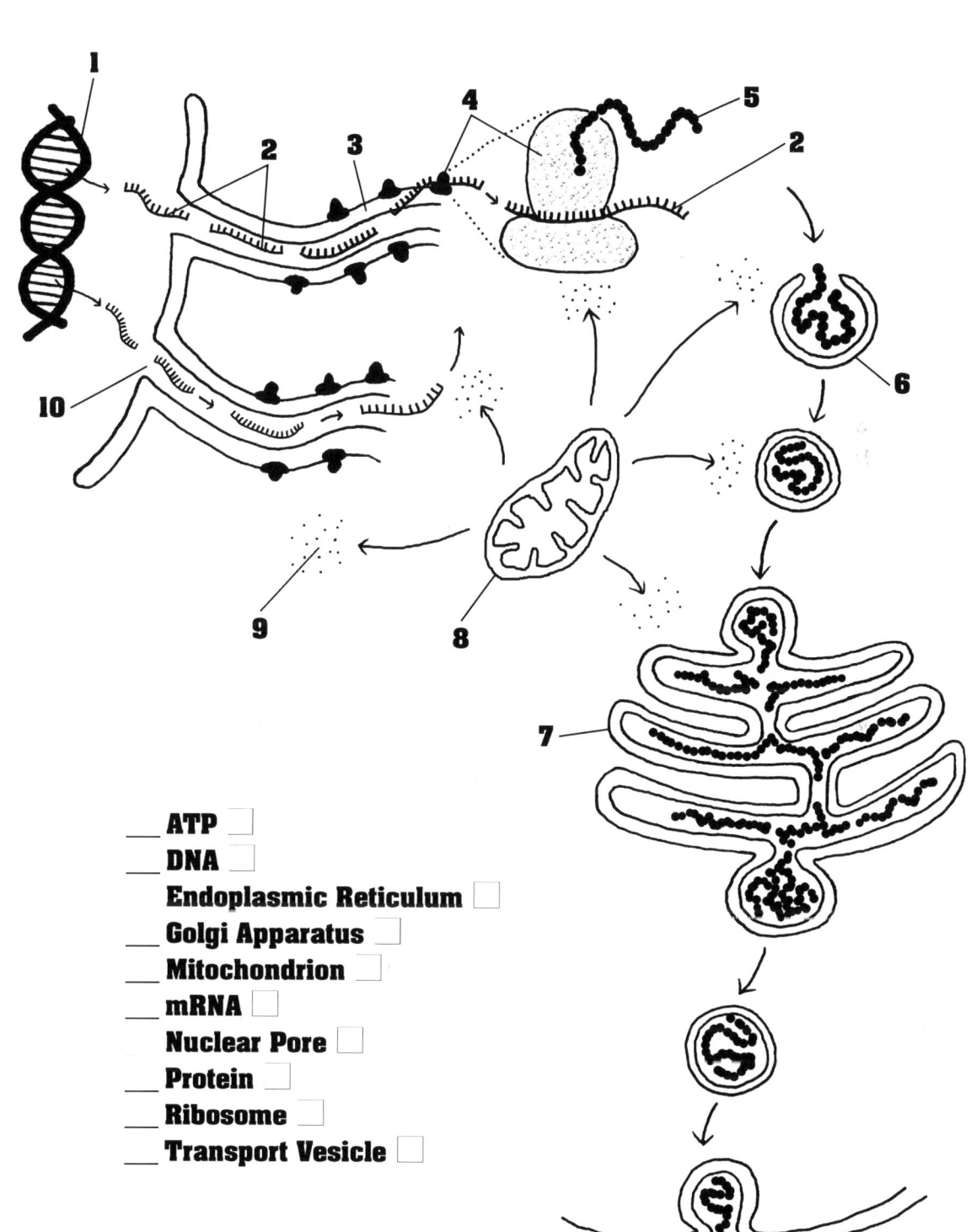

___ ATP
___ DNA
___ Endoplasmic Reticulum
___ Golgi Apparatus
___ Mitochondrion
___ mRNA
___ Nuclear Pore
___ Protein
___ Ribosome
___ Transport Vesicle

## Centrioles

A region of the cell near the nucleus contains two barrel-shaped organelles lying at right angles to each other (see page 15). The region is called the centrosome and the barrel-shaped organelles are called centrioles. Centrioles, which are composed of 27 **microtubules** (2) arranged in nine **triplet sets** (1), have the astonishing capacity of rapidly assembling (polymerizing) and disassembling (depolymerizing) microtubules, or microtubular-like structures, as the cell has need for them — and indeed most cells have many needs for them.

Microtubules, along with microfilaments and intermediate filaments constitute complex skeletal frameworks within the cytosol of the cell. At the time of cell division centrioles produce a microtubular spindle apparatus (see page 35), which appears and disappears as if by magic. Indeed the magic of polymerization and depolymerization. The magic of laying molecular, monomeric bricks at the rate of 500,000/second.

## Cilia and Flagella

Although cilia and flagella arise from basal bodies precisely like centriole basal bodies, their microtubular arrangement differs in that they have nine **doublets** (4) (rather than nine triplets) in their outer region. Additionally, they differ from centrioles in that they have two **central microtubules** (3). The doublets are adjacently in contact with each other via ATP powered **motor molecules** (5). The motor molecules are capable of "walking" one adjacent tubule past another, thus allowing for the "bending and waving" which is the hallmark of ciliary and flagellar action.

*"Beating" cilia constitute the "tracheal escalator" which, if not poisoned by nicotine, helps remove foreign particles from the lungs. Fertilized eggs are propelled down oviducts toward the uterus by the action of ciliated cells. Sperm perform their Herculean swimming feats via the locomotive power of one single flagellum!*

*Motor molecules, which achieve movements via making conformational shifts, are used many places in the body. The contraction of our skeletal muscles is the result of myosin motor molecules making conformational shifts called power strokes (see page 109). A muscle contraction is the result of multiplied millions of power strokes! And when light strikes the retina of the eye, retinal molecules respond by making conformational shifts which generate nerve impulses (see page 179). Indeed we owe much to motor molecules and the conformational shifts they perform on our behalf.*

# Centriole & Cilia

**Centriole**

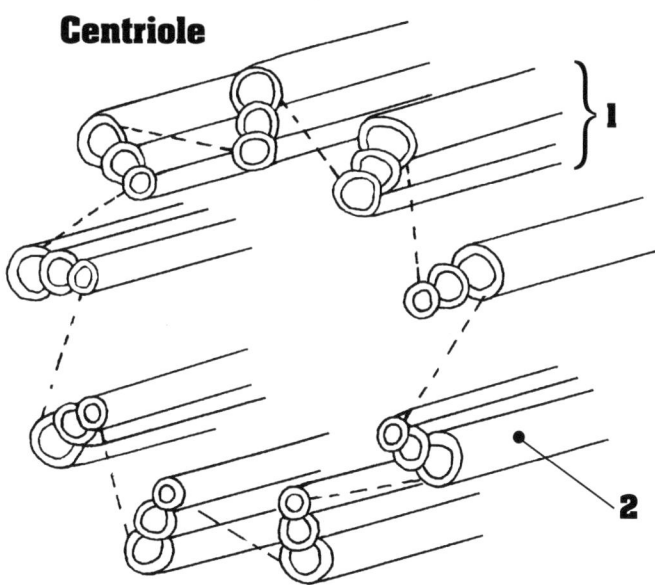

**Cilia**

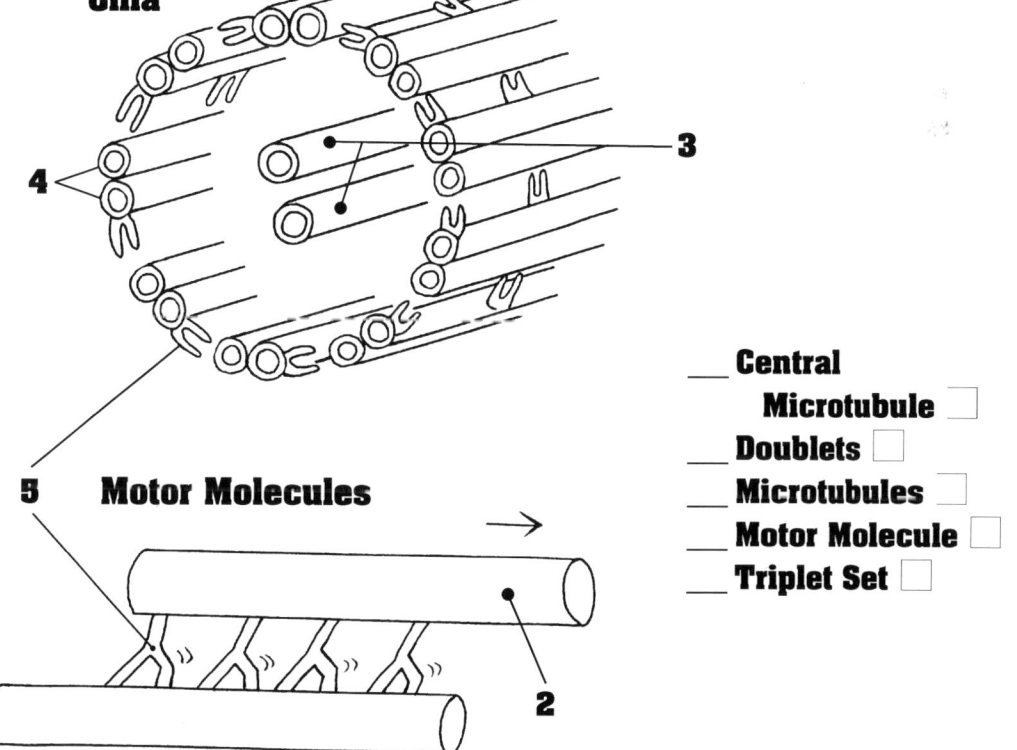

**Motor Molecules**

___ Central Microtubule
___ Doublets
___ Microtubules
___ Motor Molecule
___ Triplet Set

# DNA Structure

Elucidating the structure of the DNA (deoxyribonucleic acid) molecule is considered by many to be the most important scientific discovery of the twentieth century. Watson, Crick and Wilkins received Nobel prizes for making that discovery in 1953.

In its normal mode, the DNA molecule is a long double-helix. (Because only a short piece of a DNA molecule is presented on the opposite page, the double helix configuration is not shown.) Each of the 46 chromosomes in a human somatic cell consists of one long DNA molecule. The DNA "ladder" has two **antiparallel strands** (7) running in opposite directions. Each of the strands is comprised of alternating units of **pentose sugars** (3) and **phosphates** (1).

The "steps" of the DNA ladder are comprised of two kinds of **base pairs** (9). In one **cytosine** (5) is bonded with **guanine** (6). In the other **thymine** (8) is bonded with **adenine** (2). However, because base pairs can also be reversed — end to end — four different steps are possible. Thus, if we read the left-hand side of this short DNA fragment from top to bottom, we have ACGT.

During DNA replication the hydrogen bonds between base pairs are broken and each of the two separated strands act as a template for polymerizing a new "daughter strand." The monomers in this polymerization are **nucleotides** (4). A nucleotide is comprised of a sugar, phosphate, and base.

Note that the bases are always bonded with a pentose sugar. There is a **triple hydrogen bond** (10) between guanine and cytosine and a **double hydrogen bond** (11) between adenine and thymine.

# DNA Structure

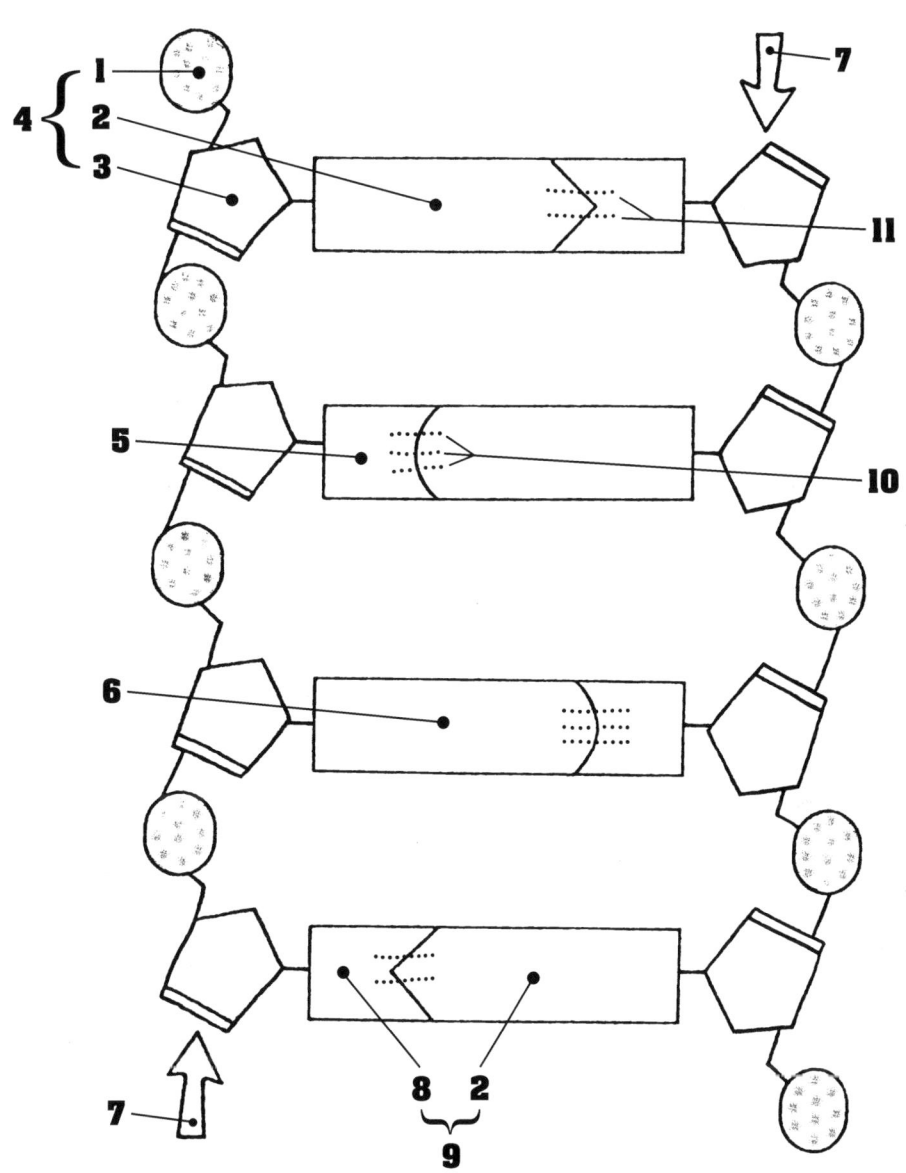

___ Adenine
___ Antiparallel Strands
___ Base Pair
___ Cytosine
___ Guanine
___ Hydrogen Bond — Double
___ Hydrogen Bond — Triple
___ Nucleotide
___ Pentose Sugar
___ Phosphate
___ Thymine

# DNA
# A Universal Code

Because the DNA (deoxyribonucleic acid) molecule is the information code for all multi-cellular, and most uni-cellular, life forms it is called a "universal code." In the beginning life forms are little more than threads of DNA. The differences between species, and between individuals within a species, are foundationally differences in the sequences of bases which comprise their DNA. Thus, in the beginning, if all species were hung out to dry on a "microtubular clothesline," they would look very much alike. The only difference, at that stage, between a tomato plant and an elephant would be a difference in the base sequences of their respective DNA molecules.

The power of DNA to create diversity lies in the fact that different sequences of DNA base pairs code for different sequences of amino acids, and hence for different proteins. Different proteins are finally reflected in the different forms and functions of life.

However, DNA, like a grand piano, has no "life of its own." It has to be "played." The "maestro that plays" DNA is the environment. Different developmental pathways are determined by different environmental influences. Consider the fact that all the diverse eukaryotic cells in our body (muscle cells, nerve cells, blood cells, bone cells, skin cells, stomach cells, etc.) have identical DNA "keyboard" codes.

Even as the diversification of early embryonic cells is determined by microcosmic, molecular, environmental influences, so the development of the child is determined by macrocosmic, environmental influences. The newborn suckles, not only at the breast of its human mother, but also at the breast of mother earth.

# DNA
# A Universal Code

In *The Dependent Gene*, David Moore, a developmental psychologist, says:

> *"Every breath we take, every meal we eat, every scene we see is assimilated into the very structures and functions of our bodies, literally becoming us. And these processes of incorporating our environments into ourselves begin at conception and continue throughout our lifetime. When we first see the light of day as newborn babies it is already impossible to identify any part of us that does not reflect the environment in which we develop from conception."*

Underlying the fact that we are creatures of habit is the more fundamental reality that we are creatures of habitat. What, and how, "we are" is always intimately intertwined with our environmental circumstances. Biologically speaking, from the cradle to the grave the symphony we call "our life" is environmentally influenced and orchestrated.

By choosing our environments we choose "who and what" we are. Although with regard to our human mother's womb we had no choice, with regard to our earth womb the choice is ours! Are you choosing wisely?

# Mitochondrion

Even as DNA serves as a "universal information code" for life, ATP serves as a "universal energy molecule" for life. All our energy requirements are met by ATP. Whenever and wherever we need energy we utilize ATP. More precisely, we obtain it from the **energy** (15) that is released from the breaking of the bond between the second and third phosphates of the ATP molecule. Metabolic events requiring only a little energy use only a few ATP molecules, while metabolic events requiring more energy use more ATP molecules.

The **ADP (adenosine diphosphate)** (2) and **P (phosphate)** (3), which result from the splitting of **ATP (adenosine triphosphate)** (1), are recycled (resynthesized) within the mitochondrion.

The energy for ATP synthesis is obtained indirectly from the combustion (decomposition) of **sugar** (9) with the aid of **oxygen** (10). The oxidative decomposition of sugar makes hydrogen available for the second phase, and yields **carbon dioxide** (5) and **water** (4) as by-products.

In the first phase sugar fragments are disassembled in **citric acid cycles** (13) within the **matrix** (14) of the mitochondrion. In the second phase, the hydrogen obtained from the decomposition of the sugar fragments in the citric acid cycle is transported by hydrogen carrier molecules to the **inner mitochondrial membrane** (8) where it is ionized. The electrons yielded in ionization are accepted by **electron (hydrogen) transport chains** (12) embedded within the membrane.

The energy from the electron flow through the electron transport chain is then used to pump protons from the matrix into the **intermembranous space** (7) between the inner and **outer mitochondrial membrane** (6) to establish a proton (electrochemical) gradient. Ultimately, energy from the proton gradient is used to power the synthesis of ATP.

Invaginations of the inner mitochondrial membrane form numerous **crista** (11), which greatly increase the surface area available for enzyme systems.

The mitochondrion is therefore an energy transformer. The energy which resides in sugar is transformed into ATP energy. For every sugar molecule that is oxidized, 32 to 38 ATP molecules are recycled (synthesized).

*The details of glucose (carbohydrate) catabolism are usually discussed in a chapter on metabolism later in the textbook, often just after the digestive system.*

*The mitochondria are exclusively maternal gifts to our human lives, being passed on from generation to generation via the eggs of the females. But wait! The plot thickens: most scientists now believe that the mitochondria were once free-living bacteria that became incorporated, as endosymbionts, into muti-cellular life to handle the numinous complexities of oxidative chemistry. Biologically, there is no such thing as an "individual." Muti-cellular life is always a symbiotic symphony of life forms. Moment by moment we are dependent upon billions of diverse wee beasties that live on us, in us, and with us! The universe is a "communiverse" — a oneness of turning. All things are a part of some grand and higher art. Your body is a "symbiotic Serengeti of wildness." You never walk alone! You never study anatomy & physiology alone!*

# Mitochondrion

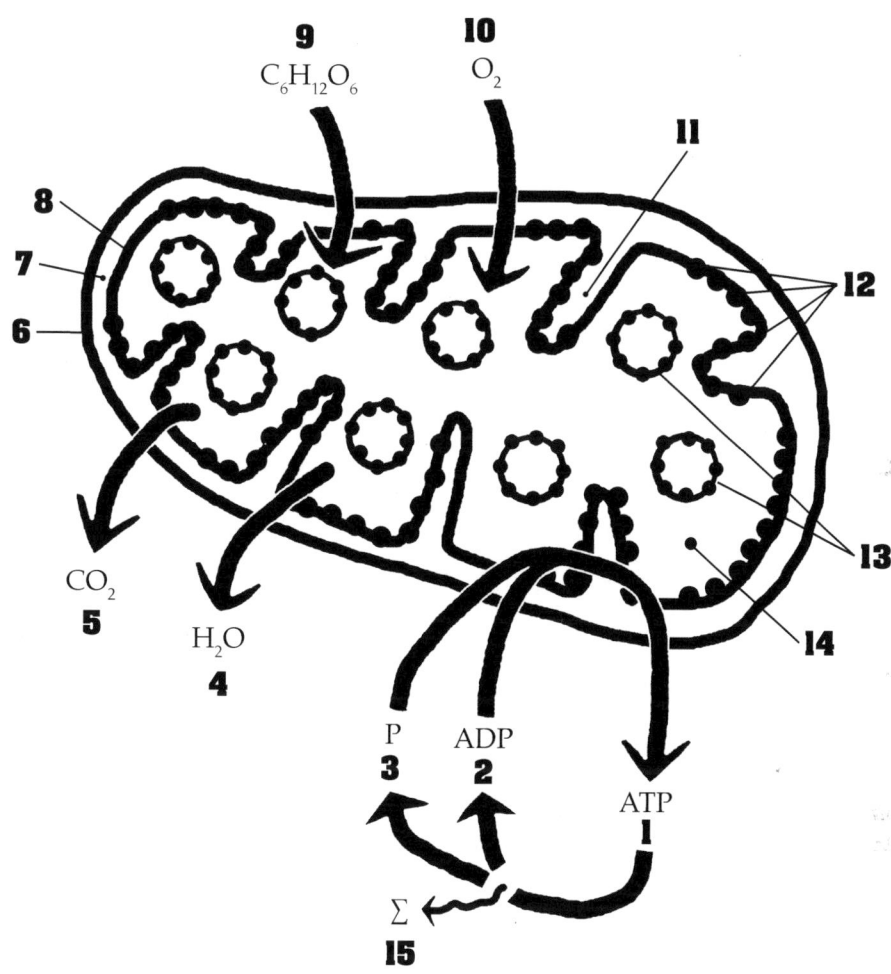

___ ADP (Adenosine Diphosphate) ☐
___ ATP (Adenosine Triphosphate) ☐
___ Carbon Dioxide ☐
___ Citric Acid Cycles ☐
___ Crista ☐
___ Electron Transport Chain ☐
___ Energy ☐
___ Intermembranous Space ☐

___ Matrix ☐
**Mitochondrial Membrane**
___ Inner ☐
___ Outer ☐
___ Oxygen ☐
___ Phosphate ☐
___ Sugar ☐
___ Water ☐

# Plasma Membrane

The plasma (cell) membrane is composed of an inner **phospholipid bi-layer** (1) and an outer **glycocalyx** (2).

**Integral** (3) and **surface** (4) **proteins** are embedded in the bi-layer. Surface proteins often serve as enzymes, or enzyme systems, whereas integral proteins often serve as channels, or ion gates, for the transport of molecular traffic into, and out of, the cell. **Cholesterol** (9) molecules, embedded within the bi-layer, act as stabilizing agents.

The phospholipid molecules have **hydrophilic phosphate heads** (8) and **hydrophobic fatty acid tails** (7).

The "fuzzy" glycocalyx layer is comprised of numerous **glycolipid** (6) and **glycoprotein** (5) molecules. The glycolipids and glycoproteins are sometimes referred to as "antenna molecules" because many of them serve as recognition and receptor sites. A sugar attached to a protein is called a glycoprotein, whereas a sugar attached to a lipid is called a glycolipid.

In the case of nerve fibers, integral proteins serve as **sodium pumps** (11). Sodium pumps "kick" **sodium ions** (10) to the outside of the membrane to establish a **sodium ion gradient** (13). The sodium ion gradient, in effect, polarizes the membrane, and creates a "mini-battery" wherein the inside of the membrane becomes -70 mv as compared with the outside of the membrane.

When **sodium ion gates** (12) open, sodium ions flow inward and the membrane is depolarized. Depolarization of the membrane is, in essence, the nature of the nerve impulse.

Cell membranes of nerve fibers are perfused with sodium pumps and sodium ion gates. Nerve impulses travel down the membrane because one sodium gate after another is opened in a domino-effect. A nerve impulse is therefore a wave of depolarization that passes down the membrane from ion gate to ion gate. Sodium pumps restore the sodium gradient after each passing impulse.

*The main event in the mitochondrion is also the creation of a "mini-battery" (between the inner and outer membranes). The "battery" is then used to power the synthesis of ATP. We could therefore make a strong case for saying we are "battery-powered." However, because we use energy from sugar to make our mitochondrial batteries, and the sugar is made via photosynthesis, we are perhaps more accurately "sun-powered."*

# Plasma Membrane

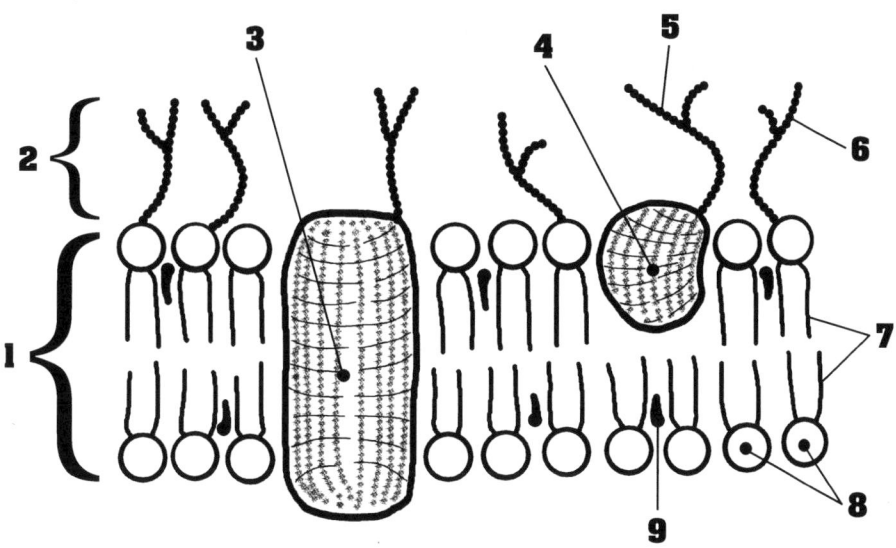

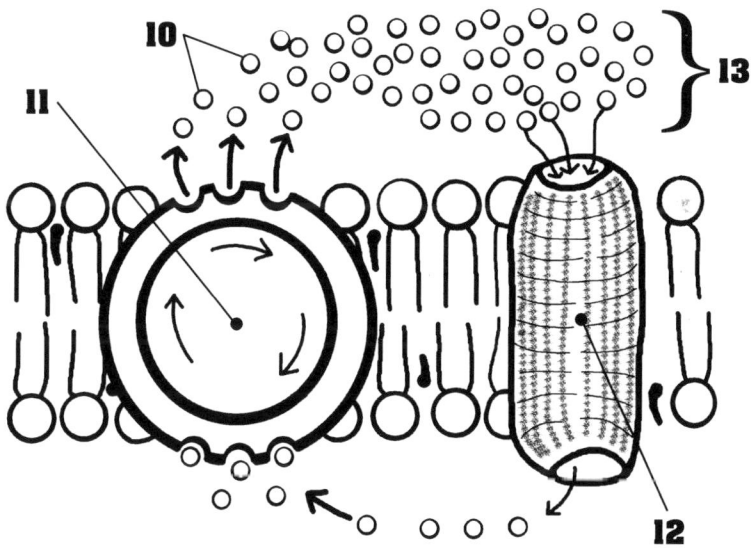

___ Cholesterol  
___ Glycocalyx  
___ Glycolipid  
___ Glycoprotein  
___ Hydrophilic Phosphate Heads  
___ Hydrophobic Fatty Acid Tails  
___ Integral Protein  

___ Phospholipid Bi-layer  
___ Sodium Ion Gate  
___ Sodium Ion Gradient  
___ Sodium Ions  
___ Sodium Pump  
___ Surface Protein

# Metaphase of Mitosis & Meiosis I

Each of our body cells has 23 maternal chromosomes (gifted to us by our mother), and 23 morphologically matching paternal chromosomes (gifted to us by our father). The 23 matching pairs are called **homologous pairs** (10). Although human somatic cells have 46 chromosomes (23 pairs), for the sake of clarity, and for the sake of preserving my sanity as an artist, I am showing cells with only six (three pairs) of chromosomes.

Prior to cell division DNA molecules replicate, condense, and become packaged into discrete bundles called **chromosomes** (7). Because the replication process doubles the DNA, each chromosome, prior to cell division, consists of two **chromatids** (6). In metaphase of mitosis chromosomes line up single file on the **equatorial plane** (8). In metaphase of meiosis I the homologous partners line up together on the equatorial plate.

Looking at a metaphase cell from a side view the chromosomes seem aligned in a straight line, but if the cell were viewed from the position of one of the poles (a polar view) the chromosomes would appear to be aligned in a circle. The spindle apparatus, which aligns the chromosomes, can be imagined as the wire skeleton of a basketball.

In metaphase of mitosis **centromeres** (9) split and the chromatids separate, with six chromatids moving toward each of the poles, thus producing two new cells, each with six chromosomes.

In metaphase of meiosis I the homologous partners separate — the centromeres do not split — with one entire chromosome, consisting of two sister chromatids, going to each of the poles. Thus each new cell will have only three — instead of six — chromosomes. Hence the first meiotic division is called a reduction division, and reduces the chromosome number by one half (from diploid to haploid). A second meiotic division (meiosis II) will separate the chromatids, as in mitosis. Meiosis only occurs in the gonads to produce eggs or sperm.

The separation of chromatids in mitosis, and the separation of homologous pairs in meiosis I is assisted and made possible by the spindle apparatus which originates from the **centrosomes** (1). Each centrosome contains two **centrioles** (2). Centrioles produce microtubules. Some microtubules reach out as **spindle fibers** (5) and physically attach to the centromeres of the chromosomes. Other microtubules, in the vicinity of the centrosome, are called **astral rays** (3). The spindle apparatus, like all other cell organelles and inclusions, is confined within the **cell (plasma) membrane** (4).

*Mitosis produces new somatic cells. For example, billions of new skin and blood cells are produced each day. Meiosis produces sex cells. A healthy adult male produces about 400 million sperm each day.*

# Metaphase of Mitosis & Meiosis I

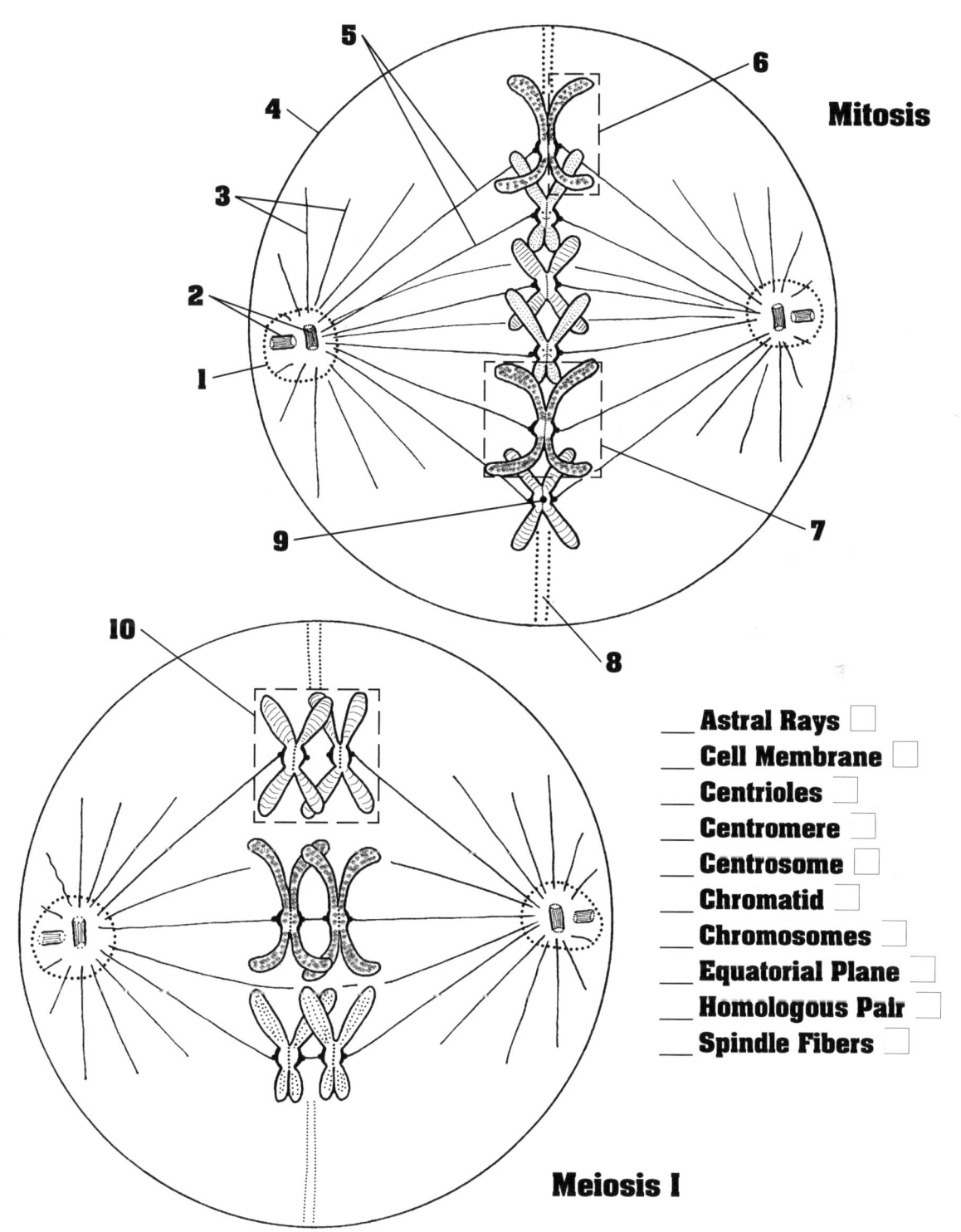

__ Astral Rays ☐
__ Cell Membrane ☐
__ Centrioles ☐
__ Centromere ☐
__ Centrosome ☐
__ Chromatid ☐
__ Chromosomes ☐
__ Equatorial Plane ☐
__ Homologous Pair ☐
__ Spindle Fibers ☐

# Tissue

The fabric of the body is comprised of four primary tissue types:

- **Covering tissue** — epitheliel
- **Supporting tissue** — connective
- **Regulatory tissue** — nervous
- **Movement tissue** — muscle

Six types of epitheliel tissue and eight types of connective tissue are considered in this exercise. (We defer muscle, nerve, bone, and blood tissue to later exercises.)

While reading the descriptions below, and perhaps looking at the tissue pictures in your textbook, you should be able to do the double matching exercise on the opposite page. First match the numbers with the tissue types, then match the same numbers with the correct tissue locations.

## Epitheliel Tissue
**Simple squamous** — flat cells which fit together like floor tiles.
**Stratified squamous** — several parallel layers of simple squamous.
**Metastasizing stratified squamous** — cancerous skin tumor.
**Cuboidal** — cube-shaped cells.
**Columnar** — tall rectangular cells.
**Ciliated columnar** — columnar cells with cilia.

## Connective Tissue
**Areolar (loose)** — a very "busy" tissue which includes three distinct fiber types and a variety of cell types.
**Adipose** — lipid-filled ringlets with nuclei residing at the outer edges.
**Reticular** — reticulated cells lying in a framework of reticular fibers.
**Dense regular** — fibroblast nuclei wedged between parallel collagen fibers.
**Dense irregular** — randomly arranged collagen fibers with fibroblast nuclei.
**Hyaline cartilage** — chondrocytes lying within lacunae "peer out" from an amorphous matrix.
**Elastic cartilage** — elastic fibers are scattered through the matrix.
**Fibrocartilage** — chondrocytes are embedded in wavy lines of collagen fibers.

*Skin cancer (epitheliel carcinoma), along with lung cancer, is becoming more and more frequent. We suspect this might have something to do with the fact that the skin and the lungs are directly exposed to our increasingly polluted environment? How's your air?*

# Tissue

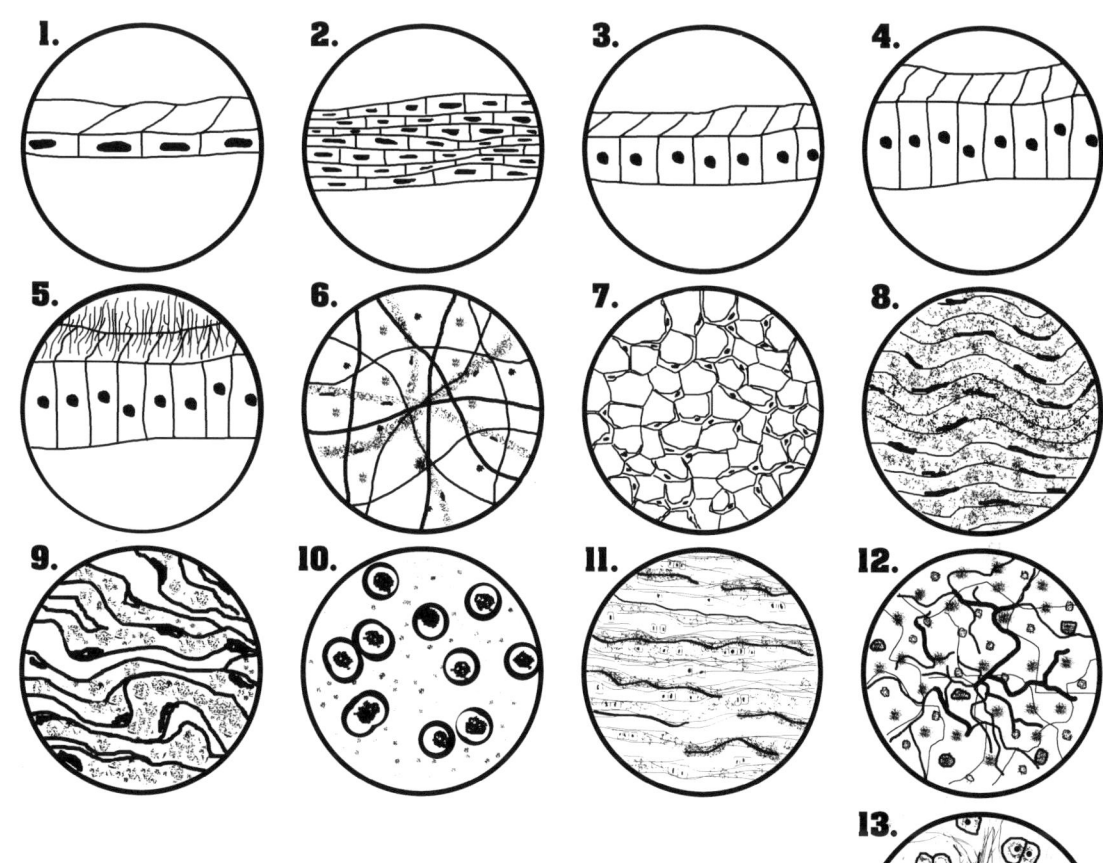

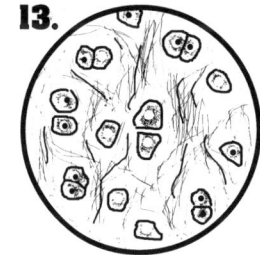

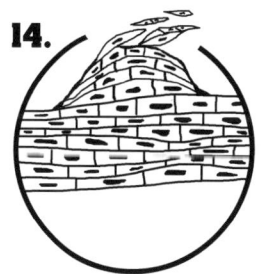

**Tissue Types**

___ Adipose
___ Areolar
___ Ciliated Columnar
___ Columnar
___ Cuboidal
___ Dense Irregular
___ Dense Regular
___ Elastic Cartilage
___ Epitheliel Carcinoma
___ Fibrocartilage
___ Hyaline Cartilage
___ Reticular
___ Simple Squamous
___ Stratified Squamous

**Tissue Locations**

___ Alveoli of Lungs
___ Epidermis
___ External Ear
___ Intervertebral Disks
___ Joint Capsules
___ Kidney Tubules
___ Lymph Nodes
___ Nose and Larynx
___ Stomach Lining
___ Tendons
___ Tracheal Lining
___ Underlies the Skin
___ Where Fat is Found

# Integumentary System

"Our Skin," which varies in depth from approximately 1.5 to 4 mm and accounts for about 7% of the body weight (about 10 pounds), is a complex organ system which performs at least the six following functions:

1. **Protection**

   - A physical barrier is provided by the continuity of the skin and by keratinized cells at its outer surface which are impermeable to most microbes.
   - Chemical barriers include:
     - A low pH "acid mantle" at the skin's surface, which inhibits bacterial growth.
     - A variety of antimicrobial substances, such as **defensin** and **cathelicidins**.
     - **Melanin pigments** (produced by melanocytes), which mitigate damage from ultraviolet light.

2. **Temperature Regulation**

   - Via constriction and dilation of dermal blood vessels the integument assists in regulating body temperature.
   - Evaporation of sweat (via sweat glands) from the skin's surface dissipates heat and effectively cools the body.

3. **Sensation**

   - Skin is richly supplied with cutaneous sensory receptors.
   - Pain stimuli, in the form of heat, cold, irritating chemicals, etc., is sensed by free nerve endings in the skin.
   - Light pressure against the skin is detected by **Meisner's corpuscles**, lying close to the surface.
   - Heavier pressure is detected by **Pacinian corpuscles**, that lie deeper in the skin.

# *Integumentary System*

**4. Metabolism**

- Skin plays a vital role in the synthesis of Vitamin D, which in turn plays a vital role in calcium metabolism.

**5. Blood Reservoir**

- Via dilation, dermal blood vessels can hold about 5% of the body's entire blood volume. Via constriction portions of the dermal blood reservoir are made available to muscles and other body organs.

**6. Excretion**

- Although most wastes are excreted in urine, limited amounts of wastes, including some toxins, are eliminated through the skin. (Wrap a heavy smoker in a white sheet and place her in a hot sauna for half an hour and you will see yellow stains (nicotine blotches) on the sheet.

*The outer epidermal layers, which are so vital in protecting the underlying cells from the deleterious effects of ultraviolet light, are all comprised of dead cells. Thus "dead" cells are as important as "live" cells in supporting and sustaining our "lives." "Dead" and "alive" have little or no meaning on the cellular and molecular level, which supports the hypothesis that, ultimately, there are no discontinuities in the universe.*

# Integument

The integument is comprised of three principal layers:

- **Epidermis** (19)
- **Dermis** (20)
- **Hypodermis** (21)

The outer epidermis is, in turn, comprised of five sub-layers. The deepest of the five epidermal layers, the **stratum basale** (9), is the generative layer. The four upper epidermal layers are transition layers, created as cells from the stratum basale migrate upward. **Dermal papillae** (16) create a wavy pattern in the stratum basale. This pattern reflects upward to the stratum corneum, creating unique individual "fingerprint" patterns.

The four transition layers above the basale are: the **stratum spinosum** (10), **stratum granulosum** (11), **stratum lucidum** (12), and **stratum corneum** (13). Thus, the five epidermal layers, from outer to inner, are: corneum, lucidum, granulosum, spinosum, and basale (CLGSB). It may help you to recall them in that order by first saying, "Cows Love Grass, Silly Bulls!"

The dermis is richly supplied with **capillaries** (7) and nerves, and thickly populated with specialized glands and sensory receptors. **Free nerve endings** (8) act as pain receptors, **Meisner's corpuscles** (14) act as light touch pressoreceptors, **Pacinian corpuscles** (23) act as heavy touch pressoreceptors, and **root hair plexuses** (4) act to sensitize hairs.

**Sebaceous (oil) glands** (15) lubricate the hair and skin and **sweat (eccrine) glands** (22) act to regulate temperature and excrete waste. The hypodermis is richly supplied with **adipose cells** (1).

Each hair has a **hair shaft** (17), and **hair root** (5), and is surrounded by a **hair follicle** (6). The base of the hair root is expanded into a **hair bulb** (2) which contains a vascularized **hair papilla** (3). Many hairs also have an associated **arrector pili muscle** (18) which, when it contracts, makes the hair "stand up" and causes "goose bumps."

# Integument

**Longitudinal Section**

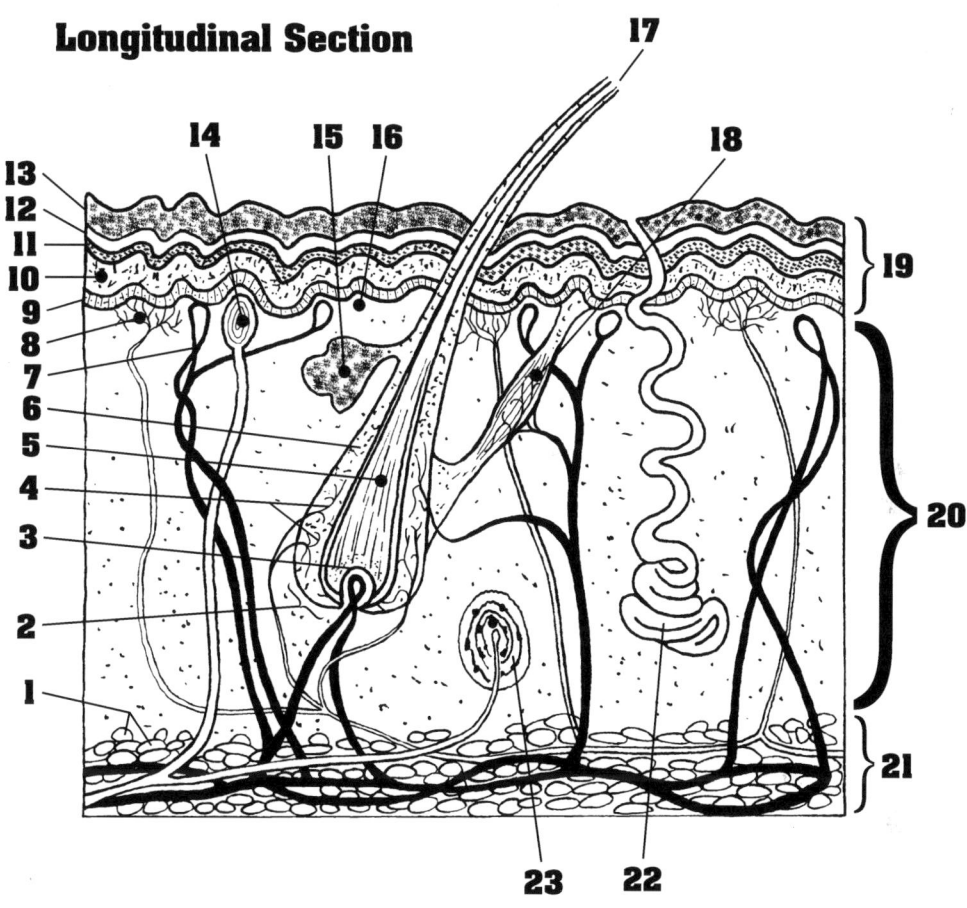

___ Adipose Cells
___ Arrector Pili Muscle
___ Capillary
___ Dermal Papilla
___ Dermis
___ Epidermis
___ Free Nerve Ending
___ Hair Bulb
___ Hair Follicle
___ Hair Papilla
___ Hair Root
___ Hair Shaft
___ Hypodermis
___ Meisner's Corpuscle
___ Pacinian Corpuscle
___ Root Hair Plexus
___ Sebaceous Gland
___ Stratum Basale
___ Stratum Corneum
___ Stratum Granulosum
___ Stratum Lucidum
___ Stratum Spinosum
___ Sweat Gland

# Hair

A **hair** (9), shown contained within a **hair follicle** (8), is comprised of three layers:

- **Medulla** (7)
- **Cortex** (6)
- **Cuticle** (5)

The hair follicle is comprised of four layers:

- **Internal epithelial root sheath** (4)
- **External epithelial root sheath** (3)
- **Glassy membrane** (2)
- Outermost **connective tissue root sheath** (1)

If the hair shaft is flat the hair is kinky or curly, if the shaft is oval the hair is wavy, and if the shaft is round the hair is straight.

The color of the hair, like the color of the skin, is determined by various proportions of melanin pigments. The pigments, which are produced by melanocytes at the base of the hair, are transferred upward to the cortical cells.

On the average hair grows at the rate of 2.5 mm per week. Old follicles age and are shed at the rate of about 90 scalp hairs per day. Hair thinning and baldness occur when hairs are not replaced as fast as they are shed.

# Hair

## Transverse Section

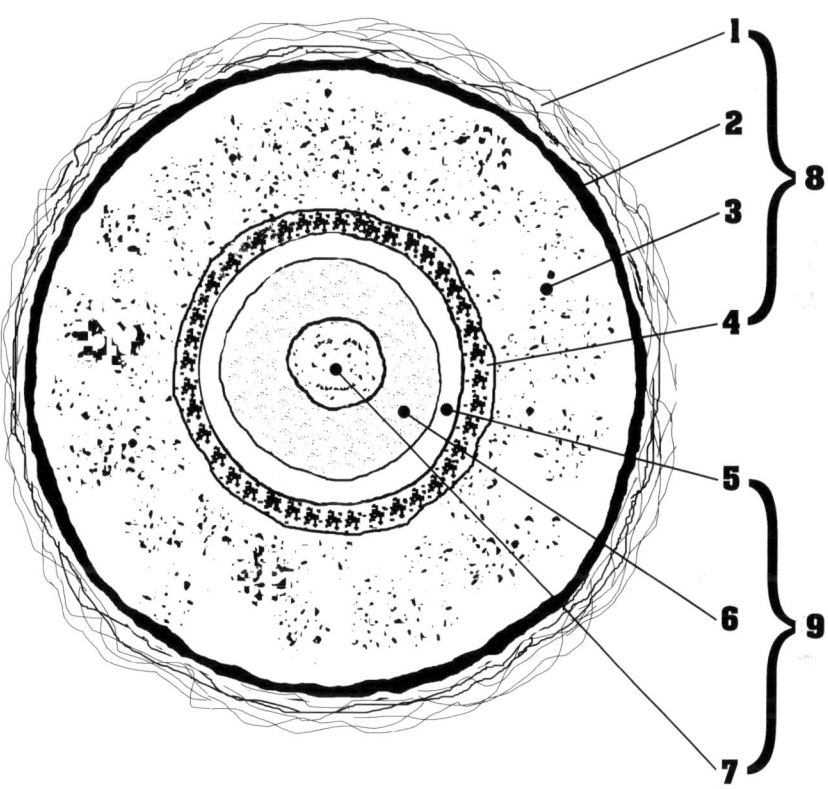

___ Cortex
___ Cuticle
___ Follicle
___ Glassy Membrane
___ Hair
___ Medulla
    Root Sheath
___   Connective Tissue
___   External Epitheliel
___   Internal Epitheliel

# Skeletal System

Bones contribute a good deal more than merely laying the foundation for shape and form:

1. **Support**

   - Large leg bones support the body trunk.
   - Vertebrae support the back and head.
   - Ribs support the thoracic wall.

2. **Protection**

   - Vertebrae protect the spinal cord.
   - The rib cage protects the heart and lungs.
   - The cranium protects the brain.

3. **Movement**

   - Muscles attach to bones via tendons and use the bones as levers to move body parts, as well as the whole body. Walking, talking, breathing, sitting, standing, and running are all bone-dependent processes.

4. **Storage**

   - Bones are one of the body's many banks, from which it takes withdrawals and to which it makes deposits. Other "body banks" include the muscles, spleen, and liver.
   - Calcium and phosphate, along with a variety of growth factors, are stored in bones.

5. **Blood Cell Formation**

   - Via a process called hemopoiesis, most blood cells are formed in the marrow cavities of certain bones. Both bone (osseous) tissue and blood (vascular) tissue are technically labeled as types of connective tissue.

# Skeletal System

The "bone marking" matching exercises below will be of help as you study individual bones in the laboratory. Use your textbook as a reference.

I. Projections that are sites for muscle and ligament attachments:

___ Crest      A. Any bony prominence
___ Epicondyle      B. Large, blunt process (on a femur)
___ Line      C. Large, rounded, rough projection
___ Process      D. Narrow ridge of bone
___ Spine      E. Raised area on or above a condyle
___ Trochanter      F. Sharp, slender, pointed projection
___ Tubercle      G. Small, rounded projection or process
___ Tuberosity      H. Very narrow ridge of bone

II. Projections that help to form joints:

___ Condyle      A. Arm-like bar of a bone
___ Facet      B. Rounded articular surface
___ Head      C. Rounded expansion on a narrow neck
___ Ramus      D. Smooth, nearly flat articular surface

III. Depressions and openings for blood vessels and nerves:

___ Fissure      A. Canal-like passageway
___ Foramen      B. Cavity within a bone
___ Groove      C. Furrow
___ Meatus      D. Narrow, slit-like opening
___ Sinus      E. Round, or oval, opening in a bone

# Osteon
## (Haversian System)

In this transverse section of compact bone we see, in the upper left hand corner, three **osteons (Haversian systems)** (1). Each osteon is composed of several layers of **concentric lamellae** (4) and each lamella is bounded by an outer ring of **osteocytes** (6).

At the center of each osteon (Haversian system) is a **central (Haversian) canal** (2) containing **blood vessels** (3).

Each osteocyte is contained within a small bony cave called a **lacuna** (5). **Protoplasmic communicating threads** (8) run through microscopic osseous channels, called **canaliculi** (7), that radiate out from each lacuna.

… And did we almost forget to mention that each osteocyte has a **nucleus** (9)?

*Central to bone growth is a process referred to as "bone remodeling." The marrow cavities of long bones are continually enlarged. As new bone cells (osteocytes) are formed by osteoblasts on the outside of the bone, old bone cells are destroyed by osteoclasts on the inside of the bone. Osteoblasts generate new bone cells, osteoclasts destroy old bone cells. The osteoblast giveth and the osteoclast taketh away! "Life" and "death," side by side, in symbiotic symphony.*

*And if you are not yet convinced of the importance of "death" to "life," we'll remind you that in your early fetal development your hands and feet were "webbed," with no separations between the fingers and the toes. The "living" cells, between your digits, were "destroyed" to free your fingers and toes.*

*What, no such thing as an "individual?" No such thing as "life" and "death? What a subversive, iconoclastic science is this anatomy & physiology!*

# Osteon
## (Haversian System)

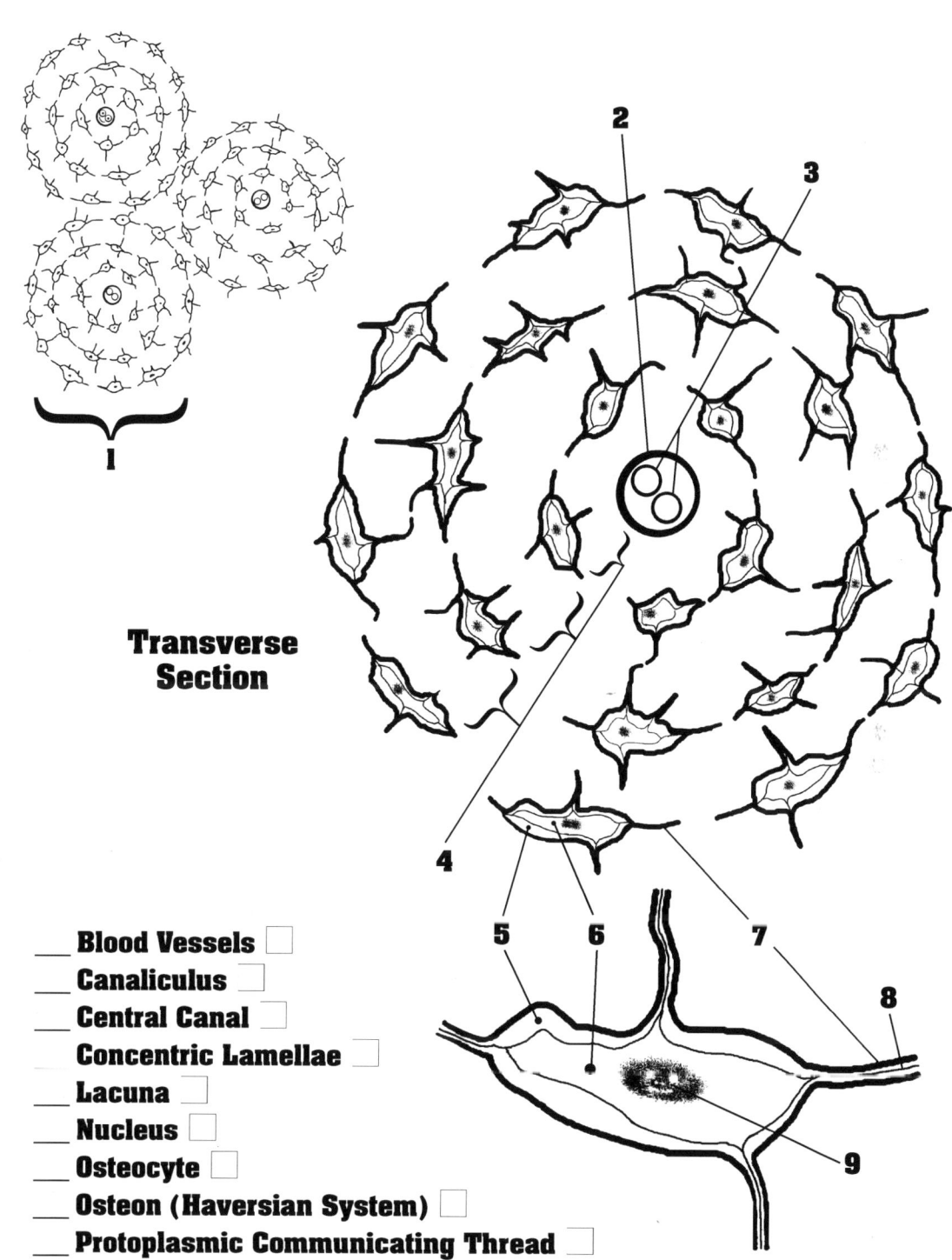

**Transverse Section**

___ Blood Vessels
___ Canaliculus
___ Central Canal
___ Concentric Lamellae
___ Lacuna
___ Nucleus
___ Osteocyte
___ Osteon (Haversian System)
___ Protoplasmic Communicating Thread

# Long Bone

Whereas **spongy bone** (9) predominates in the **epiphyses** (11), **compact bone** (6) predominates in the **diaphysis** (12). And whereas **articular cartilage** (10) covers the condylar epiphyses, **periosteum** (8) covers the diaphysis.

A **medullary cavity** (5) lined with **endosteum** (7), and containing **bone marrow** (4), runs through the center of the diaphysis.

A cartilaginous **epiphyseal plate** (1) is found in each epiphysis.

The **blood supply** (2) enters the bone through an **osteoforamen** (3).

The periosteum, endosteum, and epiphyseal plates, are all osteogenic layers containing bone-forming cells called osteoblasts, and bone-destroying cells, called osteoclasts. In a process called "remodeling," new bone cells are continuously produced via osteoblasts while old bone cells are continuously destroyed via osteoclasts.

The osteogenic layers are cartilage. Bone derives from cartilage. The early fetal skeleton is all cartilage. In a process called ossification, the cartilaginous "bone models" are replaced by bone.

*One class of vertebrates, the Chondrichthyes, which includes sharks and skates, never replaces their cartilaginous skeletons with bone, thus their bodies remain very flexible ... No shark has ever broken a bone!*

# Long Bone

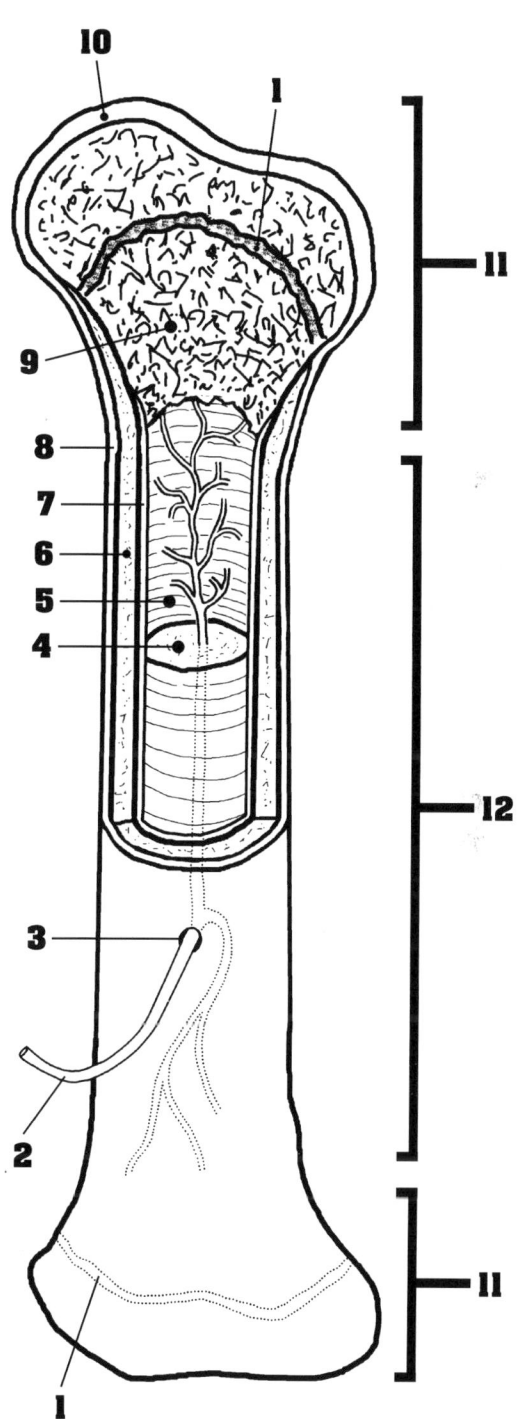

___ Articular Cartilage ☐
___ Blood Supply ☐
___ Bone Marrow ☐
___ Compact Bone ☐
___ Diaphysis ☐
___ Endosteum ☐
___ Epiphyseal Plate ☐
___ Epiphyses ☐
___ Medullary Cavity ☐
___ Osteoforamen ☐
___ Periosteum ☐
___ Spongy Bone ☐

# Infant Skull

Because the rigors of birth require a flexible skeleton, the bones of the infant cranium are loosely knit and partially separated by membranous fontanels.

Superiorly, there is an **anterior fontanel** (4) and a **posterior fontanel** (7). Laterally, there is an **anteriolateral fontanel** (11) and a **posteriolateral fontanel** (18).

The cranial bones are joined at sutures:

- **Frontal suture** (2)
  connects the two **frontal bones** (3).

- **Sagittal suture** (5)
  connects the two **parietal bones** (9).

- **Coronal suture** (1)
  connects the frontal bones with the parietal bones.

- **Lambdoidal suture** (6)
  connects the parietal bones with the **occipital bone** (8).

- **Squamosal suture** (10)
  connects the parietal bones with the **temporal bones** (17).

Other major bones of the skull are the **mandible** (15), **maxilla** (14), **zygomatic** (16), **sphenoid** (12), and **nasal** (13).

*The pliability of the infant skull in no way justifies trying to see how far you can toss your baby up into the air and catch her!*

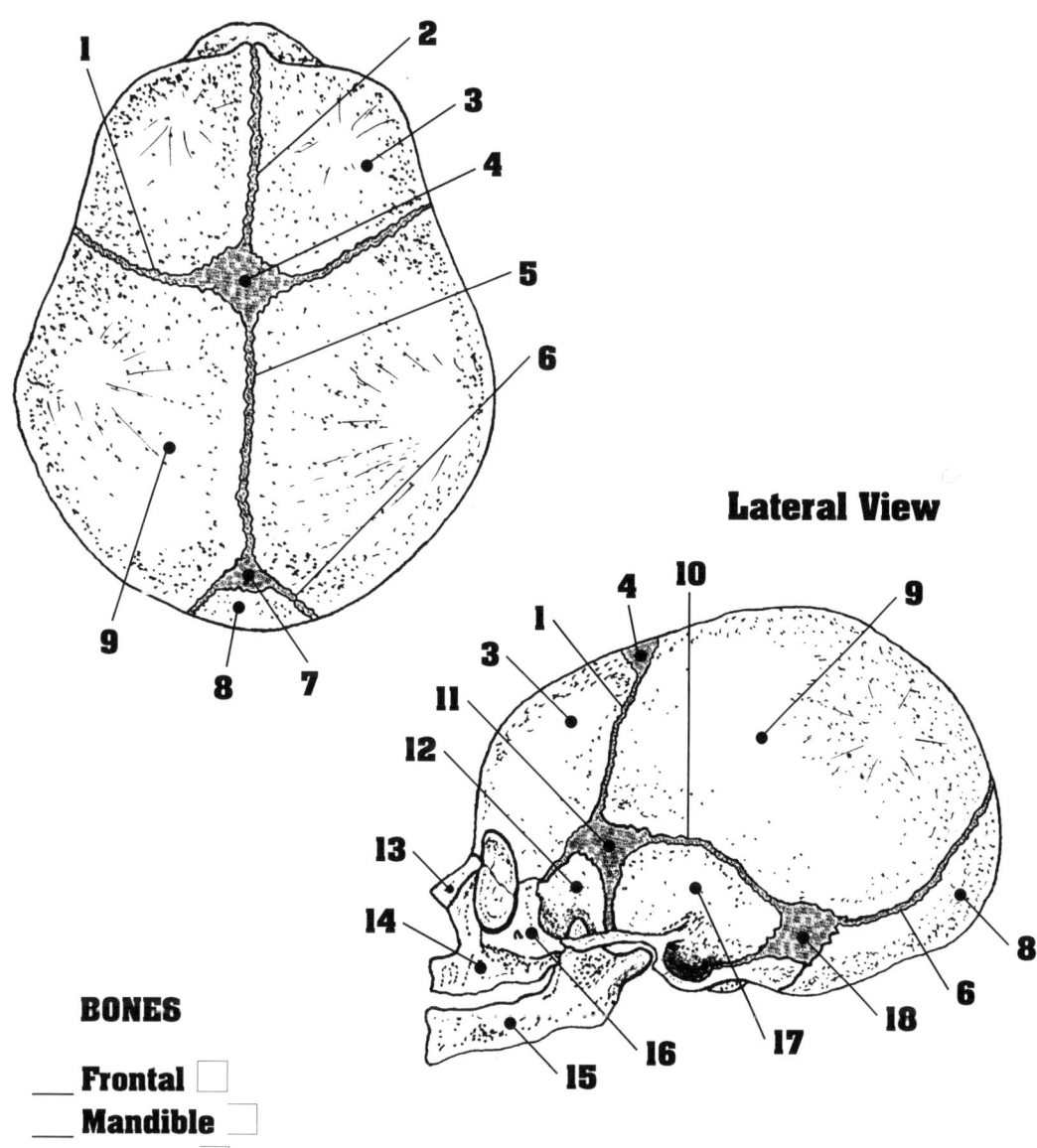

# Skull
## Lateral View

Here, on the adult skull, we see again how sutures connect bones:

- **Coronal suture** (5)
  connects the **parietal bones** (4) with the **frontal bones** (6).

- **Squamosal suture** (3)
  connects the parietals with the **temporal bones** (1).

- **Lambdoidal suture** (25)
  connects the parietal bones with the **occipital bone** (24).

A **supraorbital foramen** (8) is located on each frontal bone just above each eye socket. An **infraorbital foramen** (14) is located on each **maxilla bone** (15) just below each eye socket.

A **zygomatic foramen** (12) is located on each **zygomatic bone** (13).

An **occipital protuberance** (23) extends from the occipital bone.

A **styloid process** (20) and **mastoid process** (21) extend inferiorly from each of the temporal bones. Just superior to the styloid process an **external auditory meatus** (22) penetrates medially into the skull.

Prominent features of the **mandible** (17) are the **mandibular condyle** (19), **coronoid process** (18), and **mental foramen** (16).

A "wing" of the **sphenoid bone** (7) is seen toward the center of the cranium and the **zygomatic arch** (2) forms a "bridge" between the zygomatic bone and temporal bone.

Finally, within the eye socket, from medial to lateral, we see the **nasal** (11), maxilla, **lacrimal** (10), and **ethmoid** (9) bones.

# Skull

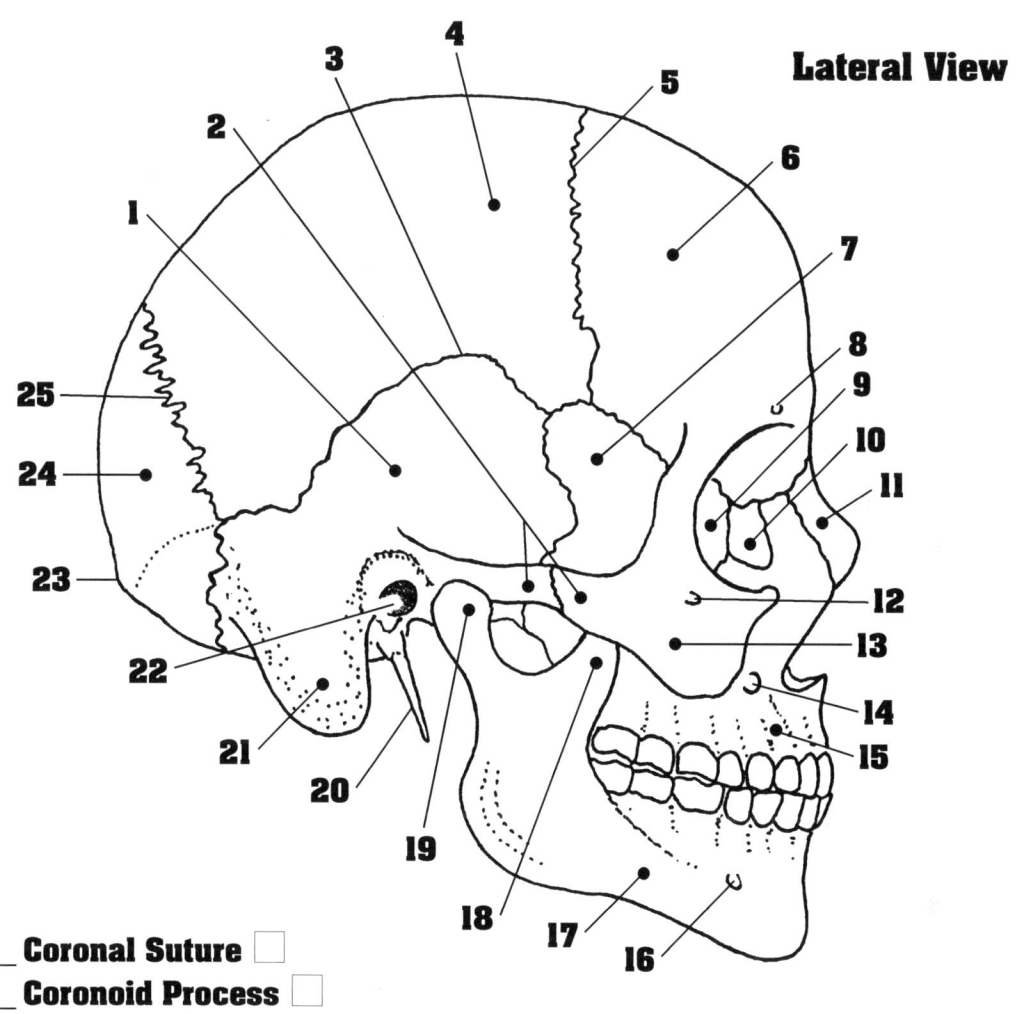

**Lateral View**

___ Coronal Suture
___ Coronoid Process
___ Ethmoid
___ External Auditory Meatus
___ Frontal
___ Infraorbital Foramen
___ Lacrimal
___ Lambdoidal Suture
___ Mandible
___ Mandibular Condyle
___ Mastoid Process
___ Maxilla
___ Mental Foramen
___ Nasal
___ Occipital
___ Occipital Protuberance
___ Parietal
___ Sphenoid
___ Squamosal Suture
___ Styloid Process
___ Supraorbital Foramen
___ Temporal
___ Zygomatic
___ Zygomatic Arch
___ Zygomatic Foramen

# Skull
## Inferior View

Anteriorly, a **transverse palatine suture** (11) connects the **palatine process of the maxilla bone** (12) with the **horizontal plate of the palatine bone** (9). The horizontal plate of the palatine bone contains the **greater palatine foramen** (10). The palatine process of the maxilla bone contains an **incisive foramen** (13).

**Nasal conchae** (8) can be seen through a large opening behind the palatine bone, and just posterior to the nasal conchae is the small **vomer bone** (7). The **foramen lacerum** (6) extends out, as an irregular fissure, from behind the vomer bone.

Laterally adjacent to the foramen lacerum are three major foramina:

- **Foramen ovale** (19) — for the passage of the mandibular branch of the trigeminal nerve.
- **Carotid foramen** (5) — a passageway for the internal carotid artery.
- **Jugular foramen** (2) — a passageway for the jugular vein.

The foramen ovale is located in the **sphenoid bone** (17), the carotid foramen in the **temporal bone** (20), and the jugular foramen in the **occipital bone** (1).

Also associated with the occipital bone are the:

- **Occipital condyle** (4)
  which articulates with the superior articular facets of the atlas.
- **Foramen magnum** (3)
  through which the spinal cord passes.
- **Hypoglossal foramen** (25)
  through which the hypoglossal nerve passes.

Features associated with the temporal bone are the **mastoid process** (24), **styloid process** (21), **external auditory meatus** (22), and two foramina: the **stylomastoid foramen** (23) at the base of the styloid process — through which the facial nerve passes, and the **mastoid foramen** (26).

The zygomatic arch is comprised of a **temporal process** (16) of the **zygomatic bone** (15), and a **zygomatic process** (18) of the temporal bone. The **maxilla bone** (14) connects with the zygomatic bone anteriorly.

# Skull

- \_\_ Carotid Foramen ☐
- \_\_ External Auditory Meatus ☐
- \_\_ Foramen Lacerum ☐
- \_\_ Foramen Magnum ☐
- \_\_ Foramen Ovale ☐
- \_\_ Greater Palatine Foramen ☐
- \_\_ Horizontal Plate of the Palatine ☐
- \_\_ Hypoglossal Foramen ☐
- \_\_ Incisive Foramen ☐
- \_\_ Jugular Foramen ☐
- \_\_ Mastoid Foramen ☐
- \_\_ Mastoid Process ☐
- \_\_ Maxilla ☐
- \_\_ Nasal Conchae ☐
- \_\_ Occipital Condyle ☐
- \_\_ Occipital ☐
- \_\_ Palatine Process of the Maxilla ☐
- \_\_ Sphenoid ☐
- \_\_ Styloid Process ☐
- \_\_ Stylomastoid Foramen ☐
- \_\_ Temporal ☐
- \_\_ Temporal Process of the Zygomatic ☐
- \_\_ Transverse Palatine Suture ☐
- \_\_ Vomer ☐
- \_\_ Zygomatic ☐
- \_\_ Zygomatic Process of the Temporal ☐

**Inferior View**

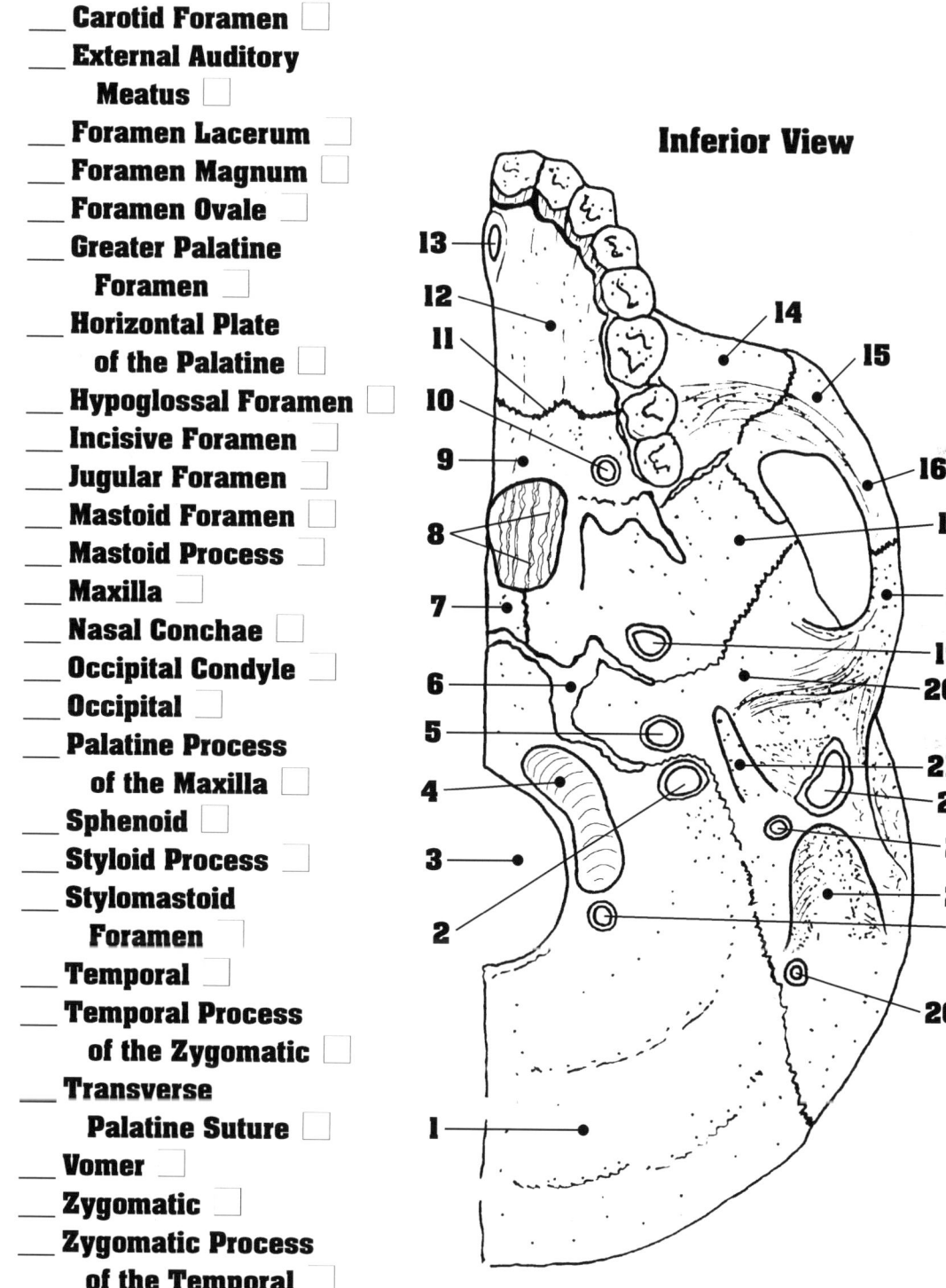

# Cranium Base

Medial to the **frontal bone** (8) the **cribiform plate** (10) of the ethmoid bone contains passageways (foramina) for olfactory nerves. The paired cribiform plates are separated by the **crista galli** (9), also on the ethmoid bone.

Just posterior to the frontal bone, the **lesser wing of the sphenoid bone** (7) forms an "upper shelf." The **greater wing of the sphenoid bone** (6) forms a "lower shelf." The **sphenoid body** (13) contains both the **sella turcica** (12) — a protective bony saddle for the pituitary gland, and the **optic foramen** (11) — through which the optic nerve passes.

Three foramina are seen in the greater wing of the sphenoid bone:

- **Foramen rotundum** (5)
  for the maxillary branch of the trigeminal nerve.

- **Foramen ovale** (4)
  for the mandibular branch of the trigeminal nerve.

- **Foramen spinosum** (3)
  through which the middle meningeal vessels pass.

The **internal auditory meatus** (15) is seen on the **temporal bone** (1).

Associated with the **occipital bone** (20) are the:

- **Foramen magnum** (18)
- **Hypoglossal foramen** (17)
- **Jugular foramen** (16)
- **Jugular groove** (19)

A portion of the **parietal bone** (2) is seen at the lateral edge of the cranium.

The **foramen lacerum** (14) is found laterally adjacent to the body of the sphenoid bone.

# Cranium Base

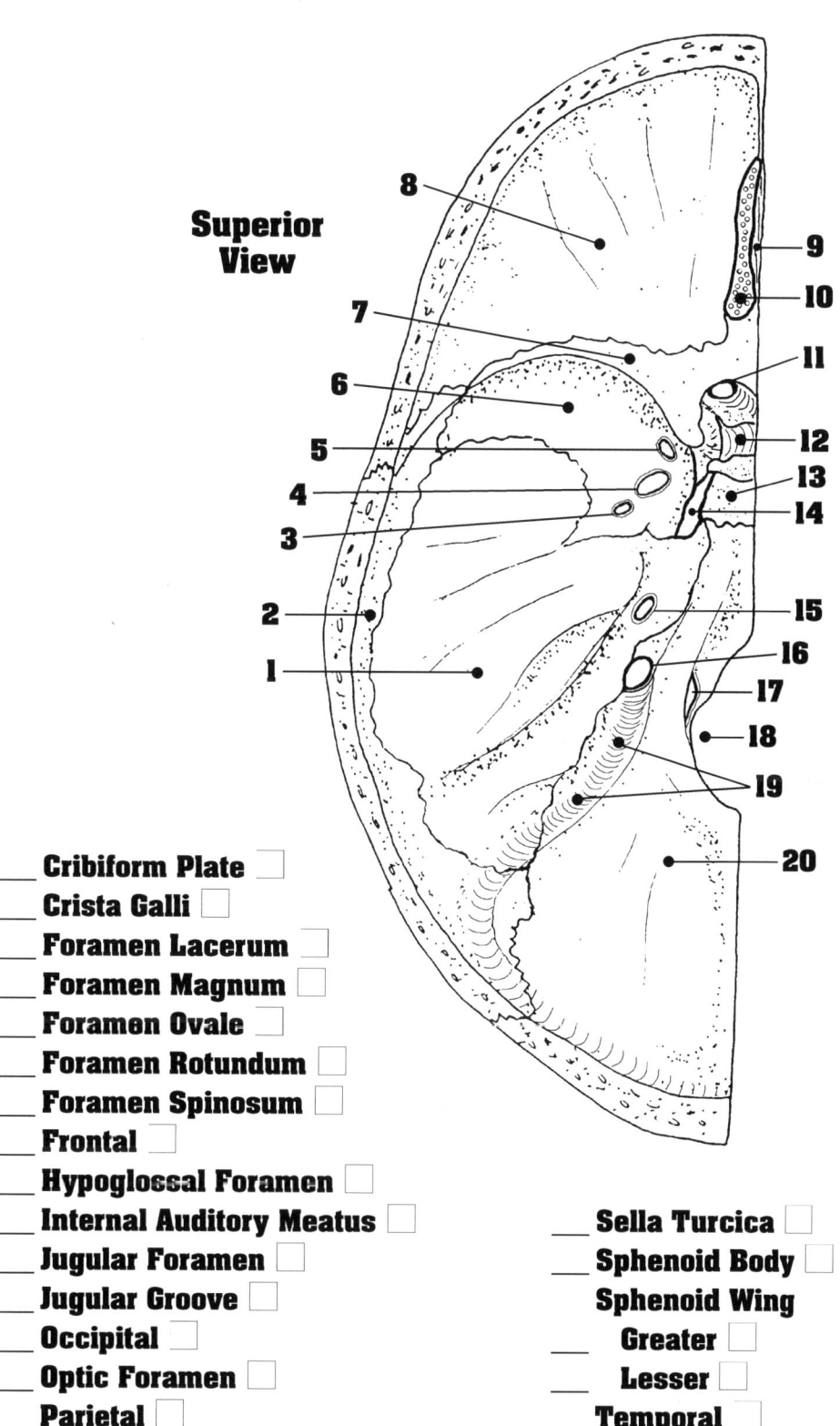

Superior View

__ Cribiform Plate
__ Crista Galli
__ Foramen Lacerum
__ Foramen Magnum
__ Foramen Ovale
__ Foramen Rotundum
__ Foramen Spinosum
__ Frontal
__ Hypoglossal Foramen
__ Internal Auditory Meatus
__ Jugular Foramen
__ Jugular Groove
__ Occipital
__ Optic Foramen
__ Parietal
__ Sella Turcica
__ Sphenoid Body
__ Sphenoid Wing
__ Greater
__ Lesser
__ Temporal

# Paranasal Sinuses

Immediately adjacent to the nasal cavity are four bones (the maxilla, frontal, ethmoid, and sphenoid) which have sinuses:

- Frontal bone — contains the **frontal sinuses** (7).
- Sphenoid bone — contains the **sphenoidal sinus** (5).
- Maxillary bones — contain the **maxillary sinuses** (11).
- Ethmoid bone — contains the **ethmoidal cells (sinus)** (6).

All the sinuses have drainage passageways into the nasal cavity.

The **orifice of the frontal sinus** (8) and the **orifice of the maxillary sinus** (10) open into the nasal cavity via a **semilunar hiatus** (9). The sphenoidal sinus has independent drainage through the **orifice of the sphenoidal sinus** (3).

The **orifice of the auditory tube** (1) is shown in the **nasopharynx** (2).

The roof of the oral cavity is comprised of the **hard** (12) and **soft** (13) **palates**.

The **sella turcica** (4) cradles the pituitary gland.

*The air quality in Riverside and San Bernardino (California) is among the worst in the U.S. More than 70% of the elementary students in those cities have some form of sinusitis. How's the air where you and your children live — or plan to live?*

# Paranasal Sinuses

**Sagittal Section**

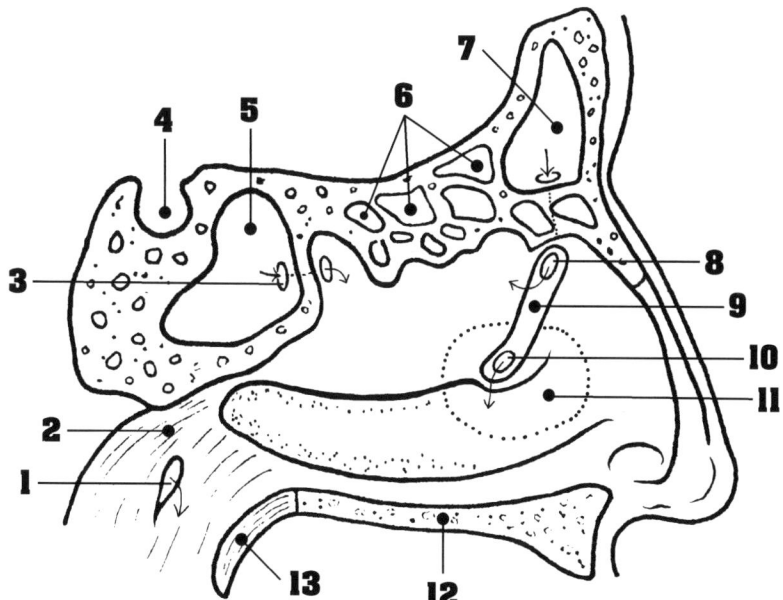

\_\_\_ **Ethmoidal Cells (Sinus)** ☐
\_\_\_ **Frontal Sinus** ☐
\_\_\_ **Maxillary Sinus** ☐
\_\_\_ **Nasopharynx** ☐
\_\_\_ **Orifice of the Auditory Tube** ☐
\_\_\_ **Orifice of the Frontal Sinus** ☐
\_\_\_ **Orifice of the Maxillary Sinus** ☐
\_\_\_ **Orifice of the Sphenoidal Sinus** ☐
    **Palate**
\_\_\_     **Hard** ☐
\_\_\_     **Soft** ☐
\_\_\_ **Sella Turcica** ☐
\_\_\_ **Semilunar Hiatus** ☐
\_\_\_ **Sphenoidal Sinus** ☐

# Mandible

The mandible has a horizontal **body** (7) and a transverse **ramus** (5). The area at the outer edge of the mandible, where the body turns upward toward the ramus, is called the **angle** (6).

Superiorly, the ramus has an anterior **coronoid process** (1) which provides attachment sites for muscles that elevate the mandible, and a posterior **condylar process** (3). The smooth, articular, superior surface of the condylar process is called the **mandibular condyle** (2). The mandibular condyle articulates with a mandibular fossa in the temporal bone above it.

A **mandibular notch** (4) is found between the coronoid and condylar processes.

**Alveolar processes** (9) extend up and between each tooth.

The **mental foramen** (8) provides a passageway for blood vessels and nerves.

# Mandible

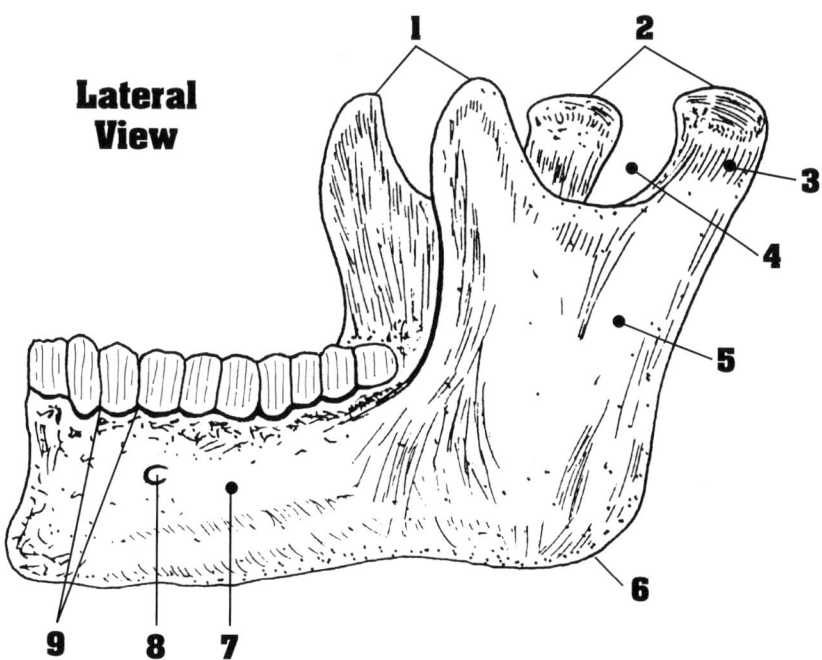

Lateral View

___ Alveolar Processes ☐
___ Angle ☐
___ Body ☐
___ Condylar Process ☐
___ Coronoid Process ☐
___ Mandibular Condyle ☐
___ Mandibular Notch ☐
___ Mental Foramen ☐
___ Ramus ☐

# Cervical & Thoracic Vertebrae

There are seven cervical vertebrae (C1 to C7), and twelve thoracic vertebrae (T1 to T12).

Seven processes extend from the posterior aspect of each vertebra:

- Two **superior articular processes** (12) extend superiorly.
- Two **inferior articular processes** (6) extend inferiorly.
- Two **transverse processes** (4) extend laterally.
- One **spinous process** (9) extends posteriorly.

**Superior articular facets** (5) on the superior articular processes articulate with **inferior articular facets** (14) on the inferior articular processes of the vertebra above.

Each vertebra has a sturdy, protective **body** (1). The vertebral arch, which extends up from the body, is comprised of two **pedicles** (2) and two **laminae** (7).

The spinal cord passes through the **vertebral foramen** (10). Spinal nerves emerge laterally through **intervertebral foramina** (15).

Cervical vertebrae have the following unique features:

- A **bifurcation** (8) on the spinous process (with the exception of C7*).
- **Transverse foramina** (3) which provide passageways for vertebral arteries and veins. All vertebrae have a large open "mouth" (vertebral foramen), but only the cervical vertebrae have "eyes" (transverse foramina) above the "mouth."
- The first two cervical vertebrae (the atlas and axis) have special features considered on another exercise.

Thoracic vertebrae have the following unique features:

- Facets for articulating with ribs. **Inferior** (16) and **superior** (11) **demifacets** on the body, and a **transverse costal facet** (13) on the transverse process.
- Relatively long, downward-pointing spinous processes.

* *The knob on the end of the spinous process of C7 can be felt by reaching around and pressing at the base of the back of the neck.*

# Cervical Vertebra

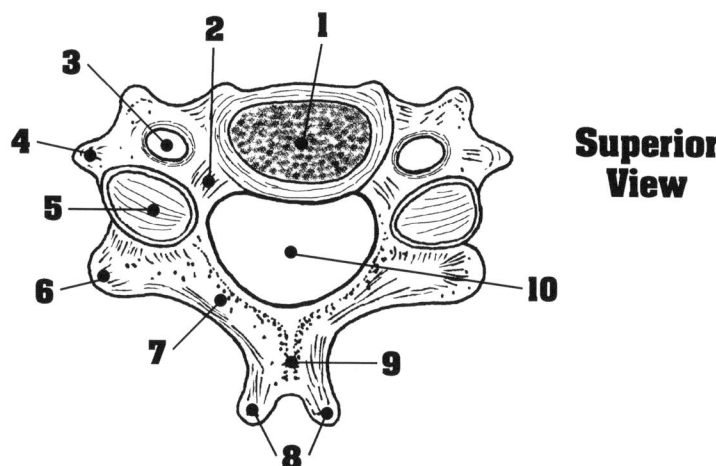

**Superior View**

# Thoracic Vertebra

**Lateral View**

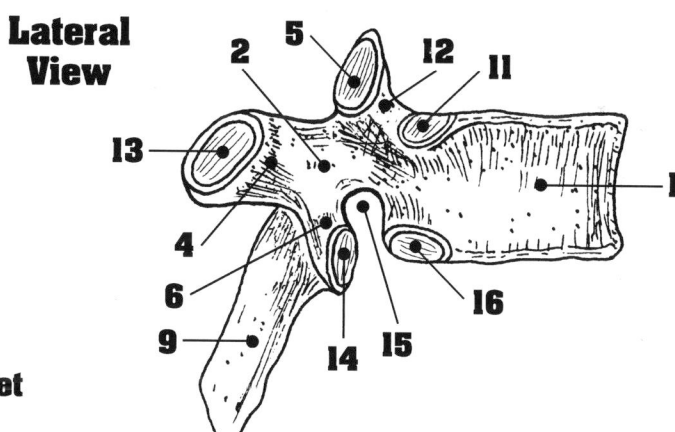

**Articular Facet**
___ Inferior ☐
___ Superior ☐

**Articular Process**
___ Inferior ☐
___ Superior ☐
___ Bifurcation ☐
___ Body ☐

**Demifacet**
___ Inferior ☐
___ Superior ☐

___ Intervertebral Foramen ☐
___ Lamina ☐
___ Pedicle ☐
___ Spinous Process ☐
___ Transverse Costal Facet ☐
___ Transverse Foramen ☐
___ Transverse Process ☐
___ Vertebral Foramen ☐

# Thoracic Vertebrae Articulations

These three articulated thoracic vertebrae not only provide an opportunity to see some features not shown on the individual vertebra, but also provide an excellent opportunity to review the processes and articulating surfaces common to all vertebrae.

Each vertebra has:

- A **body** (3)
- Two **superior articular processes** (5)
- Two **superior articular facets** (6)
- Two **inferior articular processes** (11)
- Two inferior articular facets
- Two **transverse processes** (8)
- One **spinous process** (10)

Because they articulate with the ribs, thoracic vertebrae also have:

- **Superior demifacets** (4)
- **Rib tubercle facets** (7)

An **articular joint** (9) occurs where the superior and inferior articular processes meet.

**Intervertebral discs** (1), located between the bodies of all vertebrae, act as shock absorbers and allow more flexibility in the vertebral column.

**Intervertebral foramina** (2), through which spinal nerves pass, are formed by counter-matching notches on adjacent vertebrae.

# Thoracic Vertebrae Articulations

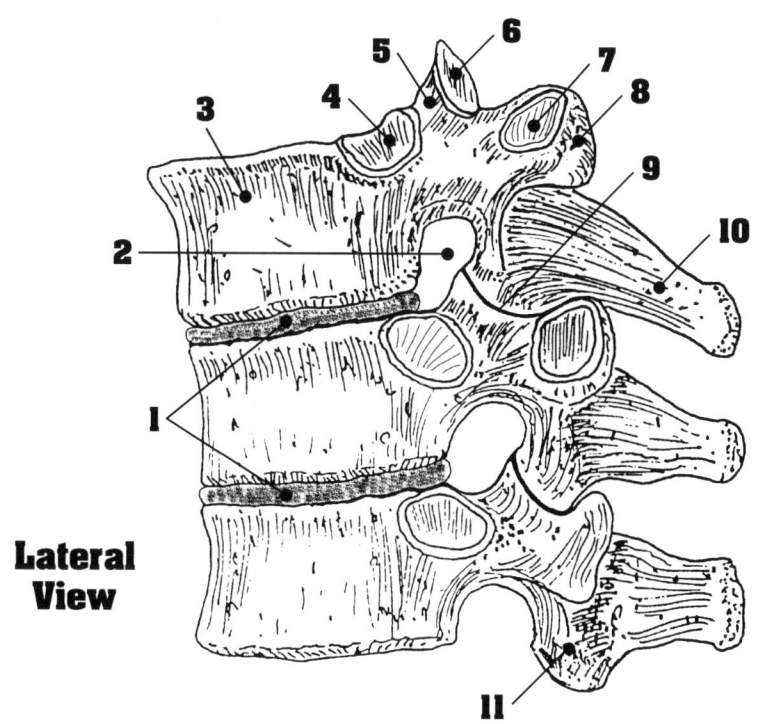

Lateral View

___ Articular Joint
    Articular Process
___    Inferior
___    Superior
___ Body
___ Intervertebral Discs
___ Intervertebral Foramen
___ Rib Tubercle Facet
___ Spinous Process
___ Superior Articular Facet
___ Superior Demifacet
___ Transverse Process

# Lumbar Vertebra

Lumbar vertebrae are distinguished from other vertebrae types by a broad, blunt **spinous process** (5), and by a significantly larger, more sturdy **body** (1).

Reviewing again features common to all vertebrae (Is not repetition the key to learning?):

- **Superior articular processes** (3)
- **Superior articular facets** (12)
- **Inferior articular processes** (7)
- **Inferior articular facets** (6)
- **Transverse processes** (4)

The **pedicles** (9) and **laminae** (8) comprise the **vertebral arch** (10). The **vertebral foramen** (11) lies within the arch.

Notches for **intervertebral foramina** (2) occur immediately posterior to the body. Intervertebral foramina provide spaces for the exiting spinal nerves.

# Lumbar Vertebra

**Lateral View**

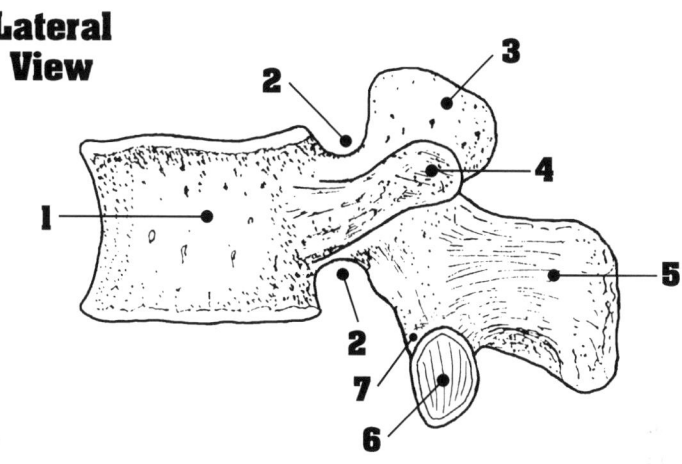

**Superior View**

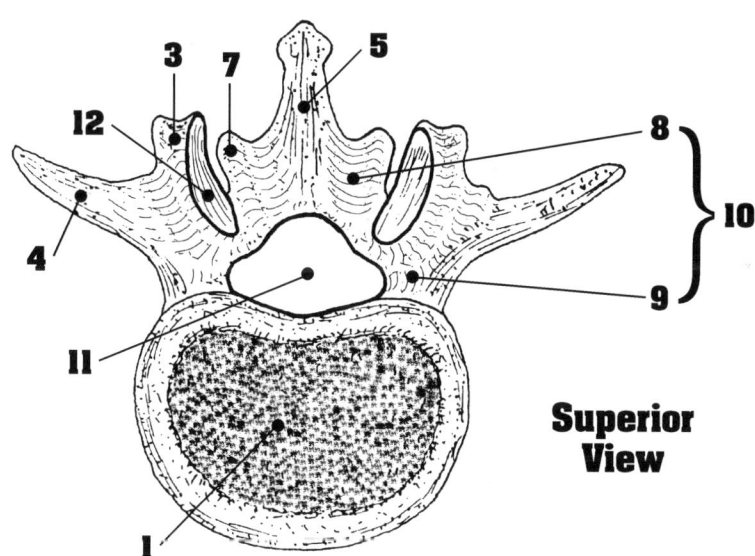

**Articular Facet**
___ Inferior
___ Superior
**Articular Process**
___ Inferior
___ Superior
___ Body

___ Intervertebral Foramina
___ Lamina
___ Pedicle
___ Spinous Process
___ Transverse Process
___ Vertebral Arch
___ Vertebral Foramen

# $C_1$ (Atlas) & $C_2$ (Axis)

Occipital condyles of the skull articulate with the **superior articular facets** (3) of the atlas. These articular surfaces are designed to allow for free movements of the skull in all directions.

To further assist, especially in the lateral movements of the skull, and to provide for the ligaments needed to support the weight of the skull during those movements, a sturdy knob called the **dens** (11) extends upward from the axis through the **anterior arch** (4) of the atlas. At the base of the dens is a **groove for the transverse ligament** (12), and on the atlas there is a **tubercle for the transverse ligament** (6) (see page 101).

As compared with other vertebrae, the atlas is unique in having an **anterior** (5) and **posterior** (9) **tubercle**, and a **groove for the vertebral artery** (7). The groove for the vertebral artery lies at the superior border of the **posterior arch** (8).

As with most other cervical vertebrae, the atlas also has **transverse processes** (1) which contain **transverse foramina** (2).

The **spinous process** (13) and the **bifurcation of the spinous process** (14) are shown on the axis.

As with all other vertebrae, the atlas and axis have a **vertebral foramen** (10), through which the _____ _____ passes?

# C₁ (Atlas) & C₂ (Axis)

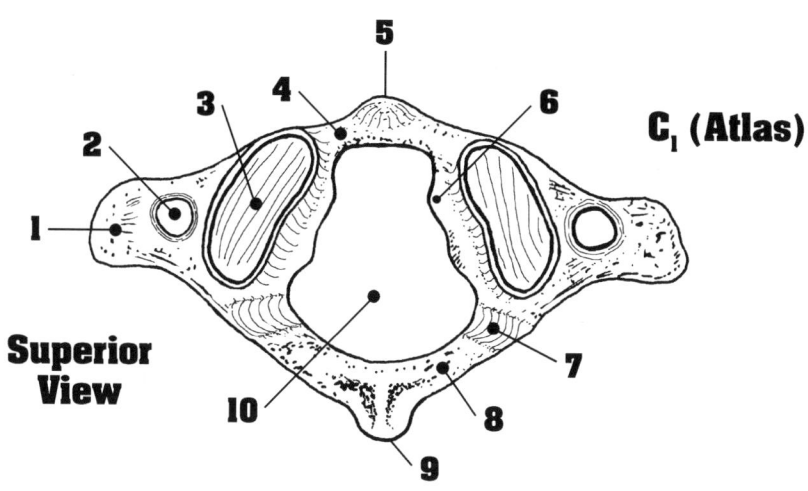

**C₁ (Atlas)** — Superior View

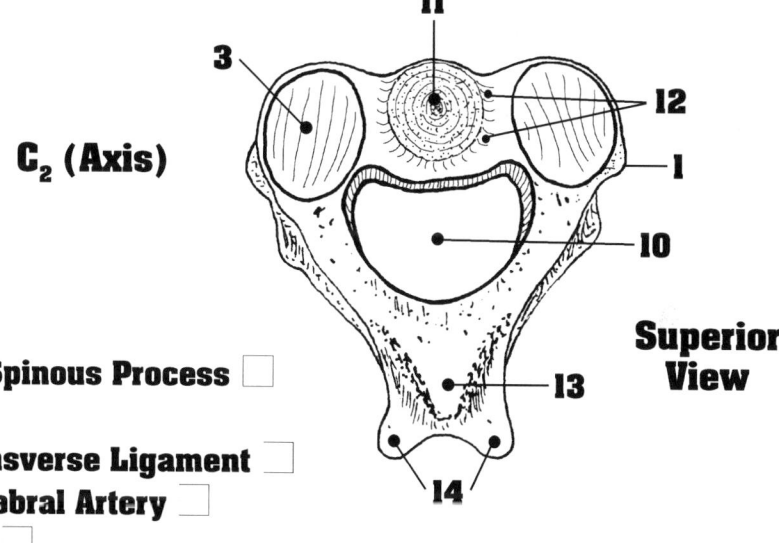

**C₂ (Axis)** — Superior View

Arch
___ Anterior
___ Posterior
___ Bifurcation of the Spinous Process
___ Dens
___ Groove for the Transverse Ligament
___ Groove for the Vertebral Artery
___ Posterior Tubercle
___ Spinous Process
___ Superior Articular Facet
___ Transverse Foramen
___ Transverse Process
Tubercle
___ Anterior
___ for the Transverse Ligament
___ Posterior
___ Vertebral Foramen

# Sacrum & Coccyx

**Transverse ridges** (2) on the anterior surface of the sacrum bear witness to the fetal fusion of five sacral vertebrae into one sacrum.

The sacrum articulates superiorly with a lumbar vertebra via three articulating surfaces: the surface of the **sacral promontory** (4) articulates with the inferior surface of the body of the lumbar vertebra immediately above it, and two **superior articular facets** (8) on two **superior articular processes** (6) articulate with the inferior articular facets of the lumbar vertebra above.

Spinal nerves enter the sacrum via a **sacral canal** (7) and emerge from the sacrum via **dorsal** (11) and **ventral** (3) **sacral foramina**.

**Auricular surfaces** (10) articulate laterally with the medial margins of the coxal (hip) bones. **Spinous tubercles** (13) arise from a **median crest** (12) on the posterior surface of the sacrum.

On the anterior surface, **sacral ala ("wings")** (5) extend outward from the promontory, and on the posterior surface, **sacral tuberosities** (9) project superiorly.

The **coccyx** (1) is comprised of four fused coccygeal vertebrae. At the coccygeosacral junction, a **sacral cornu** (14) articulates with a **coccygeal cornu** (15) to form a lateral margin of the **sacral hiatus** (16).

# Sacrum & Coccyx

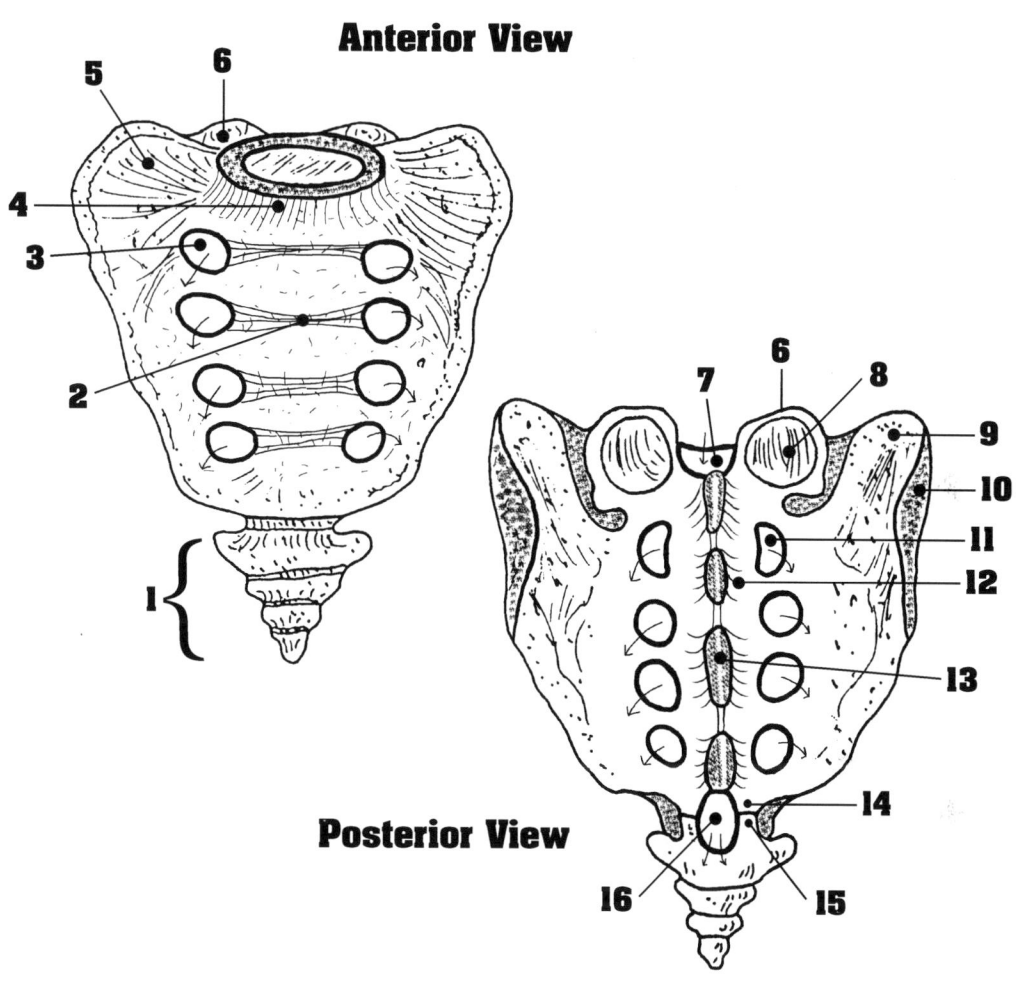

__ Auricular Surface ☐
__ Coccygeal Cornu ☐
__ Coccyx ☐
__ Median Crest ☐     __ Sacral Hiatus ☐
__ Sacral Ala (Wing) ☐     __ Sacral Promontory ☐
__ Sacral Canal ☐     __ Sacral Tuberosity ☐
__ Sacral Cornu ☐     __ Spinous Tubercle ☐
    Sacral Foramen     __ Superior Articular Facet ☐
__     Dorsal ☐     __ Superior Articular Process ☐
__     Ventral ☐     __ Transverse Ridge ☐

# Sternum

A **sternal angle** (11) separates the **manubrium (presternum)** (12) above and from the **gladiolus (mesosternum/body)** (10) below.

The superior surface of the manubrium is comprised of two **clavicular notches** (13) laterally adjacent to a central **jugular (suprasternal) notch** (14). The clavicular notches articulate with the clavicles (collar bones).

On each lateral surface of the manubrium, immediately inferior and adjacent to the clavicular notch, is a **costal (rib) notch 1** (1) which provides an articulating surface connection for the costal cartilage associated with the first rib.

The notch for the second rib — **costal notch 2** (2) — is located at the sternal angle. The next five ribs articulate with the body of the sternum as follows: **costal notch 3** (3), **costal notch 4** (4), **costal notch 5** (5), **costal notch 6** (6), and **costal notch 7** (7).

Because the first seven pairs (of the twelve pairs of ribs) have a direct connection to the sternum, they are called "true" ribs. Because the next three pairs of ribs (ribs 8, 9, and 10) have an indirect connection to costal notch 7 via longer extensions of costal cartilage, they are called "false" ribs.

The 11th and 12th pairs of ribs have no connection with the sternum and are called "floating" ribs.

A thin extension of the inferior surface of the body of the sternum called the **xiphoid process** (9) has a **xiphoid foramen** (8).

# Sternum

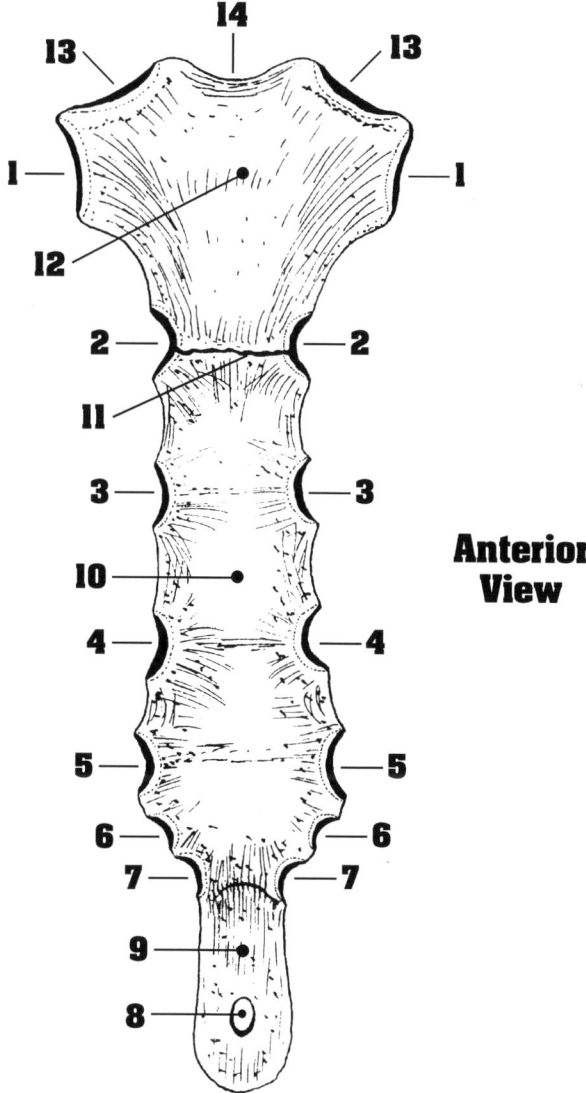

**Anterior View**

___ Costal Notch 1
___ Costal Notch 2
___ Costal Notch 3
___ Costal Notch 4
___ Costal Notch 5
___ Costal Notch 6
___ Costal Notch 7
___ Clavicular Notch
___ Gladiolus (Mesosternum/Body)
___ Jugular (Suprasternal) Notch
___ Manubrium (Presternum)
___ Sternal Angle
___ Xiphoid Foramen
___ Xiphoid Process

# Right Rib

The **vertebral end** (5) of a rib, which articulates at a juncture between two vertebrae, has a **head** (3) and **neck** (2). On the head, a **superior demifacet** (6) articulates with a superior vertebra, and an **inferior demifacet** (4) articulates with an inferior vertebra. Behind the neck an **articular tubercle** (1) provides a third point of articulation of the rib with the vertebrae. Thus the rib has a tri-faceted articular connection with the vertebrae.

Anteriorly, the rib has a **sternal end** (11) which attaches to a costal notch in the sternum via a piece of semi-flexible costal cartilage.

Each rib has a **superior border** (7), **inferior border** (10), and a **costal groove** (9). The costal groove provides a protected passageway for blood vessels and nerves.

The point of greatest curvature at the lateral margin is called the **angle** (8) of the rib.

Recall that there are seven pairs of true ribs, three pairs of false ribs, and two pairs of floating ribs.

*Reflect upon the tri-faceted articulation of the ribs with the vertebrae, which allows for the expansion and contraction of the rib cage in breathing — while at the same time providing a rigid protection for the lungs and heart.*

# Right Rib

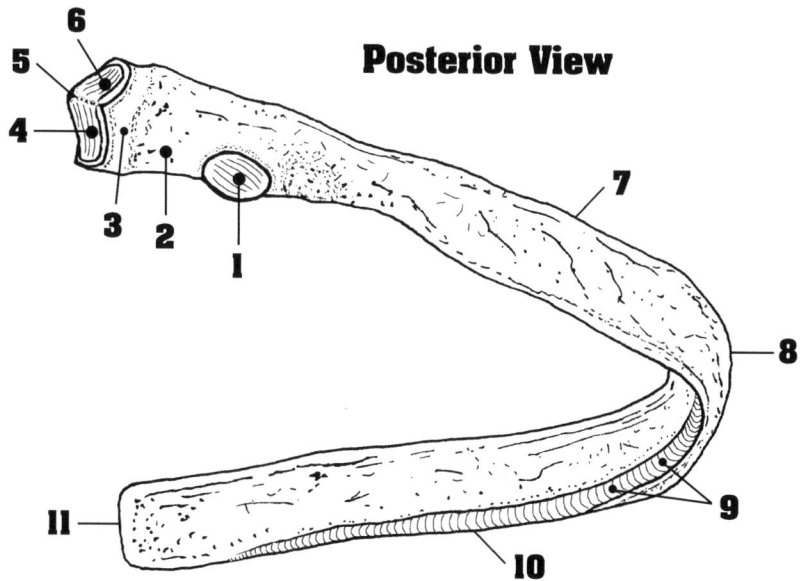

**Posterior View**

___ Angle ☐
___ Articular Tubercle ☐
**Border**
___ Inferior ☐
___ Superior ☐
___ Costal Groove ☐
**Demifacet**
___ Inferior ☐
___ Superior ☐
___ Head ☐
___ Neck ☐
___ Sternal End ☐
___ Vertebral End ☐

# Right Scapula

The **body** (10) of the scapula is bounded by three margins: a **vertebral (medial) margin** (9), an **axillary (lateral) margin** (4), and a **superior margin** (5). The vertebral margin gives way to the superior margin at the **superior angle** (6), and to the lateral margin at the **inferior angle** (11).

The superior margin has a **scapular notch** (12) and a **coracoid process** (2). The superior aspect of the axillary margin is modified into a **glenoid cavity (fossa)** (3) which articulates with the head of the humerus bone.

Posteriorly, a **spine** (13) projects laterally and gives rise to an **acromion process** (1). Above the spine there is a **supraspinous fossa** (7), and below the spine an **infraspinous fossa** (8).

*The scapula, because of its complex architecture, is sometimes referred to as a "highly irregular bone." But when we come to understand how many shoulder and upper back muscles have their origin or insertion on the scapula we are more inclined to call it "a poem in bone" (see page 119).*

# Right Scapula

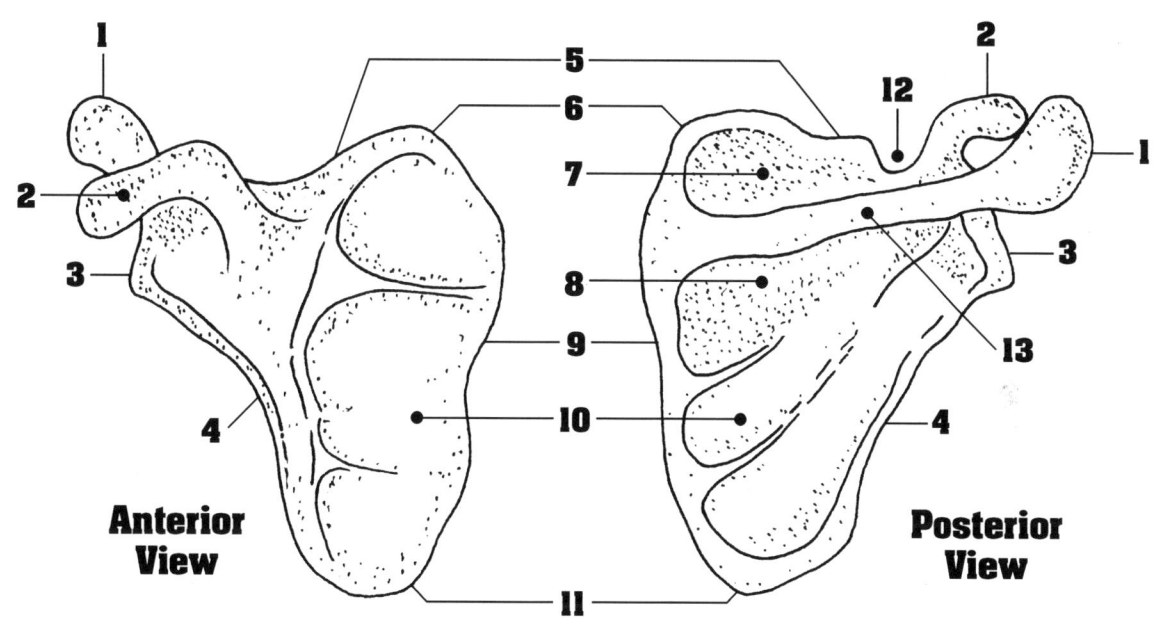

**Anterior View**

**Posterior View**

___ Acromion Process
___ Body
___ Coracoid Process
___ Glenoid Cavity (Fossa)
___ Inferior Angle
___ Infraspinous Fossa
   Margin
   ___ Axillary (Lateral)
   ___ Superior
   ___ Vertebral (Medial)
___ Scapular Notch
___ Spine
___ Superior Angle
___ Supraspinous Fossa

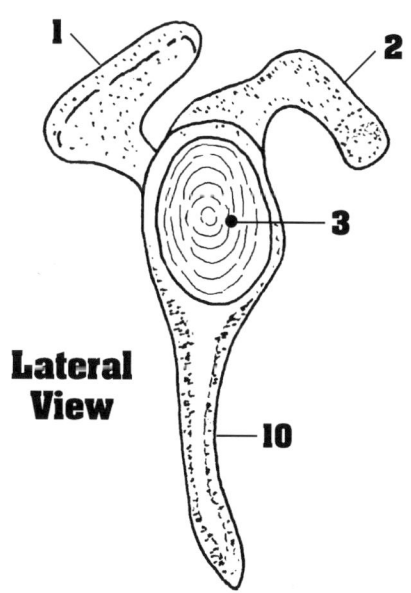

**Lateral View**

# Right Clavicle

The **sternal end** (4) of the clavicle articulates with the sternum by way of a **sternal facet** (5). The **acromial end** (3) of the clavicle articulates with the acromion process by way of an **acromial facet** (2).

A **conoid tubercle** (1) provides sites for muscle attachments.

*A superior view? Yes, that means you are looking down, from the third floor, at the right shoulder of someone passing by.*

*The inferior view? Well, you must be flat on your back, under some grating in the sidewalk, looking up at the right shoulder of someone passing over you.*

*Be warned: a tricky teacher might use this tricky bone to trick you.*

# Right Clavicle

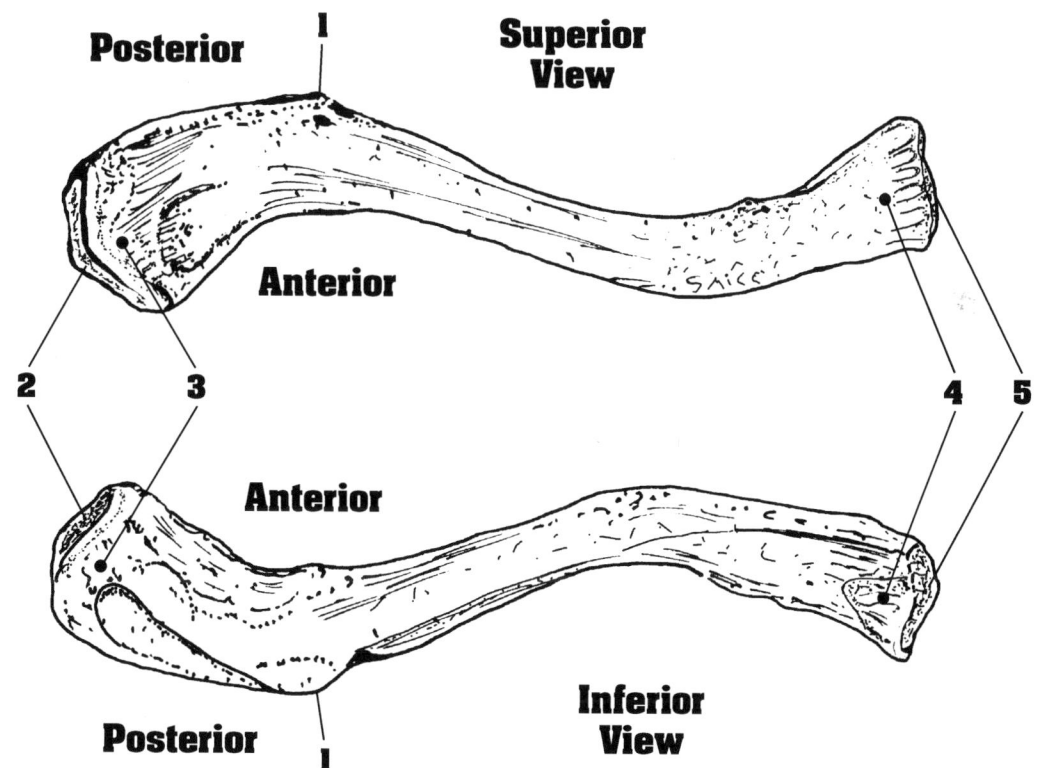

___ Acromial End
___ Acromial Facet
___ Conoid Tubercle
___ Sternal End
___ Sternal Facet

# Left Humerus

At the proximal end of the humerus, the rounded **head** (6) is separated from a **greater tubercle** (1) by an **anatomical neck** (7). There is a **surgical neck** (2) just inferior to the anatomical neck.

In the anterior view an **intertubercular groove** (12) is seen between the greater and **lesser tubercle** (8).

The rounded head of the humerus articulates with the glenoid cavity of the scapula (see page 77). The intertubercular groove provides a passageway for one of the tendons of the biceps muscle.

At the distal end of the humerus the **trochlea** (11) articulates with the proximal end of the ulna bone of the lower arm, and the **capitulum** (14) articulates with the distal end of the radius bone of the lower arm.

The **olecranon fossa** (4) which, when the lower arm is flexed, allows an inset for the olecranon process of the ulna, is located on the anterior aspect of the distal end. On the posterior aspect of the distal end of the ulna are found the **coronoid fossa** (9) and the **radial fossa** (13). When the lower arm is flexed, these fossae allow insets for the coronoid process of the ulna and the head of the radius.

A **lateral** (5) and **medial** (10) **epicondyle** are also found at the distal end.

About midway along the shaft of the humerus a small **deltoid tuberosity** (3) provides a site of attachment for the tendon of the deltoid muscle.

*Whereas condyles are smooth, rounded articulating surfaces, epicondyles are more roughened "bumps" for tendon attachments. The head of the humerus, the trochlea, and the capitulum are highly specialized condyles. As you study the proximal ends of the bones of the lower arm on the next exercise, you will want to refer back to the details of the distal end of the humerus on this exercise.*

# Left Humerus

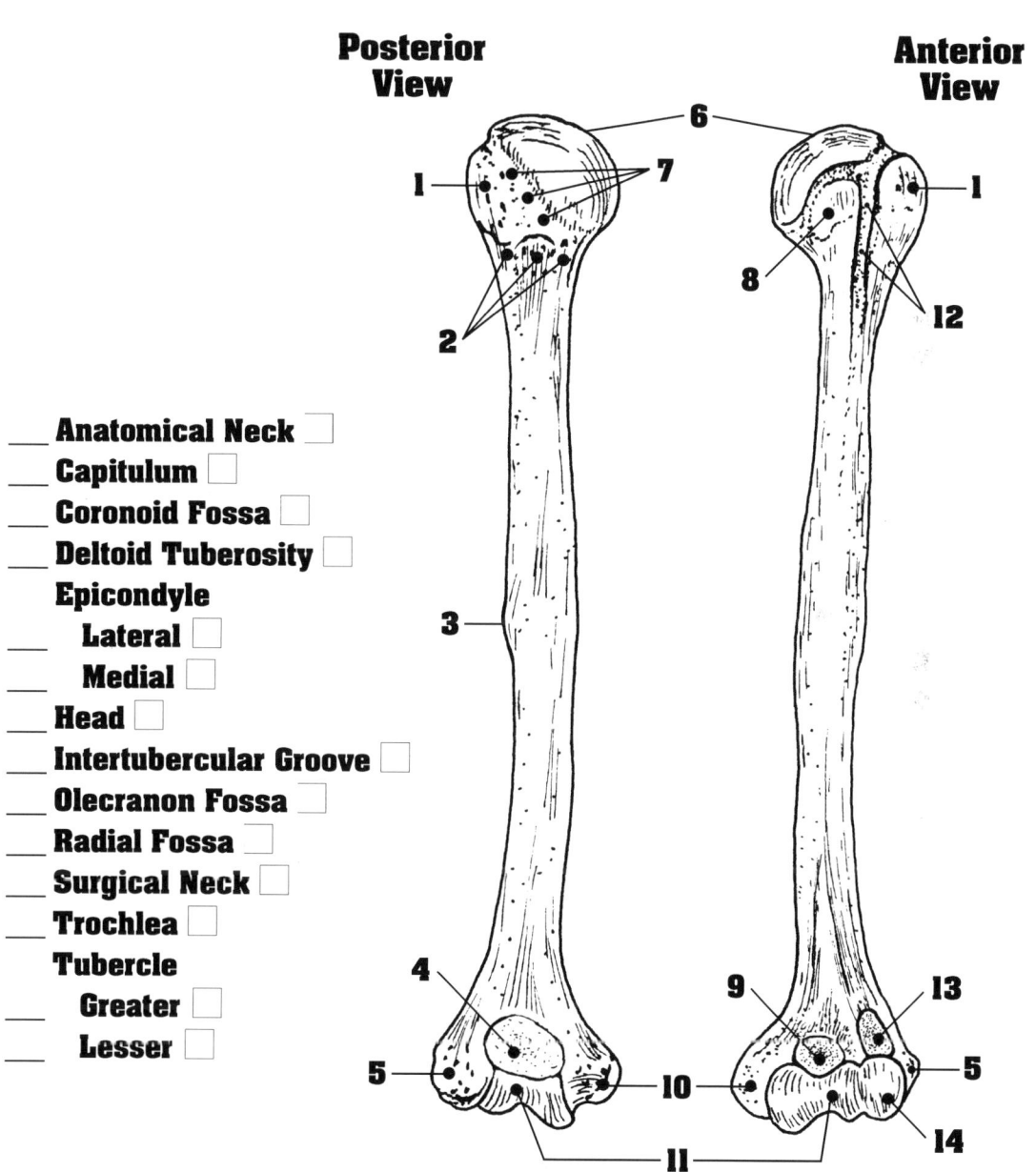

Posterior View    Anterior View

___ Anatomical Neck
___ Capitulum
___ Coronoid Fossa
___ Deltoid Tuberosity
    Epicondyle
___   Lateral
___   Medial
___ Head
___ Intertubercular Groove
___ Olecranon Fossa
___ Radial Fossa
___ Surgical Neck
___ Trochlea
    Tubercle
___   Greater
___   Lesser

# Lower Left Arm Bones

In the "anatomical position" (arms extended to the sides and "thumbs out"), the **radius** (9) is lateral to the **ulna** (2). The radius, therefore, articulates with the wrist bones (carpals) on the thumb side.

The **trochlear notch** (5) at the proximal end of the ulna articulates with the trochlea at the distal end of the humerus. When the arm is flexed the "lower jaw," or **coronoid process** (6) insets into the coronoid fossa of the humerus, and when the lower arm is extended the "upper jaw," or **olecranon process** (4) is inset in the olecranon fossa of the humerus.

And, as the lower arm is flexed, the **head of the radius** (7) slides over the capitulum of the humerus. When the arm is extended the head of the radius is inset in the radial fossa of the humerus.

When the lower arm is rotated, as when turning a doorknob, or screwdriver, the head of the radius rotates in the **radial notch** (11) of the ulna.

The **neck of the radius** (8) is immediately inferior to the head of the radius and the **radial tuberosity** (1) is inferior to the neck.

The condylar distal ends of the bones are shaped to articulate with carpal (wrist) bones and each bone has a specialized epicondyle for muscle attachments. These specialized epicondyles are the **styloid process of the radius** (10) and the **styloid process of the ulna** (3).

*Review the elbow articulations by obtaining humerus, radius, and ulna bones, and articulating them together under a beech tree at your local cemetery.*

# Lower Left Arm Bones

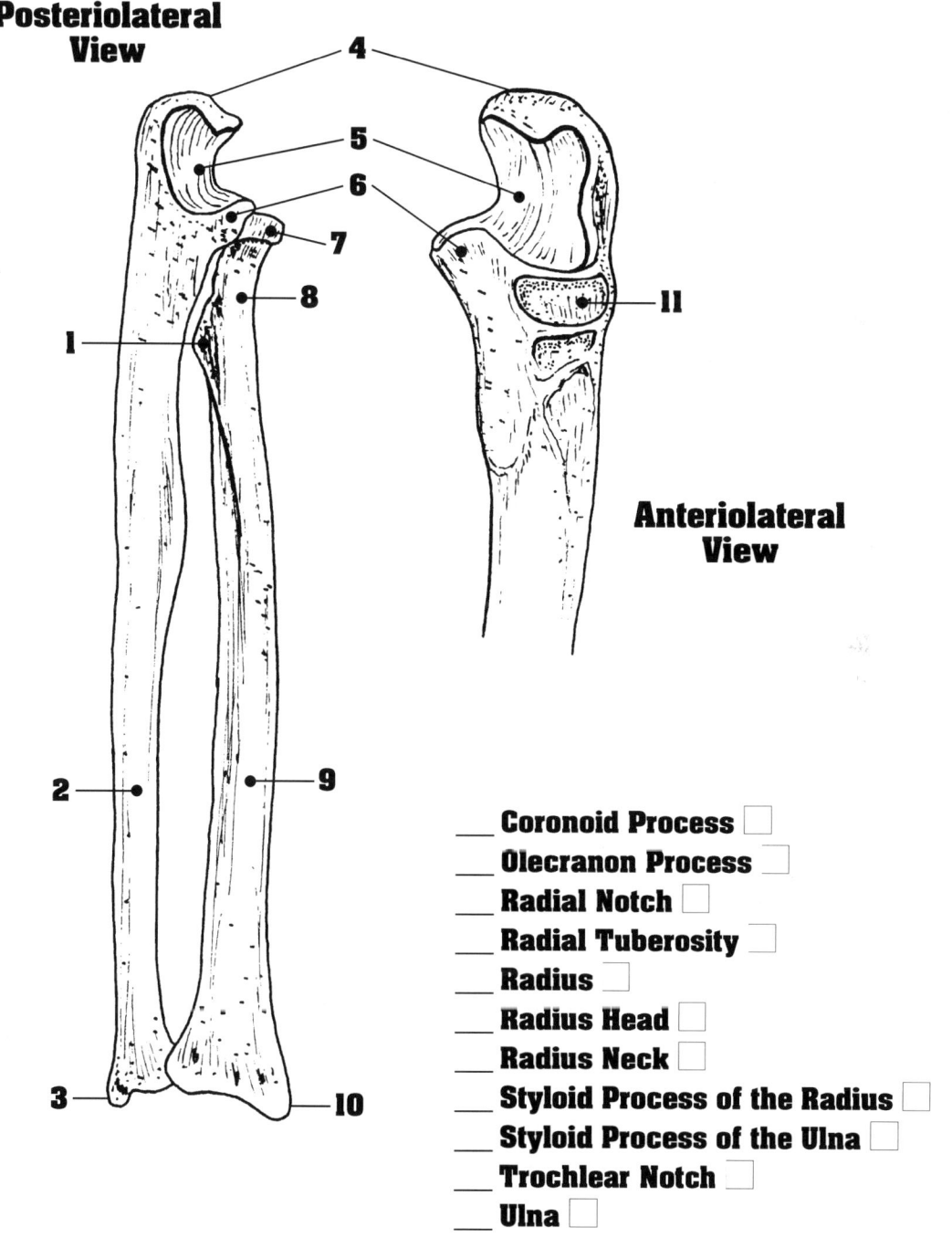

___ Coronoid Process
___ Olecranon Process
___ Radial Notch
___ Radial Tuberosity
___ Radius
___ Radius Head
___ Radius Neck
___ Styloid Process of the Radius
___ Styloid Process of the Ulna
___ Trochlear Notch
___ Ulna

# Left Wrist & Hand

The wrist is supported by eight carpal bones arranged in two rows, with four bones in each row.

Starting on the lateral (thumb side), the carpal bones in the proximal row are the:

- **Scaphoid** (6)
- **Lunate** (7)
- **Triquetrum** (8)
- **Pisiform** (9)

Starting again on the thumb side, the carpal bones in the distal row are the:

- **Trapezium** (4)
- **Trapezoid** (5)
- **Capitate** (10)
- **Hamate** (11)

The hand is supported by the **first** (3), **second** (2), **third** (1), **fourth** (13), and **fifth** (12) **metacarpals**.

Each finger is supported by a **proximal** (14), **medial** (15), and **distal** (16) **phalange**.

The thumb is supported by a proximal and distal phalange. There is no medial phalange in the thumb.

# Left Wrist & Hand

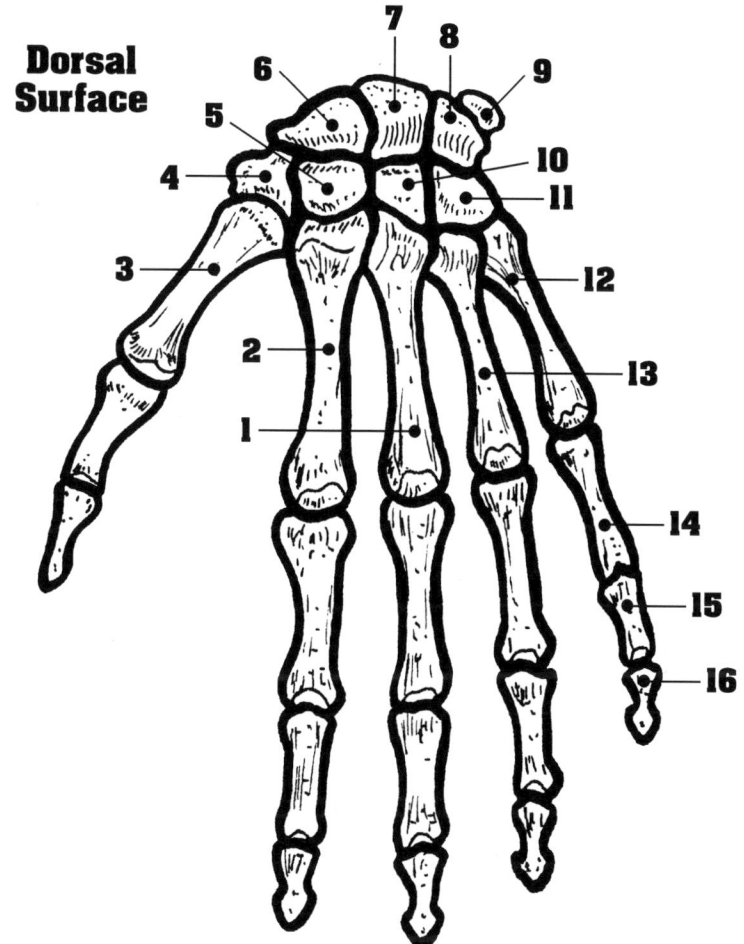

Dorsal Surface

___ Capitate
___ Hamate
___ Lunate
   Metacarpal
___  First
___  Second
___  Third
___  Fourth
___  Fifth

   Phalange
___  Distal
___  Medial
___  Proximal
___ Pisiform
___ Scaphoid
___ Trapezium
___ Trapezoid
___ Triquetrum

# Left Coxa

Each coxa (hip bone) is comprised of three fused bones: the ilium, ischium, and pubis. The three bones are fused in the region of the **acetabulum** (15) which forms a socket into which the head of the femur articulates.

The ilium comprises all the bony structure superior to the acetabulum. An **iliac crest** (7) comprises the superior border of the ilium. The posterior iliac margin has two spines: the **posterior superior iliac spine** (1) and the **posterior inferior iliac spine** (2). The anterior iliac margin also has two spines: the **anterior superior iliac spine** (9) and the **anterior inferior iliac spine** (10). An **iliac fossa** (8) is found on the medial aspect.

Each ischium has an **ischial spine** (4) and an **ischial tuberosity** (6) which, like all spines and tuberosities, provide irregular surfaces for muscle attachments.

Each pubis bone has a **superior** (11) and **inferior** (14) **pubic ramus**. The two pubic bones meet at their **symphyseal surfaces** (13).

The space between the posterior inferior iliac spine and the ischial spine forms a **greater sciatic notch** (3). A smaller space between the ischial spine and the ischial tuberosity forms the **lesser sciatic notch** (5).

The **obturator foramen** (12), a large opening surrounded by the ischium and pubis, is covered by an obturator membrane to which several muscles are attached.

# Left Coxa

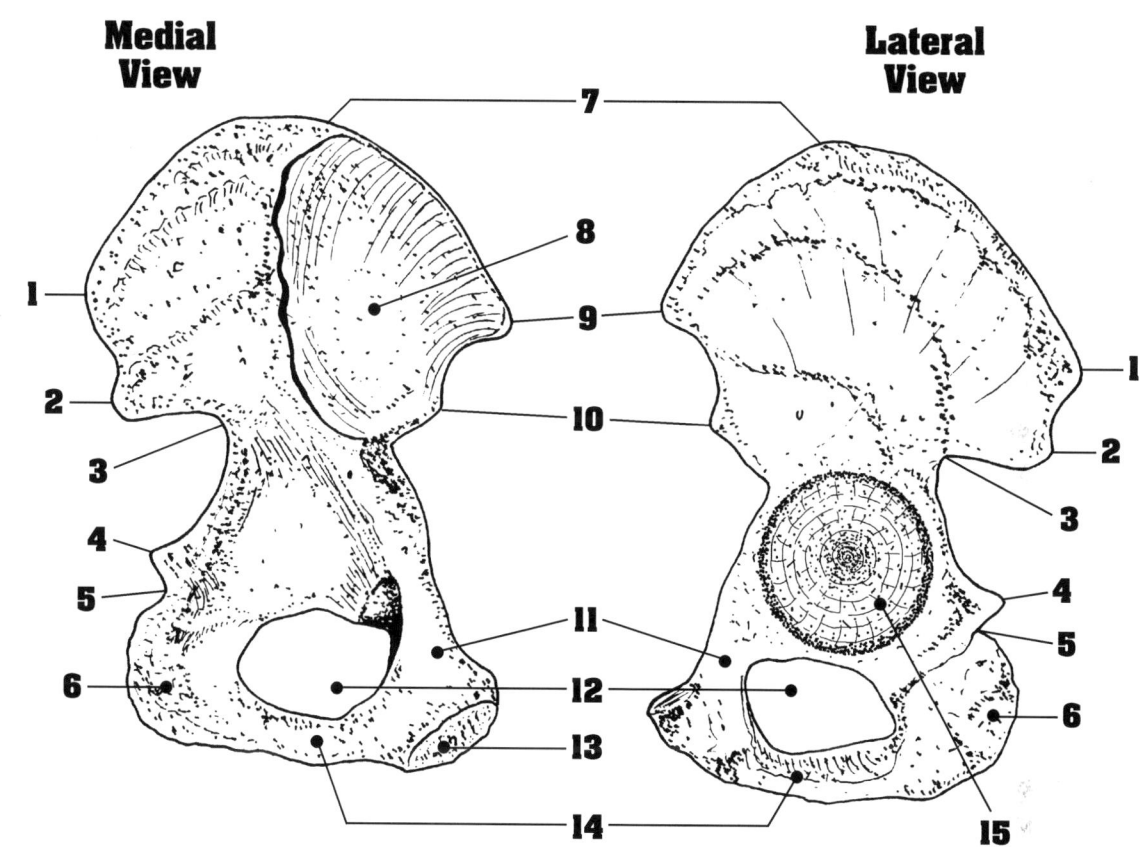

**Medial View**     **Lateral View**

___ Acetabulum  
    **Anterior Iliac Spine**  
___ Inferior  
___ Superior  
___ Iliac Crest  
___ Iliac Fossa  
___ Ischial Spine  
___ Ischial Tuberosity  
___ Obturator Foramen  

    **Posterior Iliac Spine**  
___ Inferior  
___ Superior  
    **Pubic Ramus**  
___ Inferior  
___ Superior  
    **Sciatic Notch**  
___ Greater  
___ Lesser  
___ Symphyseal Surface

# Left Femur

At the proximal end of the femur a prominent **neck** (8) supports the **head** (7). At the base of the neck, from the posterior aspect, an **intertrochanteric crest** (2) separates the **greater trochanter** (1) from the **lesser trochanter** (9). From the anterior aspect an **intertrochanteric line** (12) is seen between the greater and lesser trochanters.

At the distal end of the femur, on the posterior aspect, an **intercondylar fossa** (6) is located between the **lateral** (5) and **medial** (11) **condyles**.

**Lateral** (4) and **medial** (10) **epicondyles** are superiolateral to the condyles.

The **patellar articular surface** (13) is found at the distal end of the femur, on the anterior aspect.

The **linea aspera** (3), a roughened ridge for muscle attachments, is found on the posterior of the femur shaft.

# Left Femur

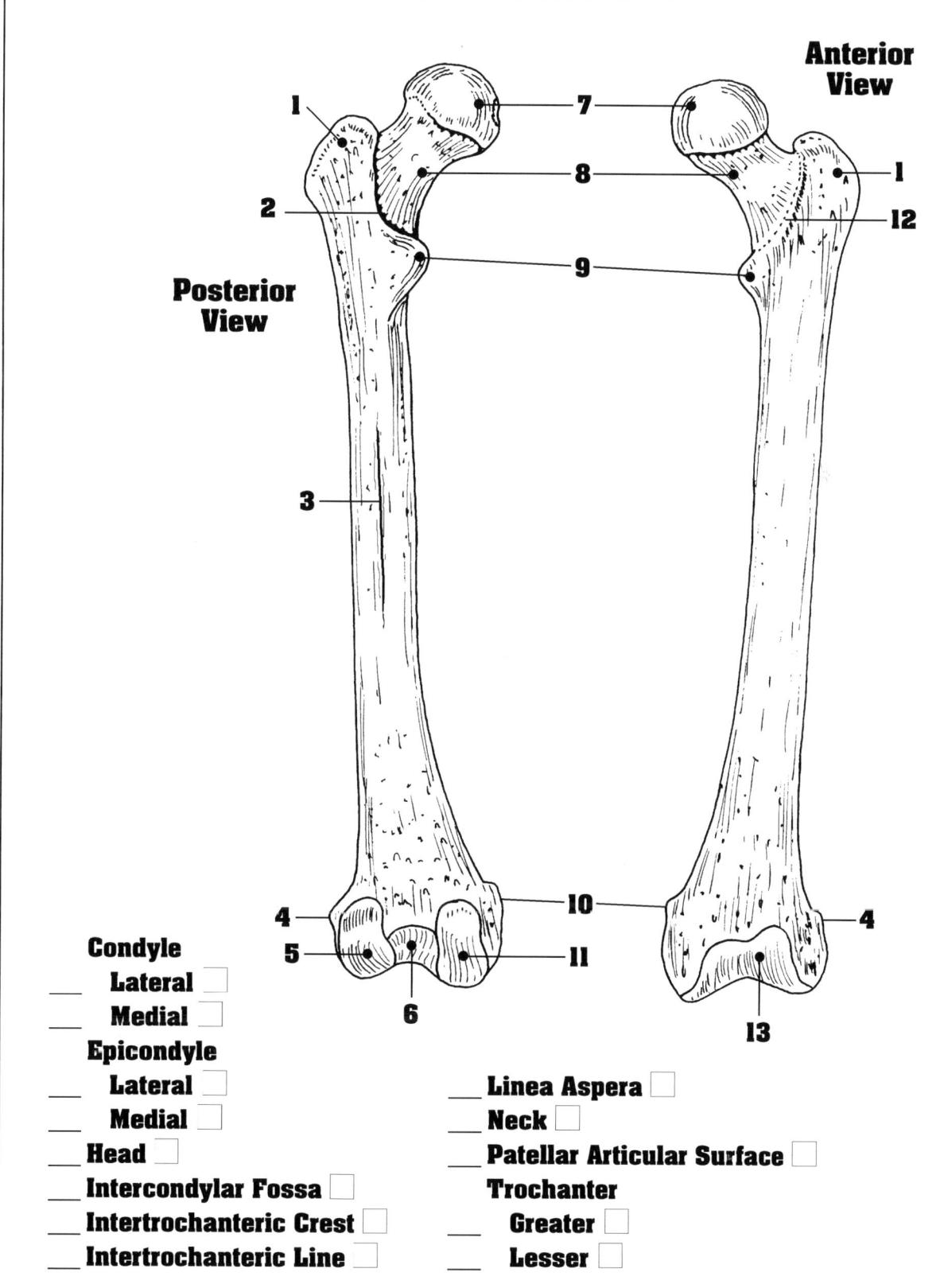

**Condyle**
- Lateral
- Medial

**Epicondyle**
- Lateral
- Medial

Head
Intercondylar Fossa
Intertrochanteric Crest
Intertrochanteric Line

Linea Aspera
Neck
Patellar Articular Surface

**Trochanter**
- Greater
- Lesser

# Lower Left Leg Bones

At the proximal end of the tibia, **lateral** (7) and **medial** (5) **condyles** are separated by an **intercondylar eminence** (6). Medially, and just inferior to the two condyles, is a **tibial tuberosity** (4), which provides an attachment point for extensor muscles in the upper leg.

The **head of the fibula** (8) articulates medially with the posteriolateral aspect of the head of the tibia. The **shaft of the fibula** (9) is lateral to the **shaft of the tibia** (3). In the area we refer to as the "shin bone," a sharp **anterior crest** (2) runs along the anterior aspect of the shaft of the tibia.

Distally, where the lower leg bones articulate with the tarsal bones of the ankle, a **medial malleolus** (1) extends from the tibia, and a **lateral malleolus** (10) extends from the fibula.

*The malleoli of the lower leg bones, like the styloid processes of the lower arm bones and the trochanters of the femur bones, are specialized epicondyles for muscle attachments.*

# Lower Left Leg Bones

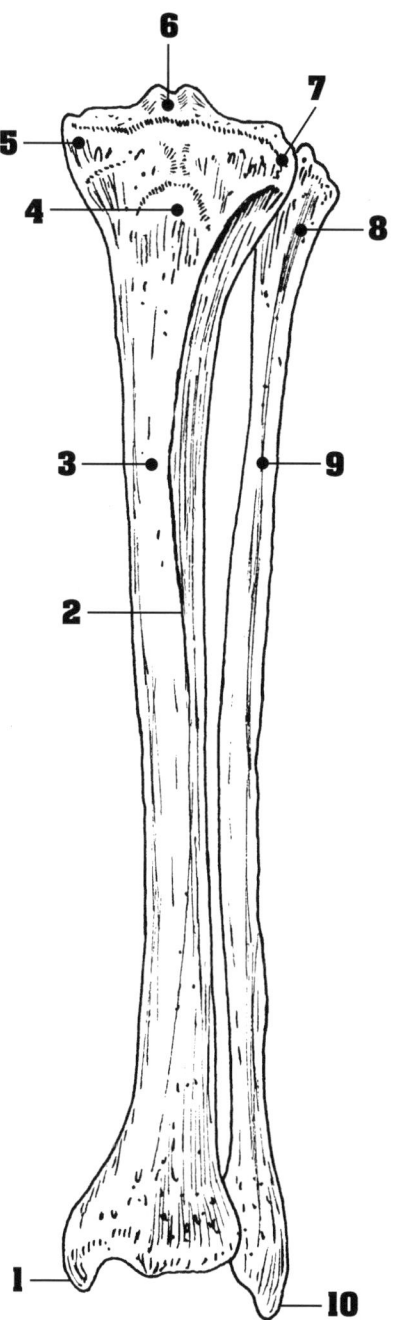

**Anterior View**

___ **Anterior Crest**
**Condyle**
___ **Lateral**
___ **Medial**
**Fibula**
___ **Head**
___ **Shaft**
___ **Intercondylar Eminence**
**Malleolus**
___ **Lateral**
___ **Medial**
___ **Tibia Shaft**
___ **Tibial Tuberosity**

# Left Ankle & Foot

There are seven tarsal (ankle) bones. The distal ends of the tibia and fibula articulate with the **talus** (7). The **calcaneus** (8), largest of the tarsal bones, forms the heel of the foot. Anterior to the talus is the block-shaped **navicular bone** (6).

The remaining four tarsal bones form a distal row which articulate with the metatarsals. From the medial, or "big toe" side, they are the **first** (4), **second** (5), and **third** (10) **cuneiform bones**, and the **cuboid bone** (9).

Just anterior to the distal row of tarsal bones are the **first** (3), **second** (2), **third** (1), **fourth** (12), and **fifth** (11) **metatarsal bones**.

Each toe is supported by a **proximal** (13), **medial** (14), and **distal** (15) **phalange**. The "big toe," like the thumb, has only a proximal and distal phalange.

# Left Ankle & Foot

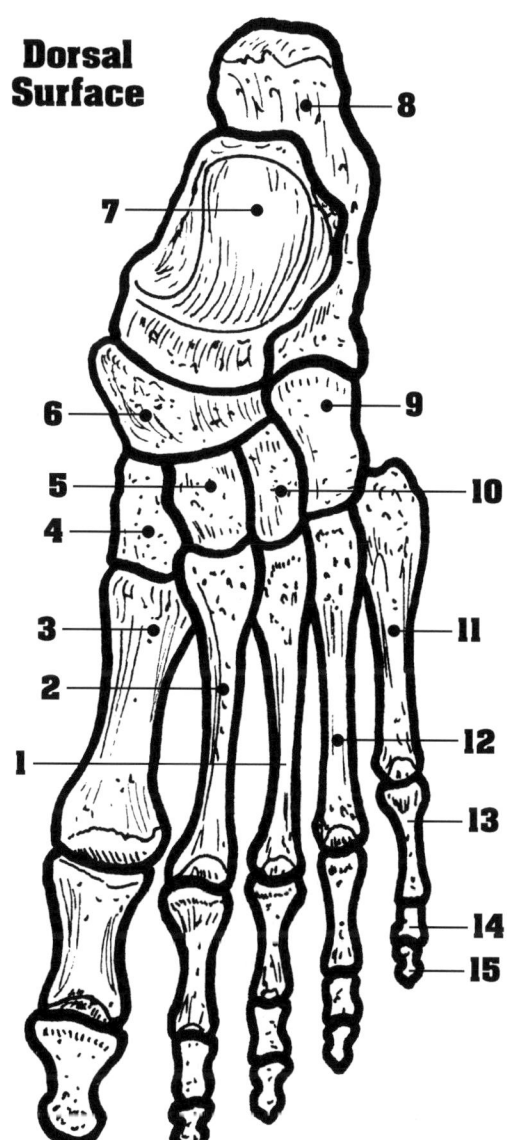

Dorsal Surface

___ Calcaneus
___ Cuboid
**Cuneiform**
___ First
___ Second
___ Third
**Metatarsal**
___ First
___ Second
___ Third
___ Fourth
___ Fifth
___ Navicular
**Phalange**
___ Distal
___ Medial
___ Proximal
___ Talus

# Articulations (Joints)

Although the joints (articulations) do not comprise a separate body system they are usually studied in a separate chapter, and we think they merit a "separate" introduction.

Although in most cases joints give our skeleton flexibility and mobility, in other cases, such as the cranium, the joints serve a protective role by disallowing flexibility and mobility.

## Classification by Function

- **Synarthroses** — immovable
- **Amphiarthroses** — slightly movable
- **Diarthroses** — freely movable

## Classification by Structure

- **Fibrous joints (sutures, syndesmoses, and gomphoses):**
  These joints have no cavity. Articulating bones are joined by tight-binding fibrous tissues. Most fibrous joints are immovable. The tightly knit wavy sutures which connect and fuse cranial bones into a single protective unit are highly specialized fibrous joints.

- **Cartilaginous joints (synchondroses and symphyses):**
  Although the cartilaginous tissues which bind the articulating bones of these joints are more flexible than the fibrous tissues of fibrous joints, these joints, like fibrous joints, have no cavity and therefore allow relatively little movement.

- **Synovial joints:**
  In synovial joints articulating bones are separated by a cavity filled with synovial fluid. Fluid-filled synovial joints allow considerable freedom of movement. Most body joints, including all limb joints are of this type.

# Articulations (Joints)

Use the pictures and descriptions in your textbook to match the types of movements with their appropriate names.

___ Abduction
___ Adduction
___ Circumduction
___ Depression
___ Dorsiflexion
___ Elevation
___ Eversion
___ Extension
___ Flexion
___ Hyperextension
___ Inversion
___ Plantar Flexion
___ Pronation
___ Protration
___ Retraction
___ Rotation
___ Supination

A. Bend the foot downward to "point the toes"
B. Bend the foot upward toward the knee
C. Bow the head
D. Distal end of the limb moves in a circle
E. Drop the jaw
F. Jut out the lower jaw
G. Look up at the stars
H. Pull jutted out lower jaw back in
I. Pull legs together after dong "the splits"
J. Raise an arm laterally
K. Raise the head (after bowing)
L. Raise the lower jaw (after dropping it)
M. Shake the head to say "no"
N. Turn a screw to the left
O. Turn a screw to the right
P. Turn the foot inward medially
Q. Turn the foot outward laterally

# Right Shoulder Joint
## (Glenohumeral)

The head of the **humerus** (1) is bound to the **scapula** (12) by a **joint (articular) capsule** (11) and by the **coracohumeral ligament** (4).

The **coracoid process** (10) of the scapula is bound to the **acromion process** (6) of the scapula by the **coracoacromial ligament** (5).

The **clavicle** (8) is bound to the acromion process by the **acromioclavicular ligament** (7), and the clavicle is bound to the coracoid process of the scapula by the **coracoclavicular ligaments** (9).

A **tendon of the biceps** (2) is anchored in the **intertubercular groove** (13) of the humerus by **transverse humeral ligaments** (3).

# Right Shoulder Joint
## (Glenohumeral)

**Anterior View**

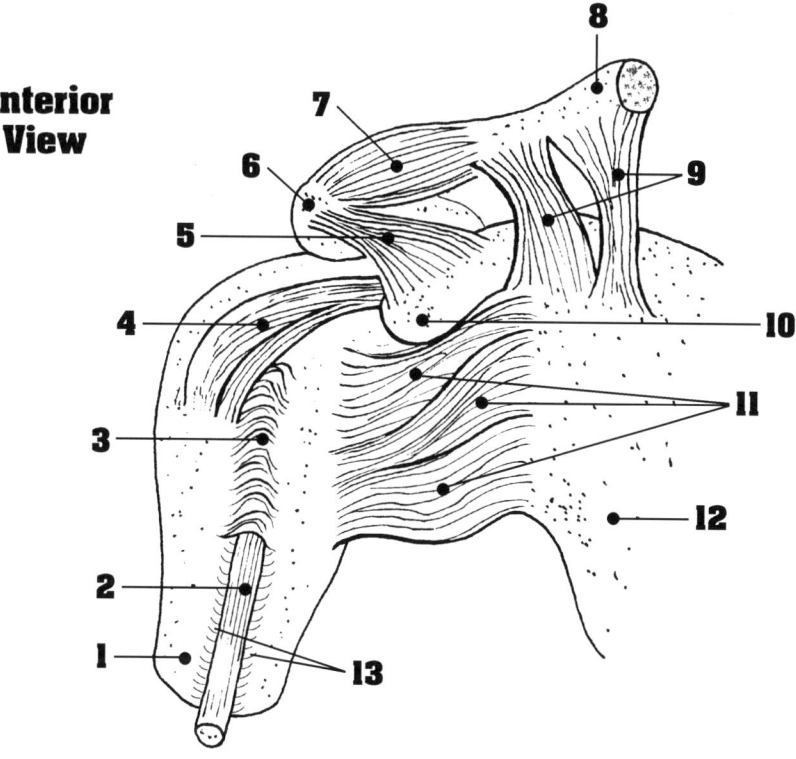

___ Acromion Process ☐
___ Clavicle ☐
___ Coracoid Process ☐
___ Humerus ☐
___ Intertubercular Groove ☐
___ Joint (Articular) Capsule ☐
Ligaments
___   Acromioclavicular ☐
___   Coracoacromial ☐
___   Coracoclavicular ☐
___   Coracohumeral ☐
___   Transverse Humeral ☐
___ Scapula ☐
___ Tendon of the Biceps ☐

© Copyright 2010 Gene Johnson

# Left Knee Joint
## (Tibiofemoral)

The outer fibrous capsule and bursae have been removed to show the inner ligament connections and configurations in this complex synovial joint.

**Medial** (5) and **lateral** (11) **condyles** of the **femur** (12) articulate with articular surfaces on the head of the **tibia** (1). However, the articulation is not direct as in most joints, for there are two protective fibro-cartilaginous menisci between the articular surfaces of the two bones: a **lateral meniscus** (9) and a **medial meniscus** (2).

The knee joint is supported laterally by a **fibular collateral ligament** (8) and medially by a **tibial collateral ligament** (3).

Three additional ligaments support the knee joint in the region of the **intercondylar groove** (13): **anterior** (10) and **posterior** (4) **cruciate ligaments** form a cross in the middle, and a **posterior meniscofemoral ligament** (14) arises from the posterior aspect of the groove.

The **fibula** (7) articulates with the posteriolateral aspect of the head of the tibia.

On the anterior aspect of the distal end of the femur the "knee cap" (patella) articulates with the **patellar articular surface** (6).

*… And so from now on, when you read the sports page, you will have a "knee up" on the public at large, for you will know what they mean when they write, "He tore his ACL," or "She tore her medial collateral ligament."*

# Left Knee Joint
## (Tibiofemoral)

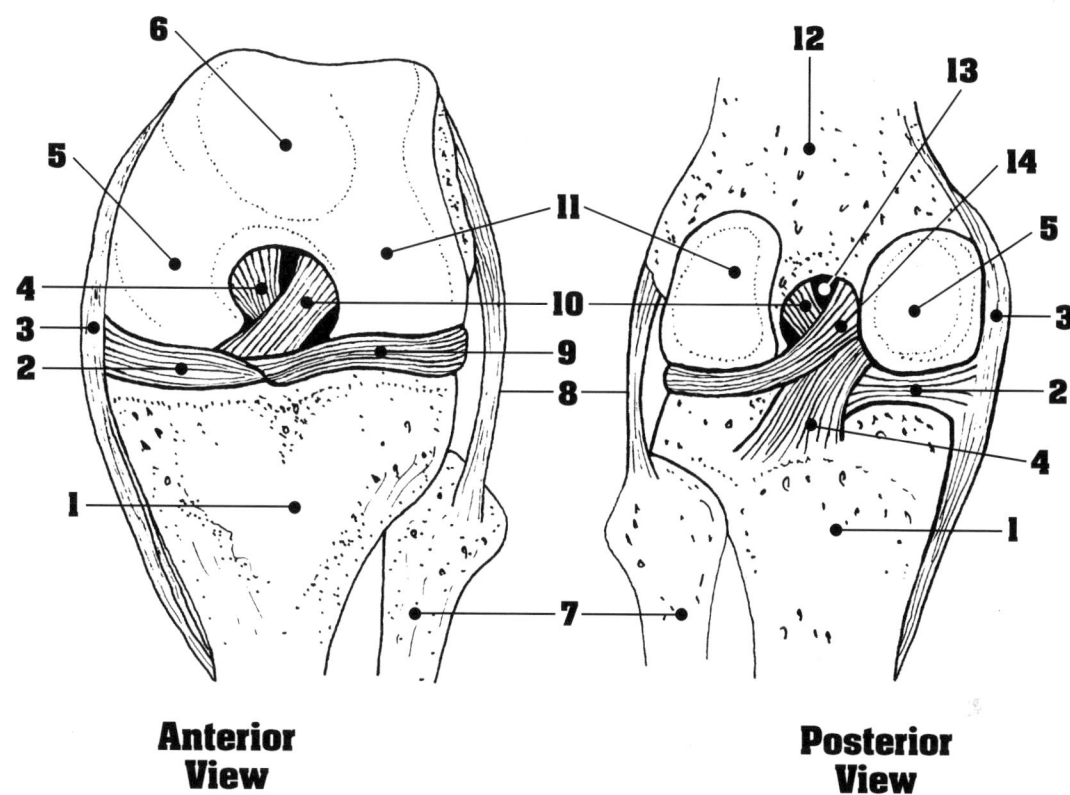

**Anterior View**

**Posterior View**

**Collateral Ligament**
- ___ Fibular
- ___ Tibial

**Condyle**
- ___ Lateral
- ___ Medial

**Cruciate Ligament**
- ___ Anterior
- ___ Posterior

- ___ Femur
- ___ Fibula
- ___ Intercondylar Groove

**Meniscus**
- ___ Lateral
- ___ Medial

- ___ Patellar Articular Surface
- ___ Posterior Meniscofemoral Ligament
- ___ Tibia

# Other Articulations

### Left Temporomandibular Joint

A **joint (articular) capsule** (3) and a **temporomandibular ligament** (2) support and bind the joint. The **mandible** (5) is stabilized medially by a **sphenomandibular ligament** (1) and posteriorly by a **stylomandibular ligament** (7).

The **styloid process** (6) and the **mastoid process** (4) are also noted.

### Atlantoaxial (Atlas-Axis) Joint

The **transverse ligament** (6) is shown lying in a groove at the base of the **dens** (1). **Alar ligaments** (2) further stabilize this remarkable pivoting joint. When you shake your head and "say no," the atlas pivots around the dens, and the transverse ligament slides back and forth in the groove at the base of the dens.

**Superior articular facets** (3) articulate with occipital condyles at the base of the skull, endowing us with the capacity to rock our heads forward and backward to "say yes." The spinal cord passes through the **vertebral foramen** (5). A vertebral artery and vein pass through the **transverse foramen** (4).

### Left Olecranal (Elbow) Joint

An **articular capsule** (2) binds and supports the articulations between the **humerus** (1), **radius** (7), and **ulna** (5). The joint is further supported and strengthened by the **ulnar collateral ligament** (4) and the **annular ligament** (9).

An **interosseous membrane** (6) binds the ulna to the radius.

The **olecranon process** (3) of the ulna forms the major osseous component of the "elbow."

The cut **tendon of the biceps** (8) is also shown.

# Left Temporomandibular Joint

___ Joint (Articular) Capsule
___ Mandible
___ Mastoid Process
___ Sphenomandibular Ligament
___ Styloid Process
___ Stylomandibular Ligament
___ Temporomandibular Ligament

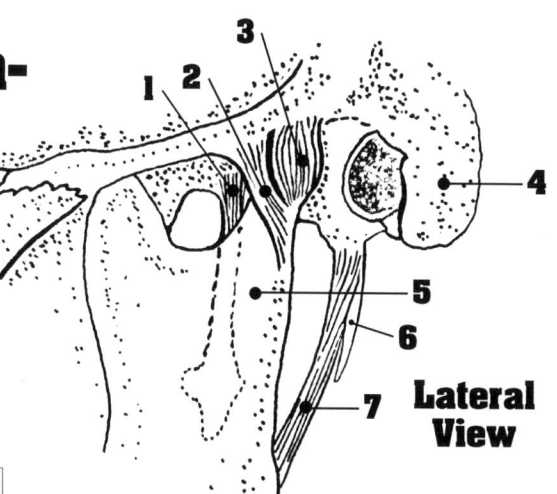

Lateral View

# Atlantoaxial Joint (Atlas-Axis)

___ Alar Ligament
___ Dens
___ Superior Articular Facet
___ Transverse Foramen
___ Transverse Ligament
___ Vertebral Foramen

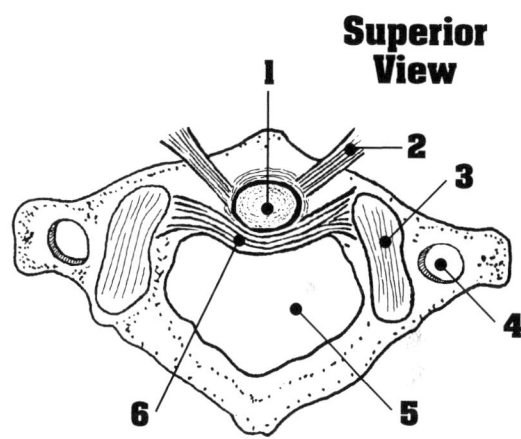

Superior View

# Left Olecranal (Elbow) Joint

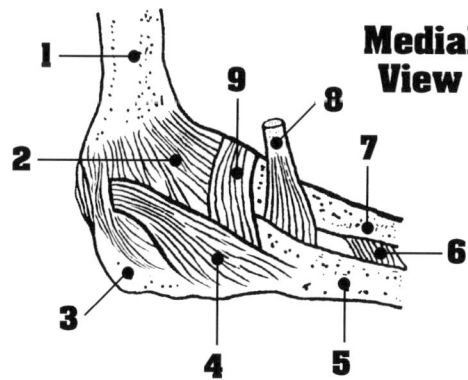

Medial View

___ Annular Ligament
___ Articular Capsule
___ Humerus
___ Interosseous Membrane
___ Olecranon Process
___ Radius
___ Tendon of Biceps
___ Ulna
___ Ulnar Collateral Ligament

# Muscular System

If you can think of your muscles, rippling under your skin, as mice scurrying along, you will be thinking as the taxonomist who gave this remarkable tissue its name: *mus*, in Latin meaning "little mouse."

**Four special properties enable muscle tissue to perform its functions:**

- **Excitability** — the ability to receive and respond to stimuli. In most cases the stimuli are chemicals called neurotransmitters.

- **Contractility** — the ability to shorten when adequately stimulated.

- **Extensibility** — the ability to be stretched or extended.

- **Elasticity** — the ability to recoil and resume shape after being stretched.

**Muscles perform five important functions:**

- **Produce movement**
- **Maintain posture**
- **Stabilize joints**
- **Generate heat**
- **Protect internal organs**

**Muscle tissue is of three types:**

- **Skeletal** — striated, voluntary
- **Smooth** — nonstriated, involuntary
- **Cardiac** — heart

Skeletal muscles come in sizes ranging from the tiny muscles within our inner ears to the large muscles of our upper legs. They also come in many shapes and patterns.

# Muscular System

Use either your textbook or your common sense to match the following names with their correct muscle patterns.

___ **Bipennate** ☐
___ **Circular** ☐
___ **Convergent** ☐
___ **Fusiform** ☐
___ **Multipennate** ☐
___ **Parallel** ☐
___ **Unipennate** ☐

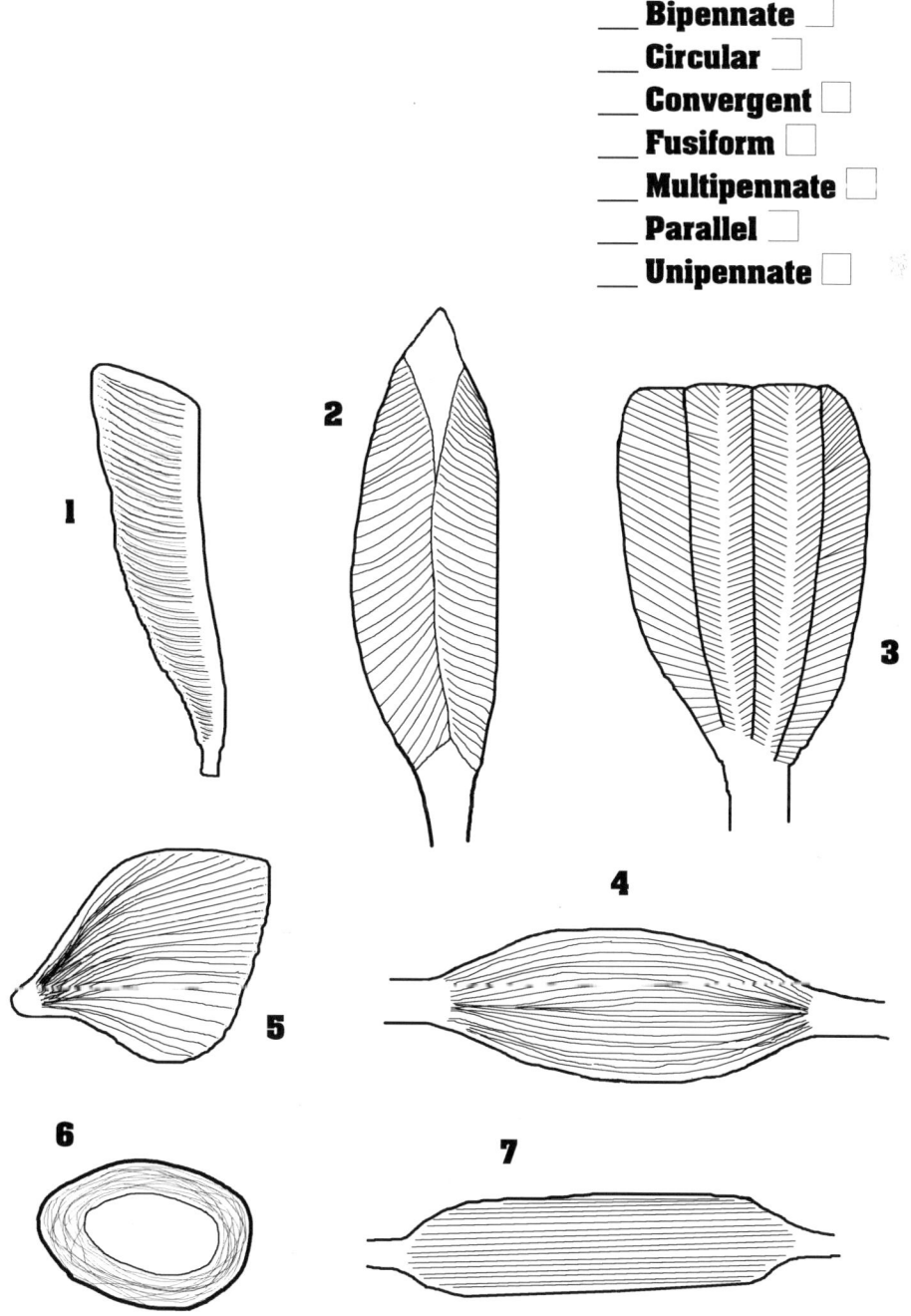

# Skeletal Muscle

A **muscle** (5) is subdivided into **fascicles** (6).
A fascicle is subdivided into **myofibers** (7).
A myofiber is subdivided into **myofibrils** (8).
A myofibril is subdivided into **myofilaments** (9).

Three connective tissue layers support muscle tissue:

A muscle is surrounded by **epimysium** (3).
A fascicle is surrounded by **perimysium** (1).
A myofiber is surrounded by **endomysium** (2).

The muscle shown has a tendinous connection (origin) on a **bone** (4).

*And all this must surely remind you of "the flea on the fly on the wart on the frog on the bump on the log on the edge of the pond!" But lo! Let us venture on, for muscles have yet smaller bumps!*

# Skeletal Muscle

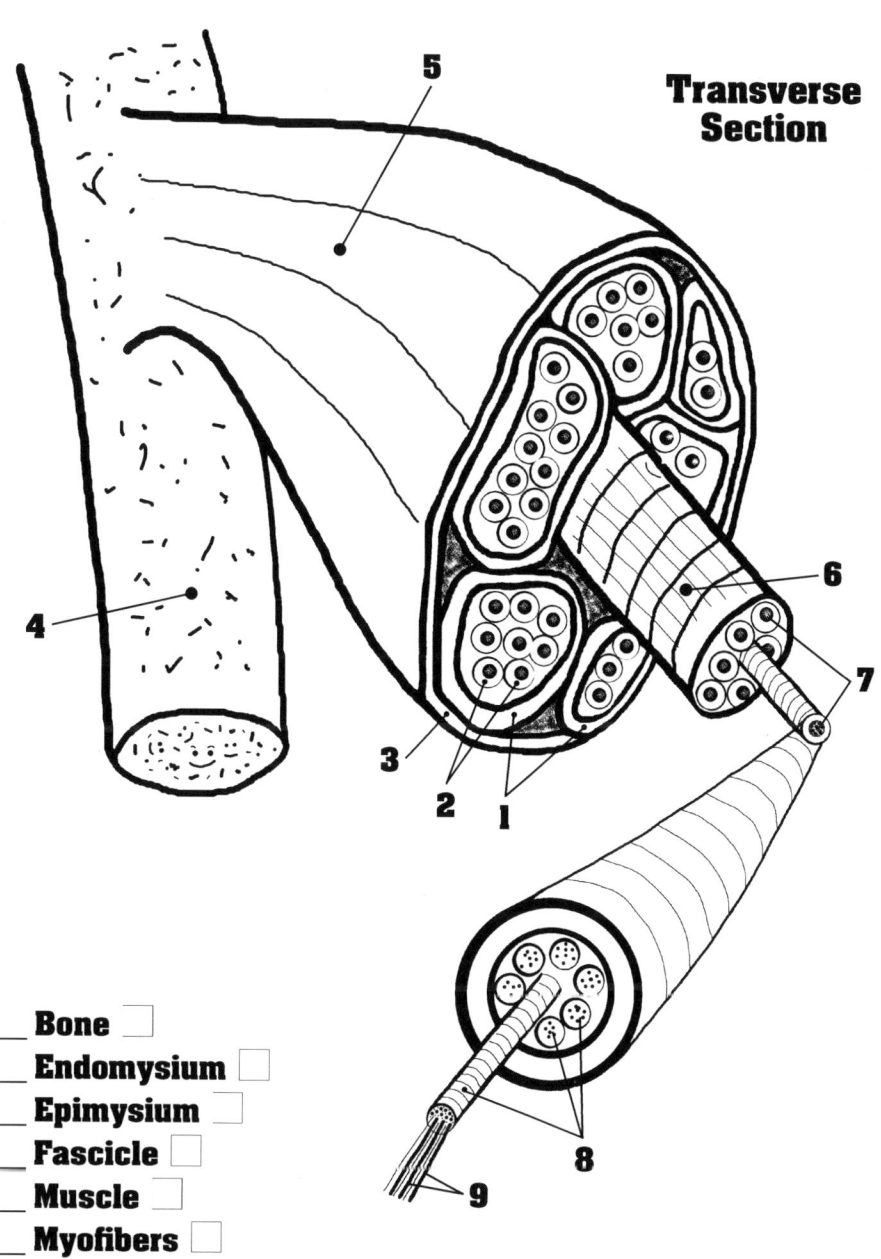

**Transverse Section**

___ Bone
___ Endomysium
___ Epimysium
___ Fascicle
___ Muscle
___ Myofibers
___ Myofibrils
___ Myofilaments
___ Perimysium

# Sarcomere

Thick and thin myofilaments, the final subdivisions on the previous drawing (and shown here at much greater magnification), are arranged in discrete units called sarcomeres. Sarcomeres are the functional units of skeletal muscle. A muscle contraction is a multiplication of sarcomeric contractions, thus, if we wish to understand how a muscle works, we must understand how a sarcomere works. Hence ...

A sarcomere is bounded at each end by a **Z disc** (4). Whereas **actin (thin myofilaments)** (7) are directly attached to Z discs, **myosin (thick myofilaments)** (8) are indirectly attached to Z discs via **myosin stabilizing filaments** (5).

Myosin (thick) myofilaments are composed of many bipolar myosin molecules. At each of its polar ends a myosin molecule has a **myosin head (cross bridge)** (2) and **myosin neck** (3). **Myosin binding sites** (1) are located on the actin myofilaments. Further details of how actin and myosin interact during a muscle contraction are shown in the next exercise ... but, we must first pay tribute to the beautiful banding patterns that are inherently a part of skeletal muscle tissue at the microscopic level — and also inherently a part of most muscle physiology tests.

At the center of the sarcomere is an **M line** (6). On either side of the M line is a short space devoid of thin myofilaments. These spaces, together with the M line, comprise the **H zone** (10).

Short spaces on either side of the Z disc that are devoid of thick myofilaments. These spaces, together with the Z disc, comprise the **I band** (9).

Myosin myofilaments, in their entirety, comprise the **A band** (11) which occupies the main central region of the sarcomere.

*We are now prepared to dig yet deeper into the manifold molecular mysteries of muscle physiology, and go on to investigate ... the "power stroke!"*

# Sarcomere

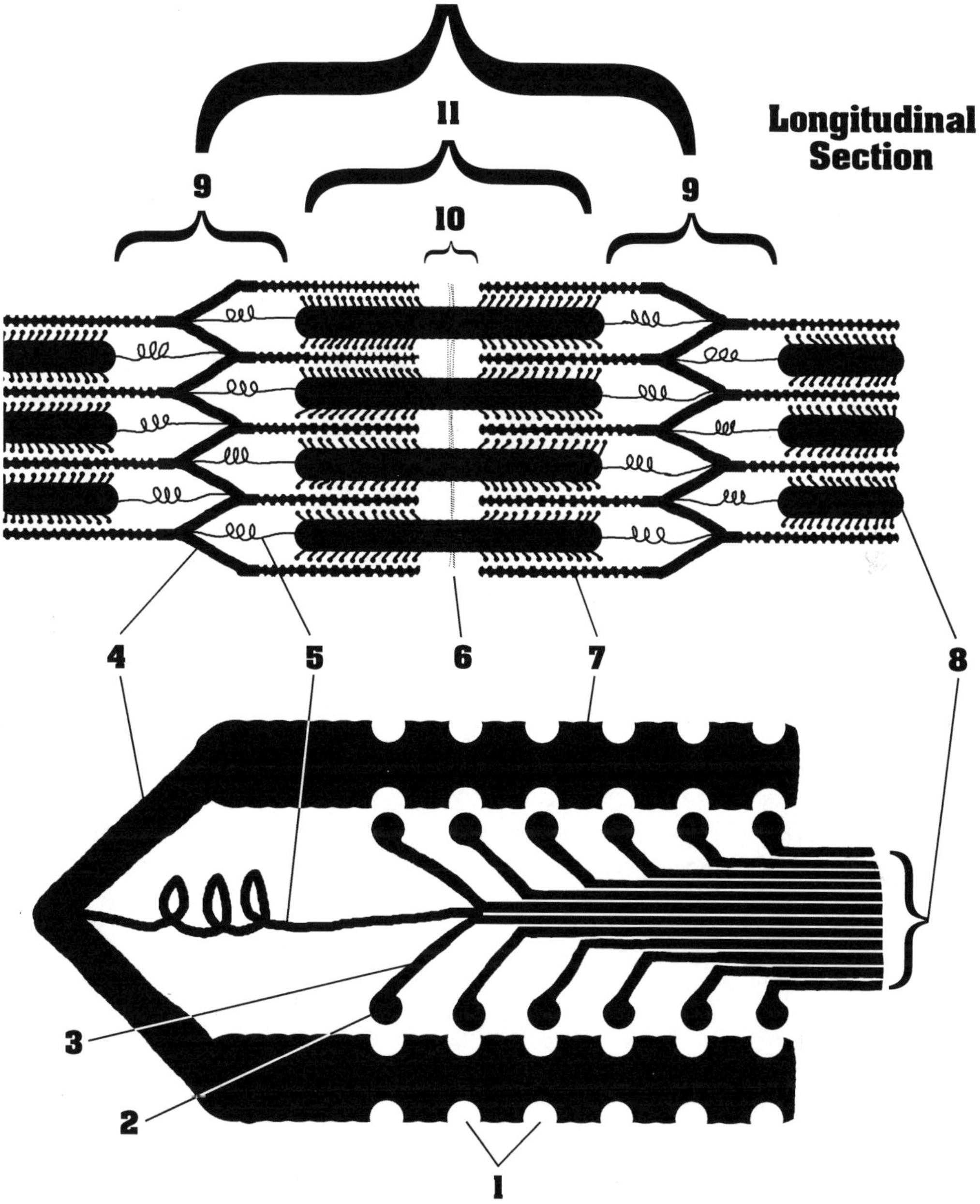

Longitudinal Section
11
9
9
10
4
5
6
7
8
3
2
1


___ A Band
___ Actin (Thin Myofilament)
___ H Zone
___ I Band
___ M Line
___ Myosin Binding Site
___ Myosin Head (Cross Bridge)
___ Myosin Neck
___ Myosin Stabilizing Filament
___ Myosin (Thick Myofilament)
___ Z Disc

# Power Stroke

"Up close and personal," we see that on each **myosin head** (10) there is an **actin binding site** (7) and an **ATP binding site** (9). We note also that there are many **myosin binding sites** (6) on the **actin** (1) myofilament.

When a muscle is relaxed the myosin binding sites on the actin are covered by a **tropomyosin thread** (5). **Troponin complexes** (4) are scattered along the tropomyosin thread.

Each troponin complex has two **calcium binding sites** (3). When **calcium ions** (2) bind to the calcium binding sites the tropomyosin thread makes a conformational shift, and the myosin binding sites are opened for engagement with actin binding sites on the myosin heads.

In the lower drawing we assume the presence of calcium ions, and we assume also that the tropomyosin thread has made a conformational shift, and that the myosin binding sites are therefore exposed. The myosin head, via its actin binding site, has engaged the myosin binding site of the actin. And when this happens **ATP** (8) undergoes **ATP hydrolysis** (13) and the **myosin neck** (11) bends forward to perform the **power stroke** (12).

As multitudinous myosin heads make their individual power strokes actin molecules are pulled (ratcheted) inward, and because the actin molecules are firmly attached to Z discs, the Z discs are also pulled inward. Thus the entire sarcomere is shortened (contracted).

*Having once engaged the actin, myosin heads will not disengage until they are reloaded with ATP. And therein lies the explanation for rigor mortis (stiff corpse). When an animal dies it runs out of oxygen, hence it runs out of ATP, hence it cannot reload the myosin heads, hence the muscles stay contracted ... until protein decomposition (putrefaction) sets in some hours later.*

*But from where did the calcium ions, which initiate the whole process, come? Ah! Let us go to the next exercise and investigate the electrochemical events that occur at a NMJ (neuromuscular junction.)*

# Power Stroke

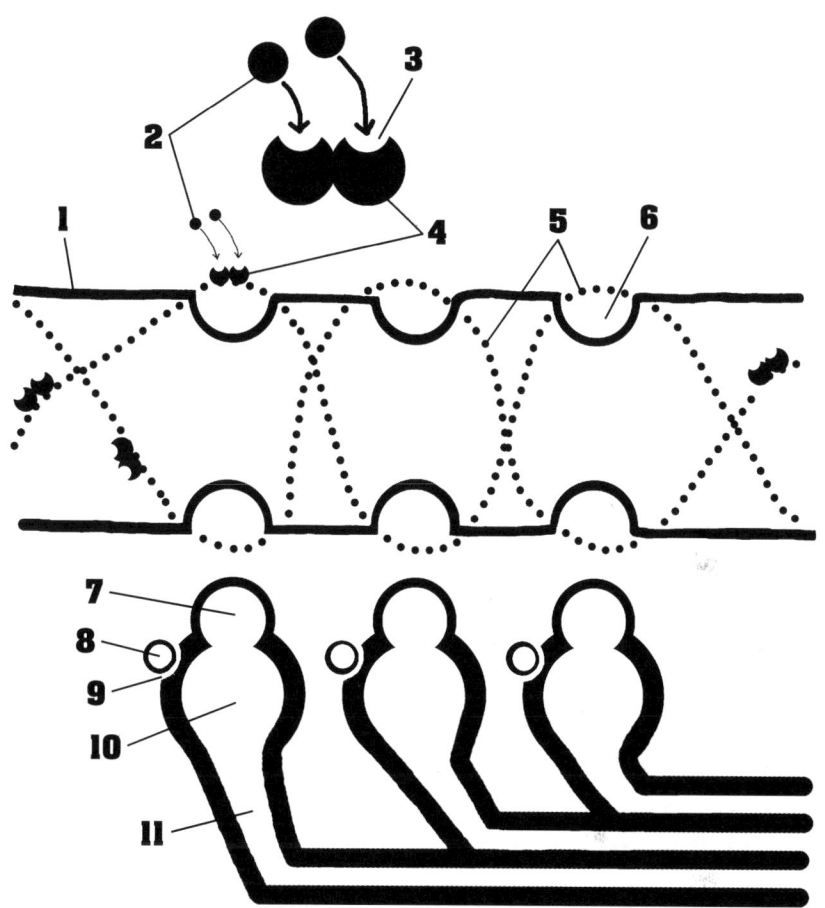

- Actin
- Actin Binding Site
- ATP
- ATP Binding Site
- ATP Hydrolysis
- Calcium Binding Site
- Calcium Ions
- Myosin Binding Site
- Myosin Head
- Myosin Neck
- Power Stroke
- Troponin Complex
- Tropomyosin Thread

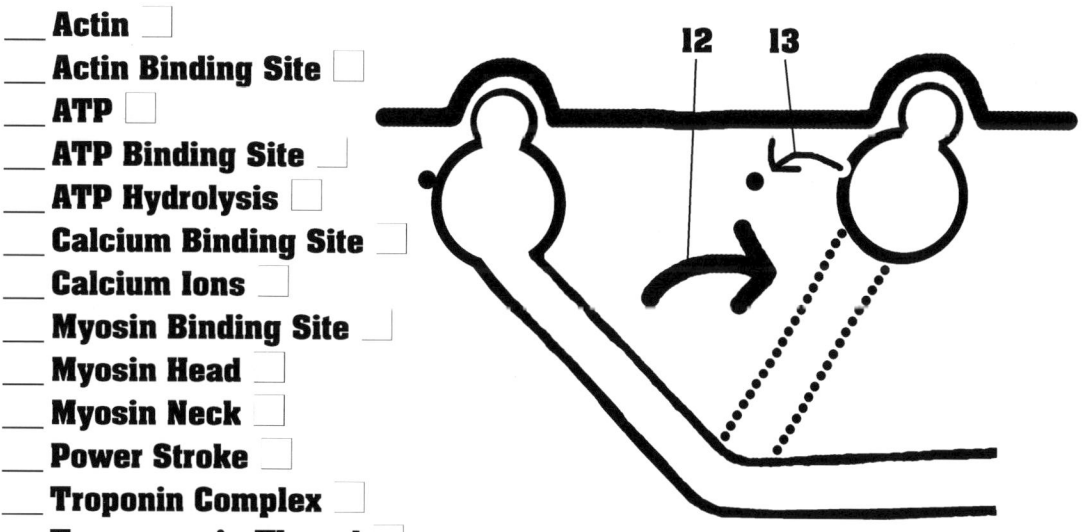

© Copyright 2010 Gene Johnson

# Neuromuscular Junction

**NAPs (nerve action potentials)** (1) travel along an **axon** (2) and arrive at the **synaptic end foot** (3) of a neuromuscular junction (NMJ).

After crossing the **synaptic cleft** (4), the action potentials travel along the **sarcolemma** (6) surrounding the muscle fibers, and as they are traveling in this region, they are called **MAPs (muscle action potentials)** (5).

Muscle action potentials then move downward via **transverse tubules** (9) into muscle fiber tissues and cause **calcium ions** (7) to be released from **terminal cisterns** (8) which lie parallel, on either side of the transverse tubule. The transverse tubule, together with the two adjacent terminal cisterns is called a **triad** (10).

Calcium released from the terminal cisterns moves out through the **sarcoplasmic reticulum** (11) into the **myofibers** (12) … and you already know what happens when it gets out there among the myofilaments that constitute the myofibers! Why, of course:

- Calcium ions attach to the calcium binding sites on the troponin complexes.

- Tropomyosin threads make conformational shifts.

- The myosin binding sites on actin are now open.

- Myosin heads engage the open myosin binding sites.

- The engagement of the myosin heads with the myosin binding sites causes ATP hydrolyses, which are immediately followed by power strokes.

*And so in ten easy steps, covered in three different drawings, you have the story of muscle contraction … and you may now be so overwhelmed with the complexities of muscle physiology, and suffering so severely from the paralysis of analysis, that you may never walk again!*

# Neuromuscular Junction

**Longitudinal Section**

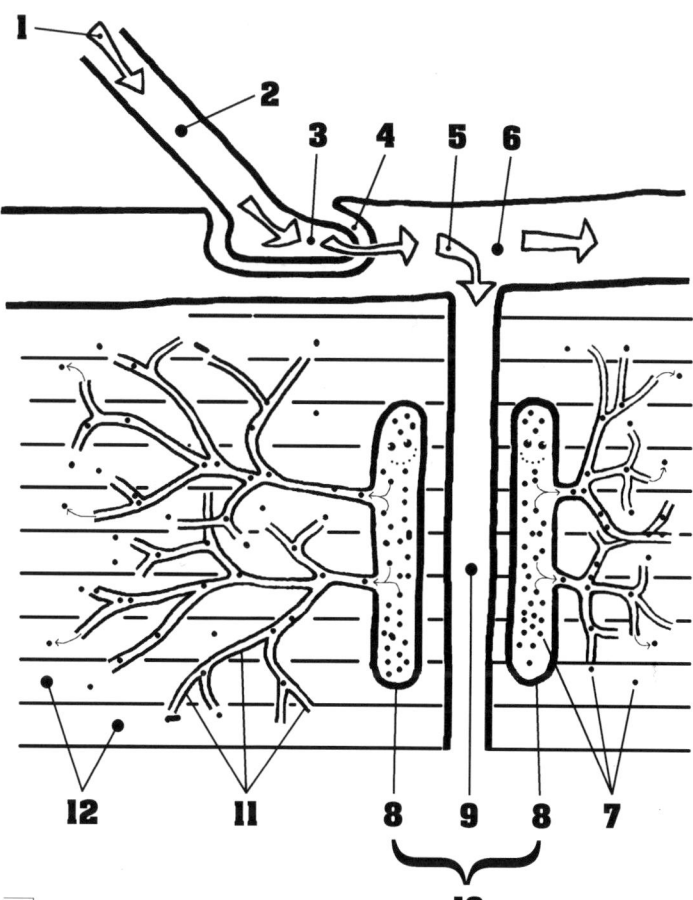

**Transverse Section Detail**

___ Axon
___ Calcium Ions
___ MAP (Muscle Action Potential)
___ Myofibers
___ NAP (Nerve Action Potential)
___ Sarcolemma
___ Sarcoplasmic Reticulum
___ Synaptic End Foot
___ Synaptic Cleft
___ Terminal Cisterns
___ Transverse Tubule
___ Triad

# Head & Neck
## Superficial Muscles

Three cranial muscles shown:

- **Frontalis** (1) — which wrinkles the forehead and raises the eyebrows.
- **Temporalis** (14) — which assists in elevating the mandible.
- **Occipitalis** (15) — which moves the scalp backward.

The **obicularis oculi** (2) closes the eye, the **obicularis oris** (9) closes and purses the lips, and the **posterior auricular** (16) adducts ("wiggles") the ear.

The following eight muscles act to move the lips or mouth:

- **Levator labii superioris** (5) — elevates the upper lip.
- **Zygomaticus minor** (6) — elevates the corners of the mouth.
- **Zygomaticus major** (7) — elevates the corners of the mouth.
- **Risorius** (8) — draws the angle of the mouth laterally.
- **Depressor labii inferioris** (11) — depresses the lower lip.
- **Mentalis** (10) — protrudes the lower lip.
- **Depressor anguli oris** (12) — draws the corner of the mouth downward and lateral.
- **Platysma** (13) — pulls the lower lip back and down and helps depress the mandible.

The **procerus** (3) elevates the nose and the **nasalis** (4) compresses the nose.

The **masseter** (18) elevates the mandible, the **trapezius** (20) draws the head back and upward, and the **sternocleidomastoid** (19) draws the head back and also assists in turning the head.

The **parotid (salivary) gland** (17), the largest of the three paired salivary glands is seen anterior to the ear.

*How mobile is your face? Read through this exercise again and perform each muscle action as it is described. If you want your face to stay smooth and shapely (like the rest of your body?), you should do some daily facial muscle exercises ... If nothing else you could at least wiggle your ears and stick out your tongue at your anatomy professor. Surely he or she would understand what you are doing and why.*

# Head & Neck
## Superficial Muscles

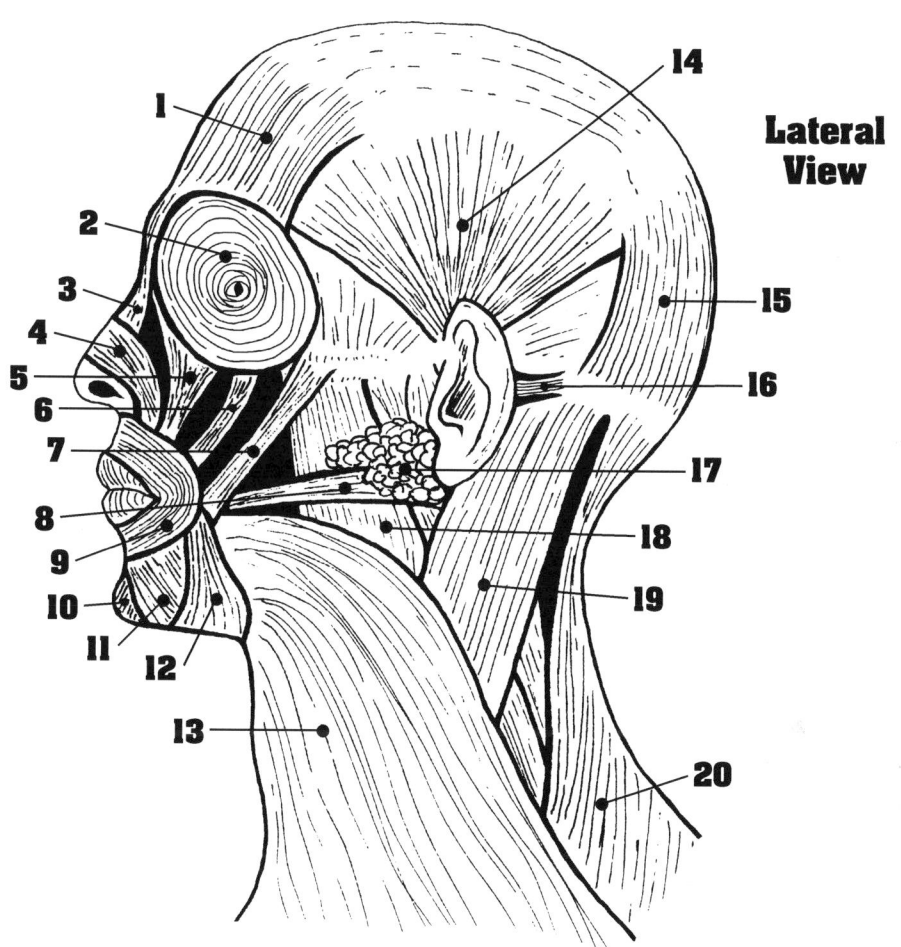

Lateral View

___ Depressor Anguli Oris
___ Depressor Labii Inferioris
___ Frontalis
___ Levator Labii Superioris
___ Masseter
___ Mentalis
___ Nasalis
___ Obicularis Oculi
___ Obicularis Oris
___ Occipitalis
___ Parotid (Salivary) Gland
___ Platysma
___ Posterior Auricular
___ Procerus
___ Risorius
___ Sternocleidomastoid
___ Temporalis
___ Trapezius
___ Zygomaticus Major
___ Zygomaticus Minor

# Neck
## Deep Muscles

First we consider the two two-bellied muscles in this drawing. The digastric muscle forms a "V" under the chin. The **digastric's anterior belly** (4) is united with the **digastric's posterior belly** (9) by an intermediate tendon. The **omohyoid's anterior belly** (2) is united with the **omohyoid's posterior belly** (19) by an intermediate tendon (hidden behind the sternocleidomastoid muscle).

The **styloglossus** (7) and **hyoglossus** (5) have their origins on styloid process (adjacent to the lower margin of the ear).

Completing the muscles in the area of the chin we see the **mylohyoid** (6) and **stylohyoid** (10).

Beneath the chin, from anterior to posterior, we see the **sternohyoid** (1), omohyoid (already mentioned), **thyrohyoid** (13), **inferior constrictor** (14), and **sternocleidomastoid** (8).

In the upper neck, posterior to the sternocleidomastoid, we see the **splenius medius** (15), **splenius capitus** (12), and the **trapezius** (16).

In the lower neck region, just posterior to the sternocleidomastoid, we see the **levator scapulae** (17) and **scalenus posterias** (18).

And finally, just above the **hyoid bone** (3), we see the **medial constrictor** (11). The inferior and medial constrictor muscles, along with the superior constrictor muscle (not shown) surround and operate the pharynx (see page 199).

*Omohyoid, stylohyoid, mylohyoid, hyoglossus, sternohyoid, thyrohyoid ... When you consider all the muscle names with "hyo" or "hyoid" in them, you can begin to appreciate how important the strange "floating" hyoid bone must be! How many muscle origins and insertions would be unattached if the hyoid bone were missing?!*

# Neck
## Deep Muscles

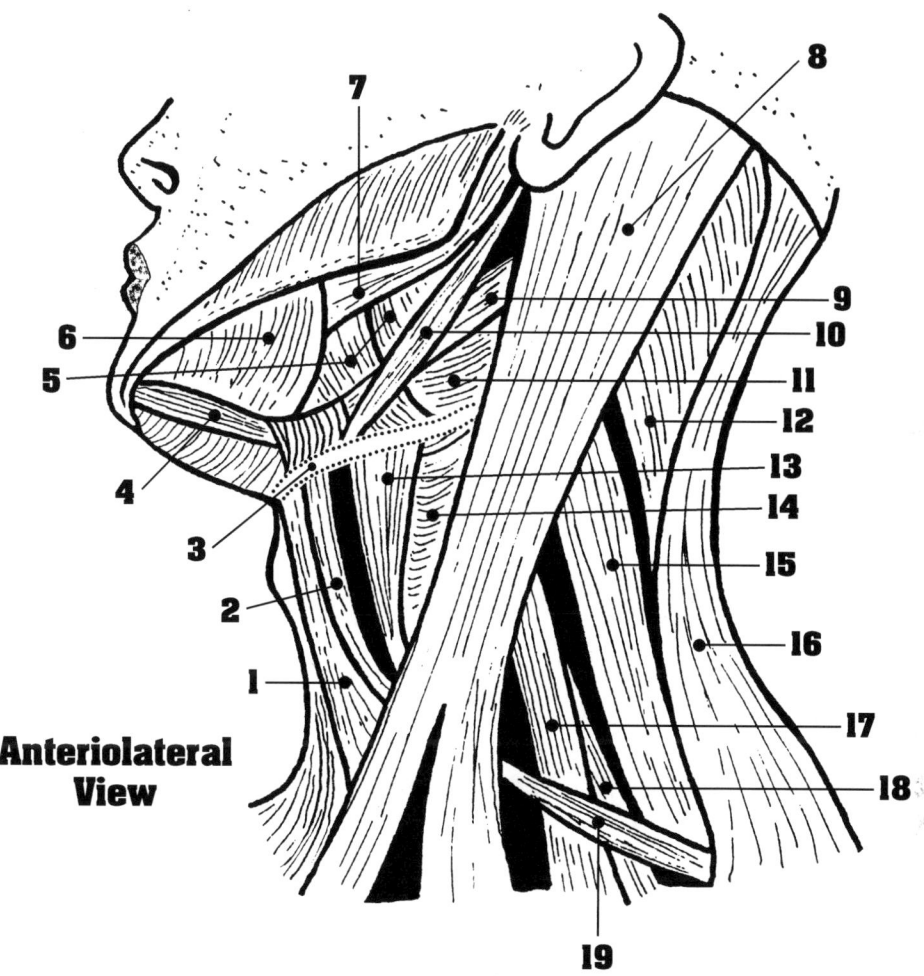

**Anteriolateral View**

___ Digastric — Anterior Belly
___ Digastric — Posterior Belly
___ Hyoglossus
___ Hyoid Bone
___ Inferior Constrictor
___ Levator Scapulae
___ Medial Constrictor
___ Mylohyoid
___ Omohyoid — Anterior Belly
___ Omohyoid — Posterior Belly
___ Scalenus Posterias
___ Splenius Capitus
___ Splenius Medius
___ Sternocleidomastoid
___ Sternohyoid
___ Styloglossus
___ Stylohyoid
___ Thyrohyoid
___ Trapezius

# Body
## Superficial Muscles

The **trapezius** (12) is a complex muscle with several origins, insertions, and functions, including drawing the head back, elevating and adducting the scapula, and bracing the shoulder. The **sternocleidomastoid** (1) which draws the head back, also assists in moving the head from side to side.

The **deltoid** (3) abducts the arm and assists in flexing and extending the humerus. The **biceps brachii** (4) flexes the forearm. The **triceps** (5) extends the forearm.

Both the **teres minor** (2), and the **infraspinatus** (13), rotate the arm laterally. The **teres major** (6) rotates the arm medially and assists in extending the humerus.

The **latissimus dorsi** (9) retracts the shoulder and extends, adducts, and rotates the humerus medially.

The **pectoralis** (7) adducts and rotates the arm medially.

The **serratus anterior** (8) pulls the scapula forward and downward.

The **external oblique** (10) compresses the abdomen and assists in its lateral rotation.

The **gluteus maximus** (14) extends and rotates the thigh laterally. The **gluteus medius** (11) abducts and rotates the thigh medially.

# Body
## Superficial Muscles

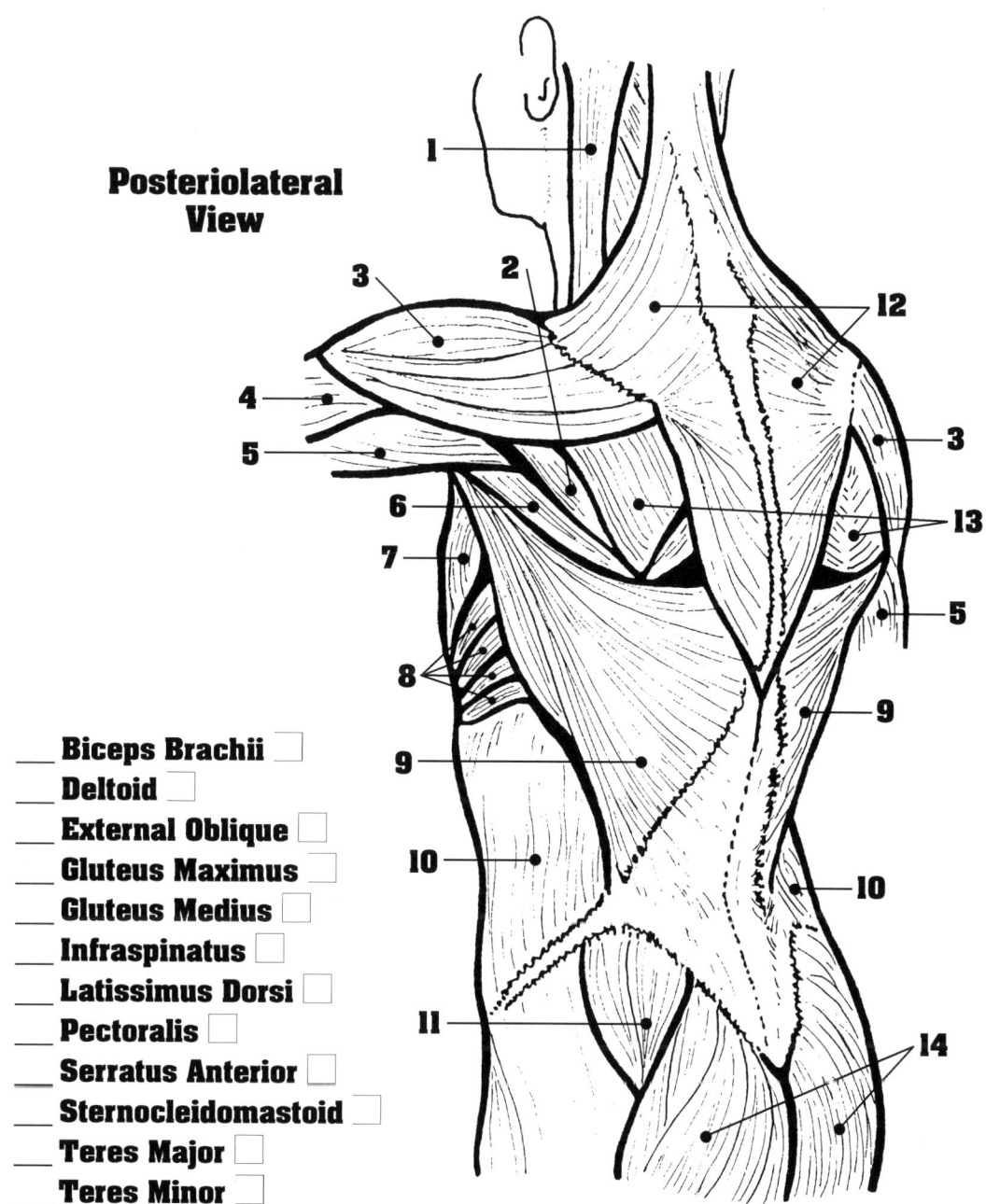

**Posteriolateral View**

___ Biceps Brachii
___ Deltoid
___ External Oblique
___ Gluteus Maximus
___ Gluteus Medius
___ Infraspinatus
___ Latissimus Dorsi
___ Pectoralis
___ Serratus Anterior
___ Sternocleidomastoid
___ Teres Major
___ Teres Minor
___ Trapezius
___ Triceps

# Shoulder & Neck
## Deep Muscles

The **spinalis capitis** (14) and **splenius capitis** (1) assist in moving the head upward and backward.

The **levator scapulae** (13) elevates the scapula and lifts the shoulders.

The **rhomboideus major** (11) and **rhomboideus minor** (12) elevate and adduct the scapula.

The **teres major** (8) adducts and rotates the arm medially. The **teres minor** (9) rotates the arm laterally.

The **supraspinatus** (2) and **infraspinatus** (10) rotate the arm laterally.

The **triceps** (7) extends the lower arm.

The skeletal elements shown are the:

- **Spine of the scapula** (3)
- **Superior margin of the scapula** (4)
- **Acromion process of the scapula** (5)
- **Proximal end of the humerus** (6)

# Shoulder & Neck
## Deep Muscles

___ Acromion Process of the Scapula ☐
___ Infraspinatus ☐
___ Levator Scapulae ☐
___ Proximal End of the Humerus ☐
___ Rhomboideus Major ☐
___ Rhomboideus Minor ☐
___ Spinalis Capitalis ☐
___ Spine of the Scapula ☐
___ Splenius Capitis ☐
___ Superior Margin of the Scapula ☐
___ Supraspinatus ☐
___ Teres Major ☐
___ Teres Minor ☐
___ Triceps ☐

**Posterior View**

# Thorax
## Deep Muscles

The **levator scapulae** (6) which acts to lift the shoulder, has its origin on the **cervical vertebrae** (5) and its insertion on the scapula.

The **pectoralis major** (9) is cut and removed to show the **pectoralis minor** (3), which pulls the scapula (and shoulder) forward and downward.

The **subscapularis** (11), just lateral to the pectoralis minor, rotates the arm medially. At the furthest lateral position, the outer edge of the **latissimus dorsi** (12) can be seen with its tendon attachment around the head of the humerus. When the latissimus dorsi contracts it adducts and rotates the **humerus** (10) medially.

The **serratus anterior** (13) pulls the scapula and shoulder downward and forward.

The **intercostals** (1) move the rib cage in the breathing process.

The external oblique is removed to show the **internal oblique** (14). Both oblique muscles compress the abdomen and assist in rotating it laterally. The **rectus abdominus** (2) flexes the vertebral column to permit the upper body to bend over and downward.

The **biceps tendon** (8) is shown near the head of the humerus. The **sternum** (4), and **clavicle** (7) are also labeled.

# Thorax
## Deep Muscles

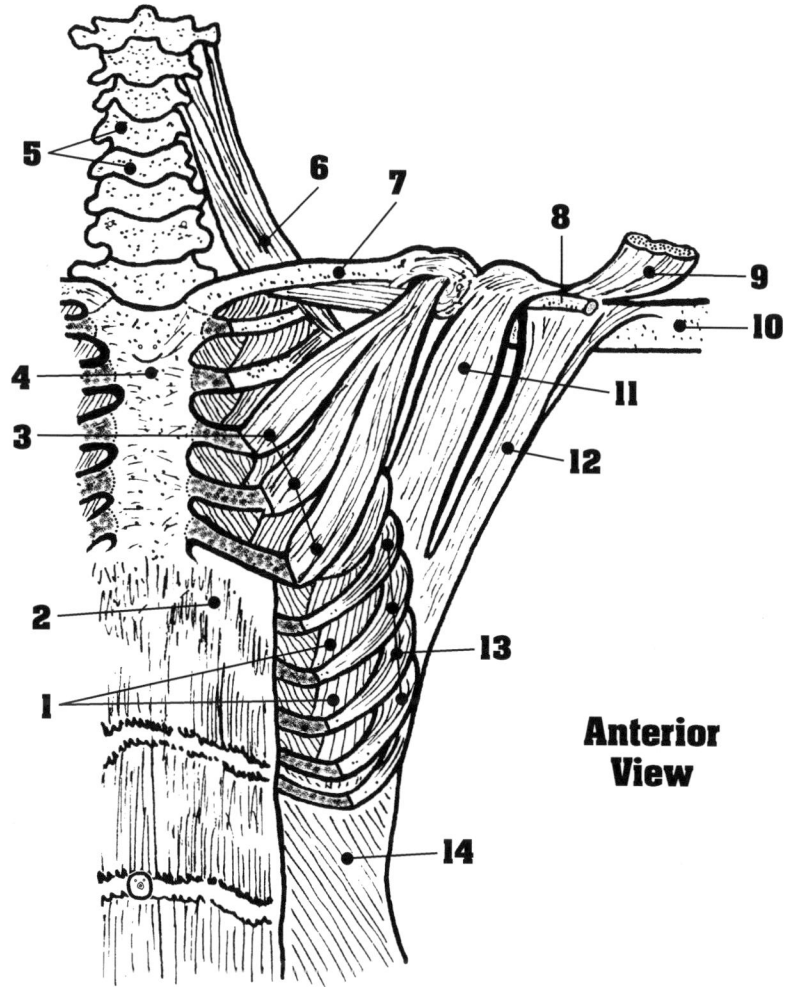

**Anterior View**

___ Biceps Tendon
___ Cervical Vertebrae
___ Clavicle
___ Humerus
___ Intercostals
___ Internal Oblique
___ Latissimus Dorsi
___ Levator Scapulae
___ Pectoralis Major
___ Pectoralis Minor
___ Rectus Abdominus
___ Serratus Anterior
___ Sternum
___ Subscapularis

# Upper Left Arm
## Deep Muscles

The shoulder girdle, comprised of the **scapula** (6) and **clavicle** (10), is knit together with several ligaments. The ligaments shown, the **coracoclavicular ligaments** (11) and **coracoacromial ligament** (13), connect with the **coracoid process** (9) of the scapula.

The **head of the humerus** (14) is indicated just inferior to the **acromion process** (12).

The **pectoralis major** (15), and **pectoralis minor** (8) have been largely removed to expose the **lateral** (3) and **medial head** (4) **of the biceps**. The biceps, which flexes the lower arm, is anchored at the shoulder girdle and inserts on the radial tuberosity.

Behind the biceps, portions of the **lateral** (16), **medial** (2), and **long** (5) **heads of the triceps** can be seen. The triceps acts to extend the lower arm.

The **brachialis** (1) and **brachioradialis** (17) also assist in flexing the lower arm.

The **coracobrachialis** (7) flexes and adducts the shoulder joint.

# Upper Left Arm
## Deep Muscles

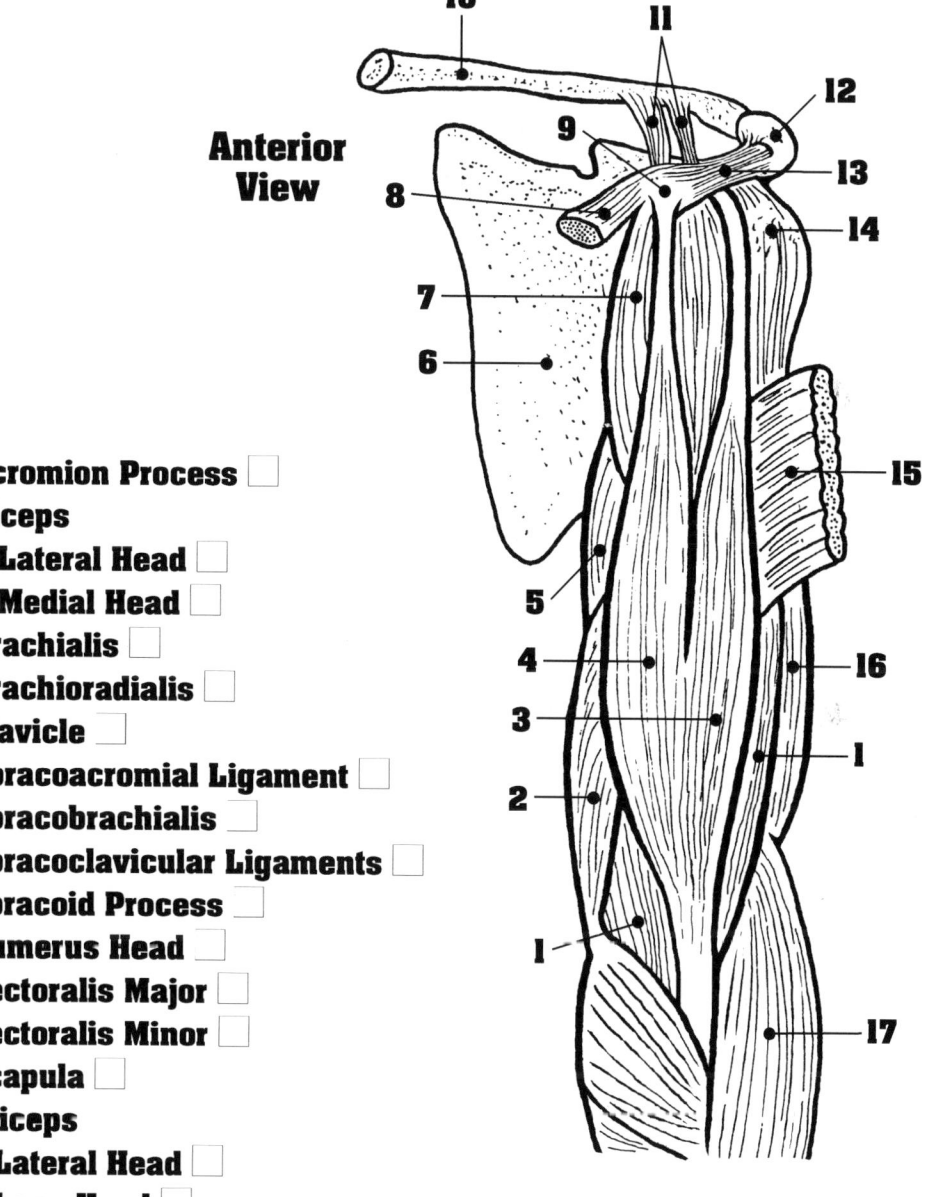

**Anterior View**

___ Acromion Process ☐
**Biceps**
___ Lateral Head ☐
___ Medial Head ☐
___ Brachialis ☐
___ Brachioradialis ☐
___ Clavicle ☐
___ Coracoacromial Ligament ☐
___ Coracobrachialis ☐
___ Coracoclavicular Ligaments ☐
___ Coracoid Process ☐
___ Humerus Head ☐
___ Pectoralis Major ☐
___ Pectoralis Minor ☐
___ Scapula ☐
**Triceps**
___ Lateral Head ☐
___ Long Head ☐
___ Medial Head ☐

# Lower Left Arm
## Anterior View

The **brachialis** (2) is seen on either side of the **biceps** (11), and the **medial head of the triceps** (1) is seen on the medial side of the upper arm.

The **medial epicondyle** (3) of the humerus is indicated on the medial side of the elbow.

From medial to lateral, we see four muscles across the middle of the forearm:

- **Flexor carpi ulnaris** (7)
  extends and adducts the wrist.

- **Palmaris longus** (6)
  assists in flexing the wrist and in stabilizing the hand.

- **Flexor carpi radialis** (5)
  flexes the wrist and adducts the hand.

- **Brachioradialis** (12)
  acts as a synergist in forearm flexion.

At the distal end of the forearm, the **flexor digitorum** (8) flexes the digits, and the **flexor pollicis longus** (9) flexes the thumb.

As they pass across the wrist, the flexor tendons are stabilized by the **flexor retinaculum** (10).

The **pronator teres** (4) pronates the arm.

# Lower Left Arm

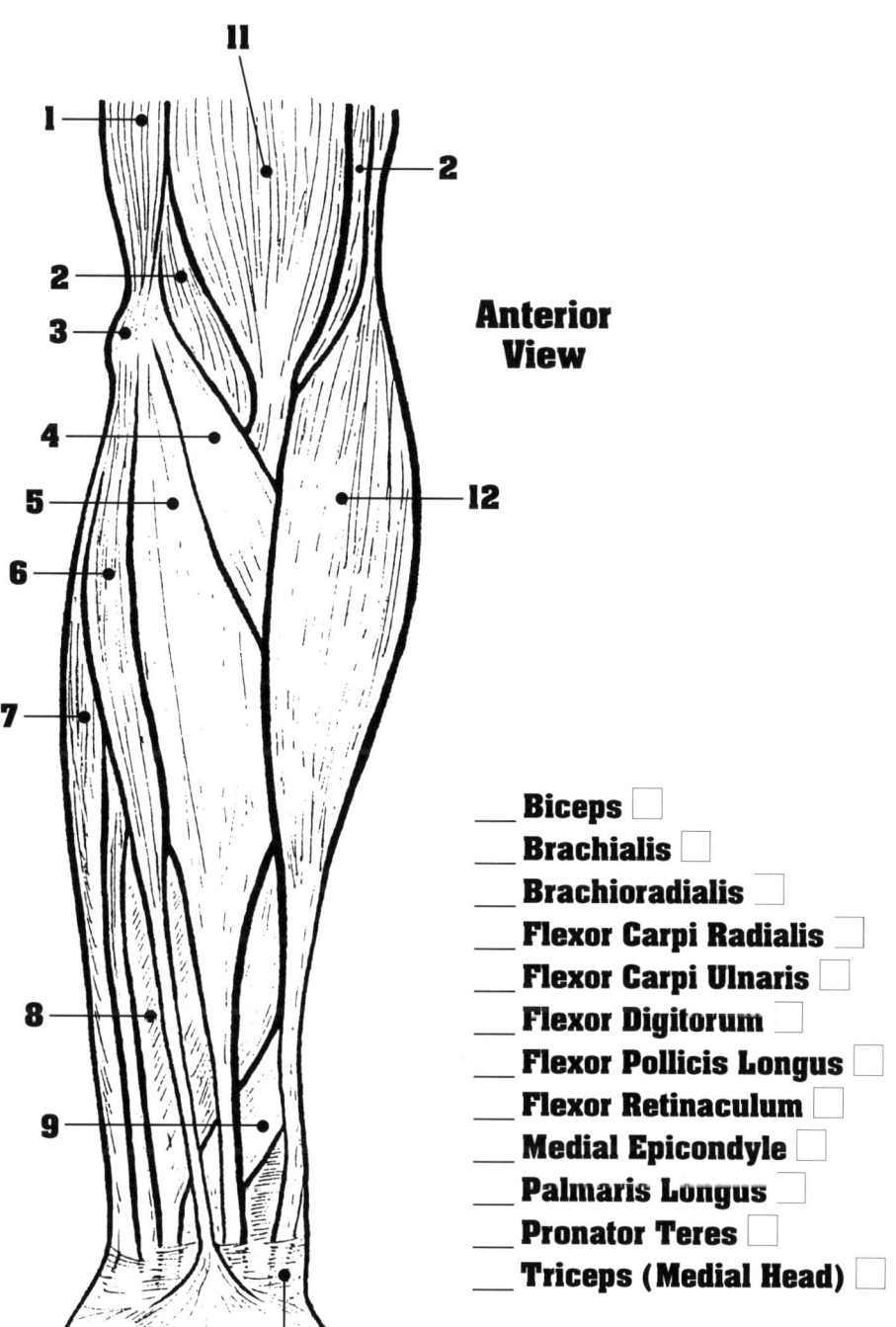

**Anterior View**

___ Biceps
___ Brachialis
___ Brachioradialis
___ Flexor Carpi Radialis
___ Flexor Carpi Ulnaris
___ Flexor Digitorum
___ Flexor Pollicis Longus
___ Flexor Retinaculum
___ Medial Epicondyle
___ Palmaris Longus
___ Pronator Teres
___ Triceps (Medial Head)

# Lower Left Arm
## Posteriolateral View

Six muscles, which originate in the upper arm or shoulder, insert on the lower arm bones:

- **Triceps** (12) — extends the lower arm.
- **Biceps brachii** (1) — flexes the lower arm.
- **Brachialis** (2) — flexes the lower arm.
- **Brachioradialis** (3) — flexes the lower arm.
- **Extensor carpi radialis longus** (4) — extends and abducts the wrist.
- **Anconeus** (13) — assists in pronating the arm.

The five major muscles of the forearm originate at the proximal ends of the radius and ulna and insert on the carpals, metacarpals, or phalanges, thus acting to move the wrist, hand, and fingers. From lateral (the thumb side) to medial (the little finger side) they are the:

- **Extensor carpi radialis brevis** (5) — assists the extensor carpi radialis longus in extending and abducting the wrist.
- **Extensor digitorum** (6) — extends the wrist and fingers and can also abduct the fingers.
- **Extensor digiti minimi** (7) — extends the fifth digit and the wrist.
- **Extensor carpi ulnaris** (8) — extends and adducts the wrist.
- **Flexor carpi ulnaris** (14) — extends and adducts the wrist.

Just superior to the wrist are two smaller muscles which act to move the thumb:

- **Abductor pollicis longus** (9) — abducts and extends the thumb.
- **Extensor pollicis brevis** (10) — extends the thumb.

The numerous tendons that run through the wrist are supported and stabilized by the **extensor retinaculum** (11).

# Lower Left Arm

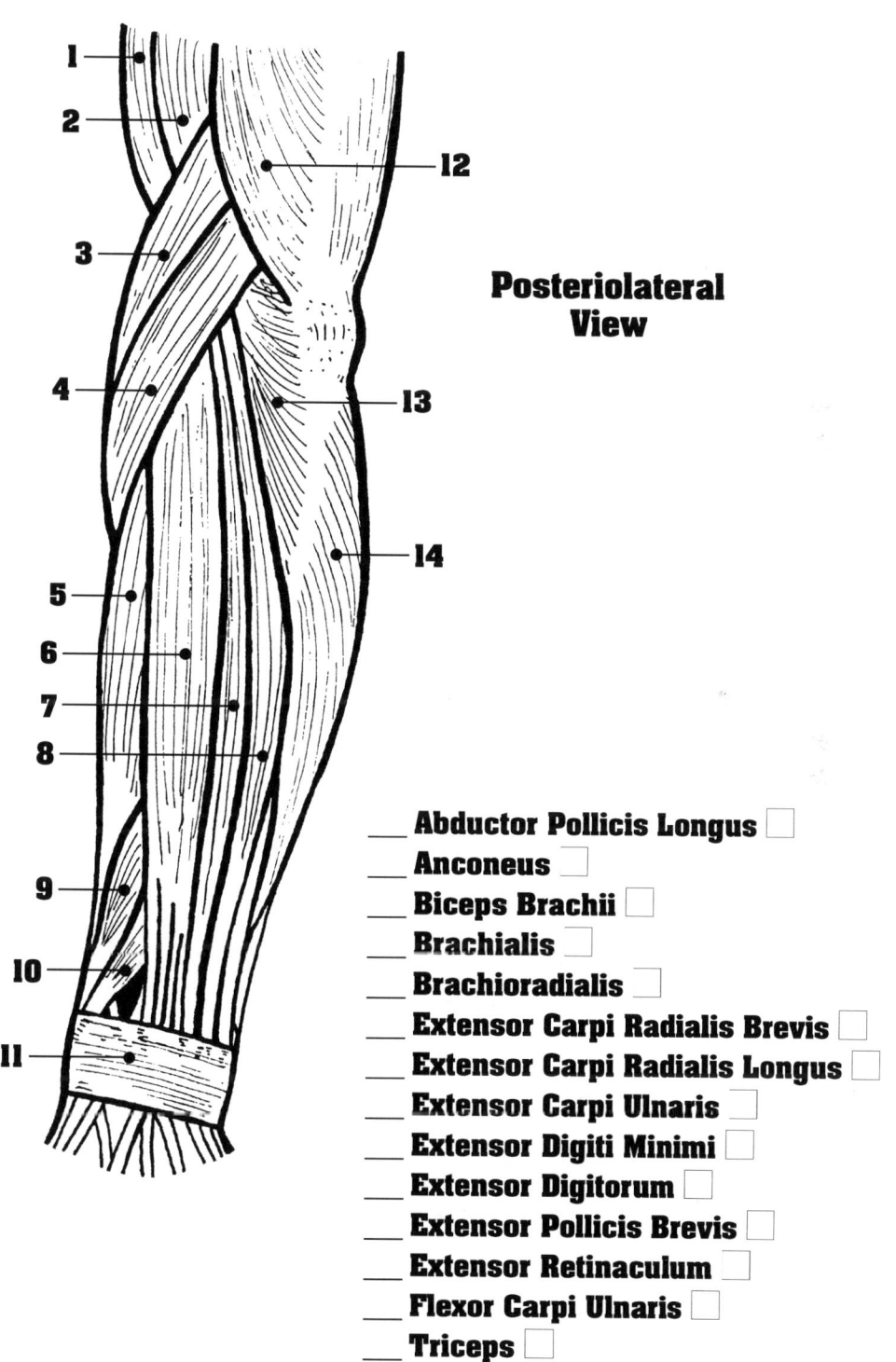

**Posteriolateral View**

___ Abductor Pollicis Longus ☐
___ Anconeus ☐
___ Biceps Brachii ☐
___ Brachialis ☐
___ Brachioradialis ☐
___ Extensor Carpi Radialis Brevis ☐
___ Extensor Carpi Radialis Longus ☐
___ Extensor Carpi Ulnaris ☐
___ Extensor Digiti Minimi ☐
___ Extensor Digitorum ☐
___ Extensor Pollicis Brevis ☐
___ Extensor Retinaculum ☐
___ Flexor Carpi Ulnaris ☐
___ Triceps ☐

# Right Leg

Movements of the thigh are accomplished largely by muscles anchored to the pelvic girdle. Inferior to the **iliac crest** (11), four muscles act to move the thigh:

- **Gluteus maximus** (9) — extends and rotates the thigh.
- **Gluteus medius** (10) — abducts and rotates the thigh.
- **Tensor fasciae latae** (12) — flexes and adducts the thigh.
- **Sartorius** (13) — flexes and laterally rotates the thigh.

Four upper leg muscles are shown. A small distal portion of the **semimembranosus** (5) is seen in the popliteal region (behind the knee). In the upper leg the **biceps femoris** (7) is at the posterior surface, and the **rectus femoris** (14) is at the anterior surface. The **iliotibial tract** (8) — which, although not muscle tissue, works in conjunction with muscles — is seen in the middle, overlying the **vastus lateralis** (6), which can be seen on either side of the tract.

Through the middle of the lower leg, from posterior to anterior, are five muscles:

- **Gastrocnemius** (4)
- **Soleus** (3)
- **Peroneus longus** (2)
- **Extensor digitorum longus** (16)
- **Tibialis anterior** (15)

The **peroneus brevis** (17) and the **Achilles tendon** (1) are seen at the distal end of the lower leg.

# Right Leg

___ Achilles Tendon
___ Biceps Femoris
___ Extensor Digitorum Longus
___ Gastrocnemius
___ Gluteus Maximus
___ Gluteus Medius
___ Iliac Crest
___ Iliotibial Tract
___ Peroneus Brevis
___ Peroneus Longus
___ Rectus Femoris
___ Sartorius
___ Semimembranosus
___ Soleus
___ Tensor Fasciae Latae
___ Tibialis Anterior
___ Vastus Lateralis

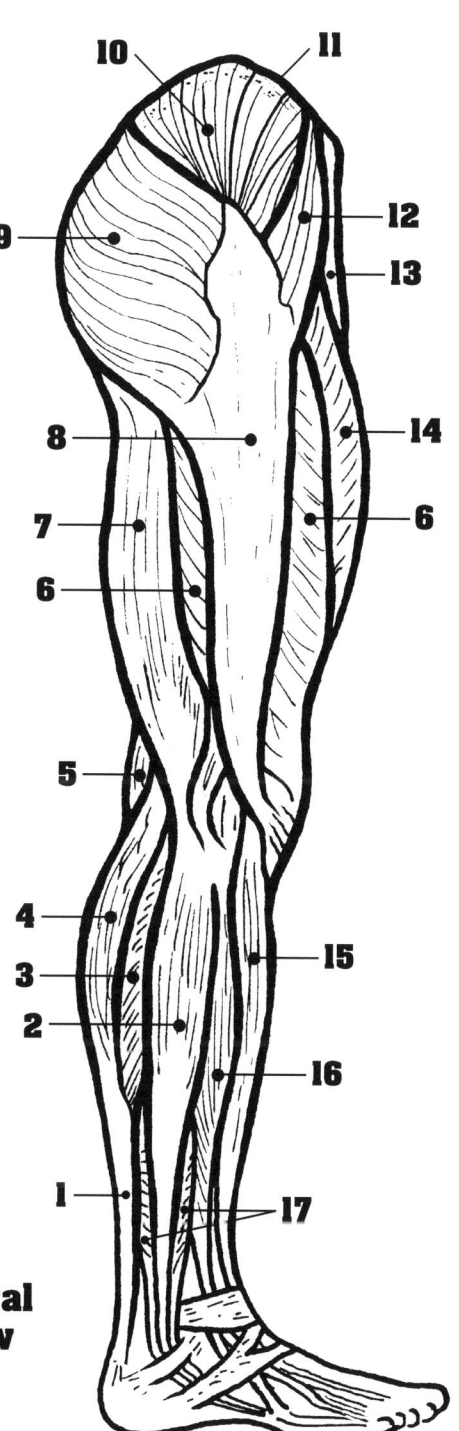

**Lateral View**

# Upper Left Leg
## Anterior View

Muscles of the upper leg perform the following actions:

- **Psoas major** (1) and **iliacus** (2)
  flex and rotate the thigh laterally.

- **Gluteus medius** (10) and **tensor fascia latae** (11)
  abduct the thigh.

- **Adductor magnus** (6), **adductor longus** (4) and **pectineus** (3)
  adduct the thigh.

- **Sartorius** (12)
  rotates the thigh laterally while rotating the leg medially.

- **Gracilis** (5)
  adducts the thigh and rotates the leg at the knee.

- **Rectus femoris** (13), **vastus lateralis** (15), and **vastus medialis** (7)
  extend the leg at the knee.

Also shown are the **iliac crest** (9), **iliotibial tract** (14), **tendon of the rectus femoris** (16), **tendon of the sartorius** (8), and **patellar ligament** (17).

# Upper Left Leg

___ Adductor Longus
___ Adductor Magnus
___ Gluteus Medius
___ Gracilis
___ Iliac Crest
___ Iliacus
___ Iliotibial Tract
___ Patellar Ligament
___ Pectineus
___ Psoas Major
___ Rectus Femorus
___ Sartorius
___ Tendon of the Rectus Femoris
___ Tendon of the Sartorius
___ Tensor Fascia Latae
___ Vastus Lateralis
___ Vastus Medialis

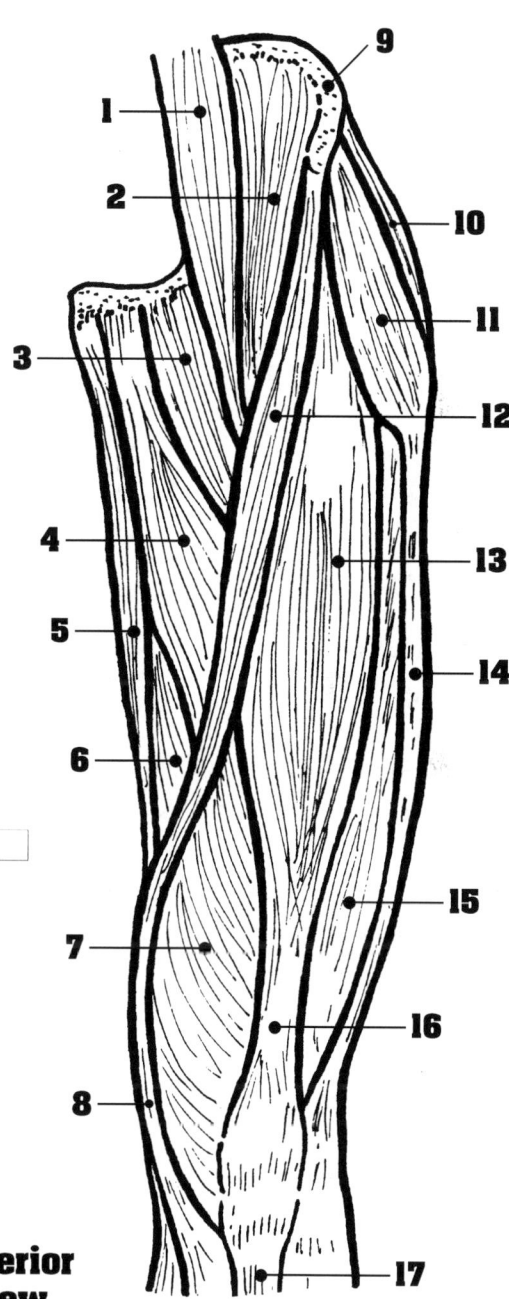

**Anterior View**

# Upper Left Leg
## Posterior View

The **gluteus maximus** (7) is largely removed to show seven underlying muscles:

- **Gluteus medius** (10) — abducts and medially rotates the thigh.

- **Piriformis** (9), **superior gemellus** (8), **obturator internus** (14), **inferior gemellus** (13), and **quadratus femoris** (6) — all rotate the thigh laterally.

- **Adductor magnus** (5) — adducts and laterally rotates the thigh.

From lateral to medial, across the middle of the upper leg the:

- **Vastus lateralis** (4) — extends the knee.

- **Biceps femoris** (3), **semitendinosus** (16), and **semimembranosus** (17) — extend the thigh and flex the knee.

- **Gracilis** (15) — adducts the thigh and medially rotates the leg.

The distal end of the **sartorius** (18) is seen medially near the knee. In the popliteal region, behind the knee, the **plantaris** (2) is seen near its point of origin. The plantaris has a very long, slender tendon which inserts on the calcaneus (heel) bone.

The proximal ends of the double-bellied **gastrocnemius** (1) appear just below the knee.

Also indicated are the **iliac crest** (11) and the **sacrum** (12).

# Upper Left Leg

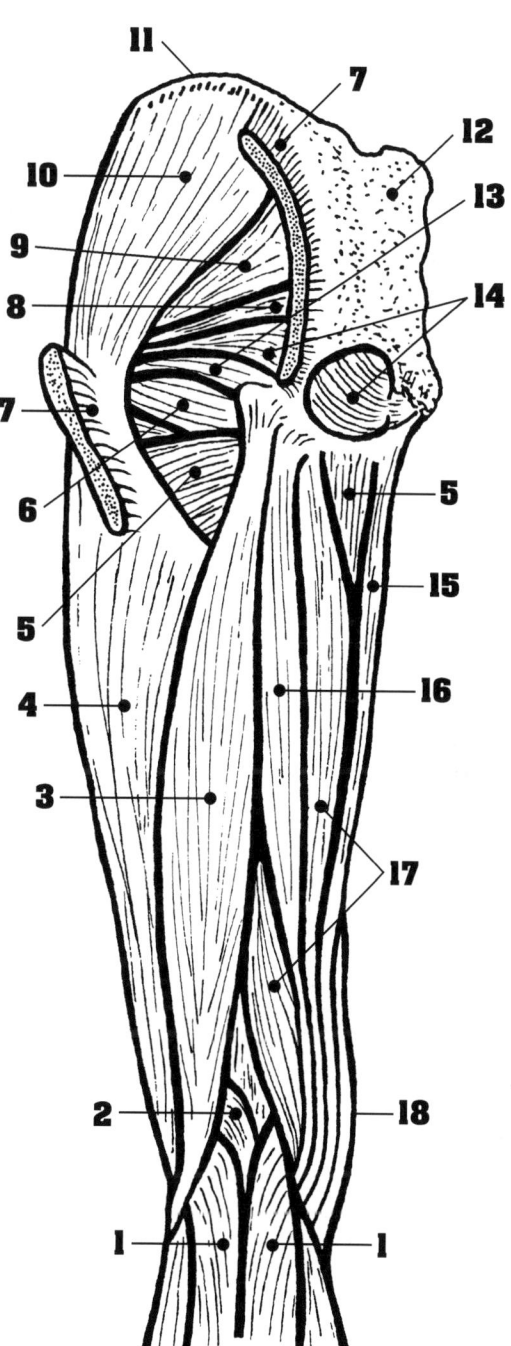

___ Adductor Magnus
___ Biceps Femoris
___ Gastrocnemius
  Gemellus
  ___ Inferior
  ___ Superior
___ Gluteus Maximus
___ Gluteus Medius
___ Gracilis
___ Iliac Crest
___ Obturator Internus
___ Piriformis
___ Plantaris
___ Quadratus Femoris
___ Sacrum
___ Sartorius
___ Semimembranosus
___ Semitendinosus
___ Vastus Lateralis

**Posterior View**

# Lower Left Leg

The **medial malleolus** (5), at the distal end of the tibia establishes the medial side of the leg.

From medial to lateral the:

- **Gastrocnemius** (2)
  plantar flexes the foot.

- **Tibialis anterior** (7)
  dorsiflexes the foot.

- **Extensor digitorum longus** (8)
  dorsiflexes the foot and extends the toes.

- **Peroneus longus** (6)
  plantar flexes and everts the foot.

Toward the distal end of the lower leg the:

- **Soleus** (3)
  plantar flexes the ankle.

- **Extensor hallucis longus** (9)
  extends the great toe and dorsiflexes the foot.

- **Peroneus brevis** (4)
  plantar flexes and everts the foot.

The **patellar ligament** (1) runs vertically on the anterior surface of the knee.

A **superior** (10) and **inferior** (11) **extensor retinaculum** stabilize the tendons at the distal end of the leg.

# Lower Left Leg

___ Extensor Digitorum Longus
___ Extensor Hallucis Longus
    Extensor Retinaculum
___    Inferior
___    Superior
___ Gastrocnemius
___ Medial Malleolus
___ Patellar Ligament
___ Peroneus Brevis
___ Peroneus Longus
___ Soleus
___ Tibialis Anterior

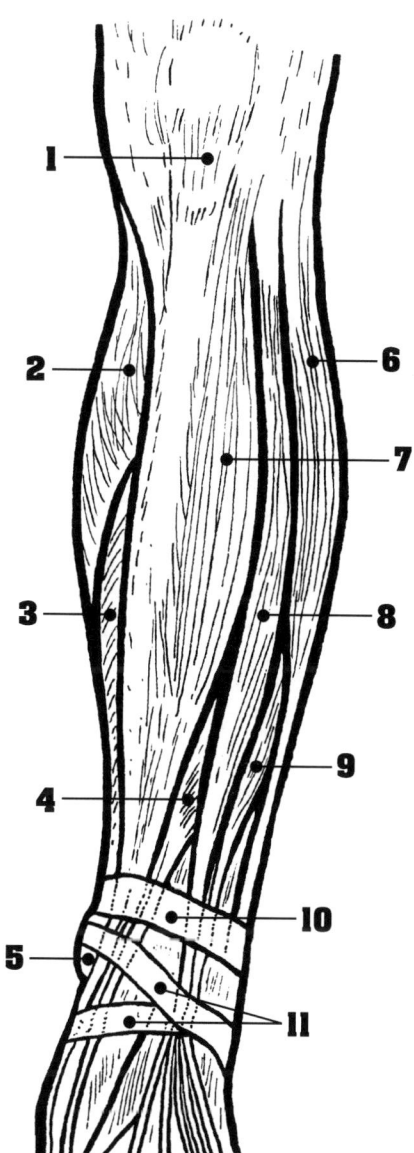

**Anterior View**

# Extrinsic Tongue Muscles

The **thyroid cartilage** (16) and the **cricoid cartilage** (1) are the two major components of the larynx. The **first ring of the trachea** (18) is shown just inferior to the larynx. Immediately superior to the larynx is the **hyoid bone** (14).

A **thyrohyoid membrane** (15) is found between the thyroid cartilage and the hyoid bone. A **cricothyroid membrane** (2) is found between the thyroid cartilage and the cricoid cartilage. The **first tracheal membrane** (17) is found between the cricoid cartilage and the first tracheal ring.

The **styloid process** (10) of the temporal bone, seen just posterior to the tongue, provides a point of origin for three muscles:

- **Styloglossus** (9) — acts to retract the tongue posteriorly.
- **Stylohyoid** (11) — inserts on the hyoid bone and helps to manipulate the "voice box."
- **Stylopharyngeus** (12) — assists in elevating the larynx.

The following three muscles act to move the **tongue** (6):

- **Genioglossus** (7) — depresses and protracts the tongue.
- **Hyoglossus** (13) — depresses the sides of the tongue.
- **Palatoglossus** (8) — elevates and retracts the tongue.

The **geniohyoid** (3) elevates the hyoid bone.

A **tooth** (5) is shown in an alveolar socket of the **mandible** (4).

*When muscles are given names relating to the bones to which they are attached, the origin is given first and the insertion is given last. For example the styloglossus has its origin on the styloid process and its insertion on the tongue (glossus). The hyoglossus has its origin on the hyoid bone and its insertion on the tongue (glossus).*

# Extrinsic Tongue Muscles

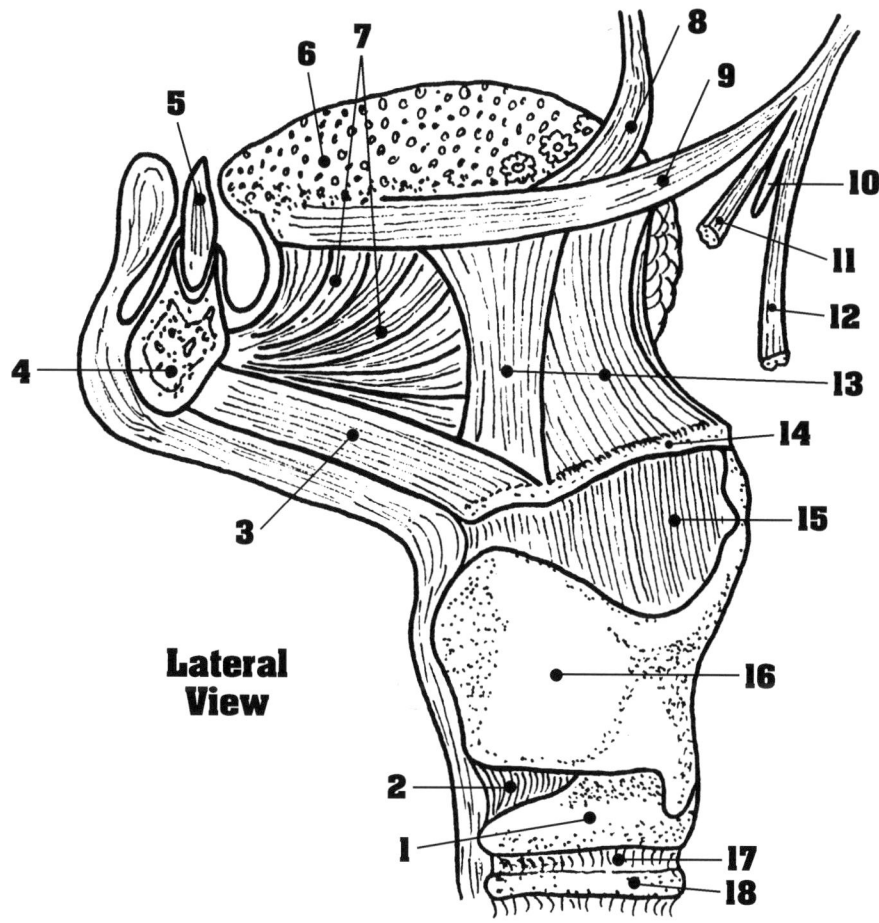

Lateral View

___ Cricoid Cartilage
___ Cricothyroid Membrane
___ First Ring of the Trachea
___ First Tracheal Membrane
___ Genioglossus
___ Geniohyoid
___ Hyoglossus
___ Hyoid
___ Mandible
___ Palatoglossus
___ Styloglossus
___ Stylohyoid
___ Styloid Process
___ Stylopharyngeus
___ Thyrohyoid Membrane
___ Thyroid Cartilage
___ Tongue
___ Tooth

# Extrinsic Left Eye Muscles

Six extrinsic eye muscles act to rotate the **eyeball** (11). The rotational direction is, in each case, indicated by the name of the muscle:

- **Superior rectus** (9) — directs the eye superiorly.

- Inferior rectus (not shown because it is obscured by the optic nerve and superior rectus) — directs the eye inferiorly.

- **Lateral rectus** (8) — directs the eye laterally.

- **Medial rectus** (4) — directs the eye medially.

- **Superior oblique** (3) — directs the eye inferio-laterally.

- **Inferior oblique** (10) — directs the eye superio-laterally.

The **levator palpebrae superioris** (5), which raises the **superior palpebra (upper eyelid)** (12), is cut away to show the superior rectus beneath.

The **superior oblique tendon** (2) runs through a connective tissue pulley called the **trochlea** (1), which is located on the **frontal bone** (13). The **common annular tendon** (7) binds all the muscles at their sites of origin on the **sheath of the optic nerve** (6).

*Three cranial nerves, the abducens, oculomotor, and trochlear, are involved in controlling the extrinsic eye muscles. Keeping the eyeballs properly aligned and coordinated is a major neuromuscular challenge. Should there be any failure in this complex coordination, a condition called strabismus (crossed eyes) can occur.*

# Extrinsic Left Eye Muscles

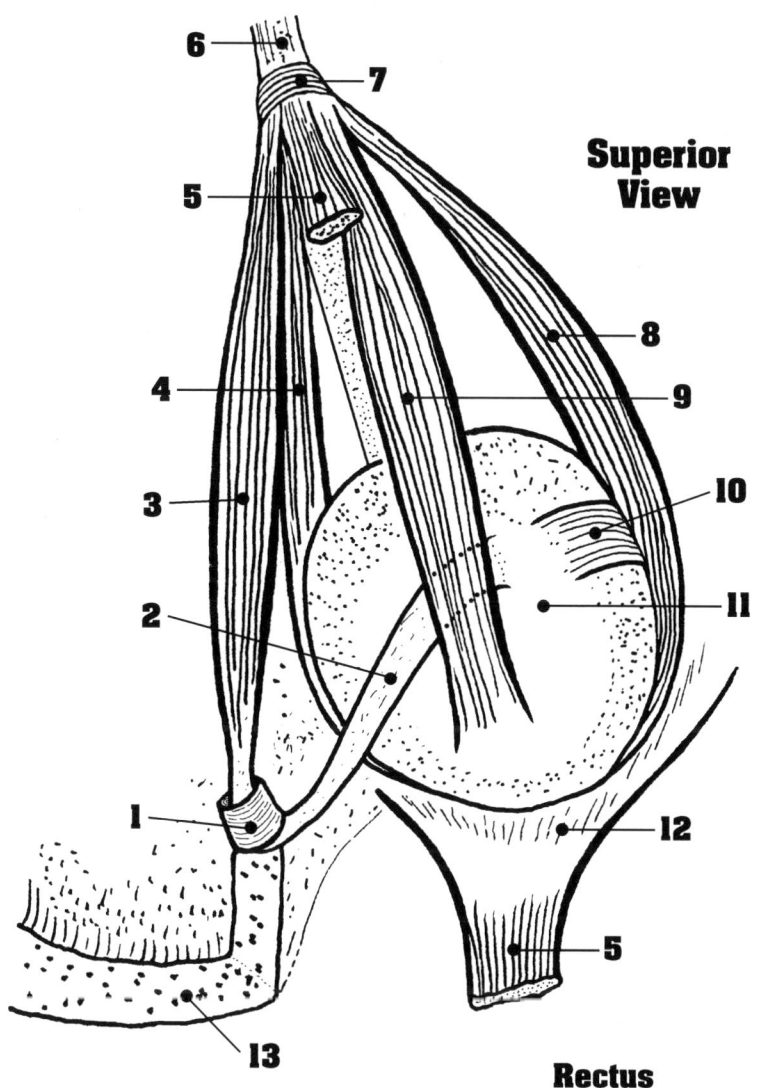

**Superior View**

___ Common Annular Tendon
___ Eyeball
___ Frontal Bone
___ Inferior Oblique
___ Levator Palpebrae Superioris

**Rectus**
___ Lateral
___ Medial
___ Superior
___ Sheath of the Optic Nerve
___ Superior Oblique
___ Superior Oblique Tendon
___ Superior Palpebra
___ Trochlea

# Nervous System

Two of the body's eleven systems (the nervous system and the endocrine system) are devoted to regulating metabolic activities. Although the two systems are, in many cases, intimately integrated, and often referred to as the **neuroendocrine system**, they do, in general differ in their primary modes of communication and regulation. Whereas the nervous system sends electrochemical signals via nerve fibers, the endocrine system sends chemical messages (endocrines) via the blood. Because movement through blood is slower than movement through nerve fibers the nervous system has the ability, in most cases, to act more quickly. However the regulatory effects of the endocrine system are generally longer lasting. Many aspects of the metabolism are co-regulated, either simultaneously, or sequentially, by both systems.

The nervous system is equipped with millions of **sensory receptors** which continuously monitor changes both inside and outside the body. Receptors respond to changes with volleys of electrochemical signals (nerve impulses) collectively referred to as **sensory input**. In a process called **sensory integration** the sensory input is processed in **interpretive centers** within the brain and spinal cord (central nervous system). In response to integration volleys of nerve impulses, collectively called **motor output**, are sent out to effector organs.

Thus the nervous system operates via a five-part sequence:

- **Sensory reception** — via sensory receptors.

- **Sensory input** — via sensory nerves.

- **Sensory integration** — via the brain and spinal cord.

- **Motor output** — via motor nerves.

- **Motor responses** — via motor effector organs.

# Nervous System

While the brain and spinal cord constitute the **central nervous system (CNS)**, the nerves which carry nerve impulse to and from it, are collectively referred to as the **peripheral nervous system (PNS)**. While the **sensory (afferent) nerves**, which carry impulses toward the central nervous, comprise the **sensory division** of the peripheral nervous system, the **motor (efferent) nerves**, which carry impulses away from the brain and spinal cord comprise the **motor division** of the peripheral nervous system.

The motor division is divided into the **autonomic nervous system** and the **somatic nervous system**. The autonomic nervous system is further subdivided into **sympathetic** and **parasympathetic divisions**. The following matching exercise will help you associate the above divisions of the nervous system with their major functions:

**Divisions**

___ **Autonomic**
___ **Central (CNS)**
___ **Motor**
___ **Parasympathetic**
___ **Peripheral (PNS)**
___ **Sensory**
___ **Sympathetic**
___ **Somatic**

**Functions**

A. **Carries impulses away from CNS**
B. **Carries impulses toward CNS**
C. **Carries motor & sensory impulses**
D. **Facilitates "fight or flight"**
E. **Facilitates "rest and repose"**
F. **Involuntary motor responses**
G. **Processes and interprets "input"**
H. **Voluntary motor responses**

# Neuron

Whereas **dendrites** (5) carry action potentials toward **nerve cell bodies** (4), **axons** (3) carry actions potentials away from nerve cell bodies.

The **axon hillock** (7) acts as an integrative center where incoming action potentials are calculated to determine the frequency of outgoing action potentials.

The axon of the neuron shown here is wrapped with **Schwann cells (neurolemmocytes)** (2). Collectively, Schwann cells constitute what is called a myelin sheath. The spaces between Schwann cells are called **nodes of Ranvier** (1). Myelinated fibers carry impulses at a faster rate because **saltatory action potentials** (8) jump from node to node rather than having to travel continuously along the axon, as is the case in unmyelinated fibers.

**Terminal end filaments** (10) with **synaptic end knobs** (9) diverge at the end of the axon.

… And the cell body, of course, has a **nucleus** (6).

# Neuron

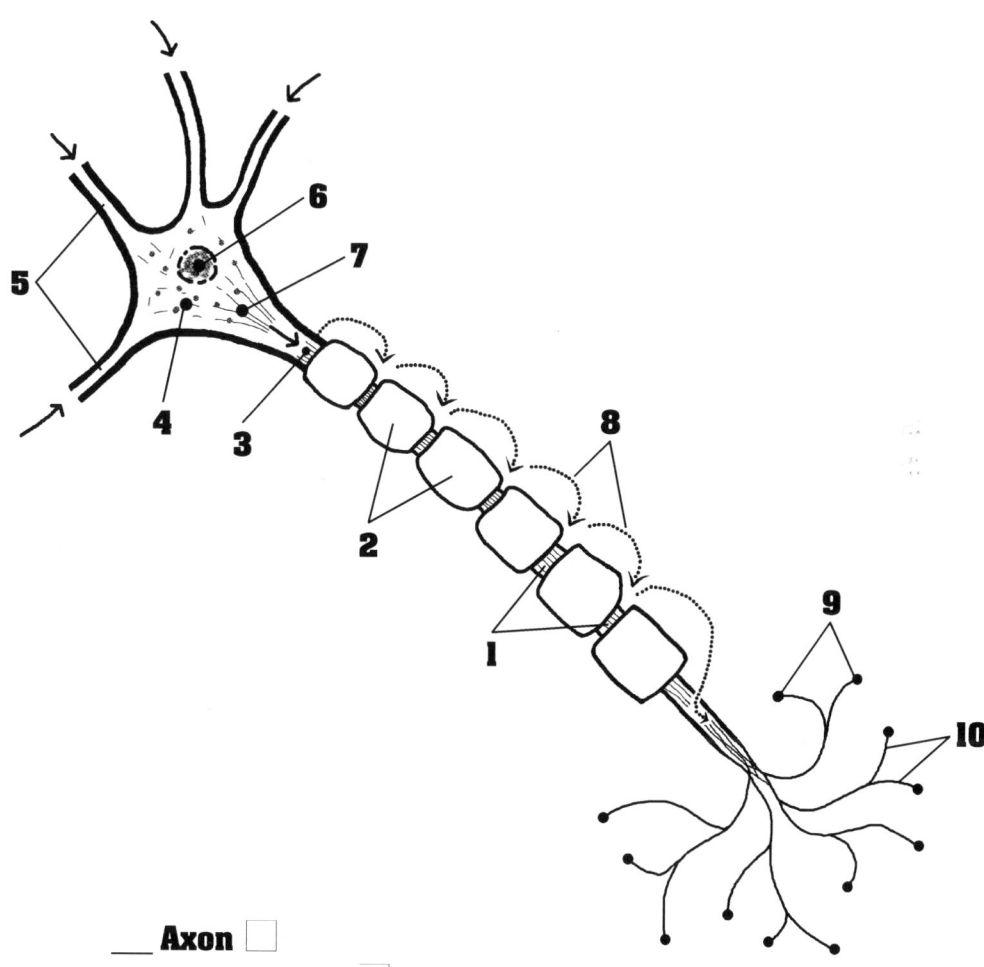

___ Axon
___ Axon Hillock
___ Dendrites
___ Nerve Cell Body
___ Nodes of Ranvier
___ Nucleus
___ Saltatory Action Potentials
___ Schwann Cells (Neurolemmocytes)
___ Synaptic End Knobs
___ Terminal End Filaments

# Action Potential

The nature of a nerve impulse is best understood in conjunction with an oscilloscope "picture" showing the changes in voltage which occur as the action potential passes the points of electrode recorders. The drawing to the right shows the typical "spike" produced by the electrical events (depolarizations and repolarizations).

Previous to a stimulus the **resting membrane potential** (1) is -70 mv. After a **stimulus** (2), there is a **lag phase** (4) of about .2 ms before **sodium gates open** (3) to allow the passage of sodium ions across the membrane. Depolarization proceeds relatively slowly until the **threshold** (5) of -55 mv is reached. If the stimulus is not great enough the depolarization will not reach the threshold and will result in an **aborted depolarization** (6).

If, however, the stimulus is strong enough to cause sufficient depolarizations so as to reach the threshold, additional sodium gates will open, and a self-propagating wave of depolarization will be started. The wave of depolarization, lasting about .5 ms, will then proceed until the membrane is not only depolarized, but also reversely polarized to a point of +30 mv, at which point the **sodium gates close** (7), and the **potassium gates open** (8). With sodium ions no longer flowing in, and with potassium ions now flowing out — both ions carrying positive charges — a repolarization of the membrane is quickly achieved in about .5 ms. Just as the wave of depolarization overshoots into reverse polarization so also does the wave of repolarization "undershoot" into **hyperpolarization** (9) before settling at a **post AP resting potential** (10) of -70 mv. The nerve fiber is now ready to "fire" again.

An action potential, then, is comprised of a wave of depolarization (a 100 mv shift from -70 mv to +30 mv), followed by a wave of repolarization (a 100 mv shift from +30 mv to -70 mv) — and all this achieved in less than 5 ms! If there were no "limiting factors" this would mean nerve impulses could pass as frequently as 200/second! There are, however, limiting factors. It is generally estimated that APs occur at frequencies between about 10 to 100/second. The variation in the speed at which APs travel (.1 to 100 meters/second) is perhaps even greater than the variation of their frequencies. The speed at which an impulse travels depends primarily on the diameter of the nerve fiber through which it passes and the degree of myelination ("insulation") which surrounds the nerve.

# Action Potential

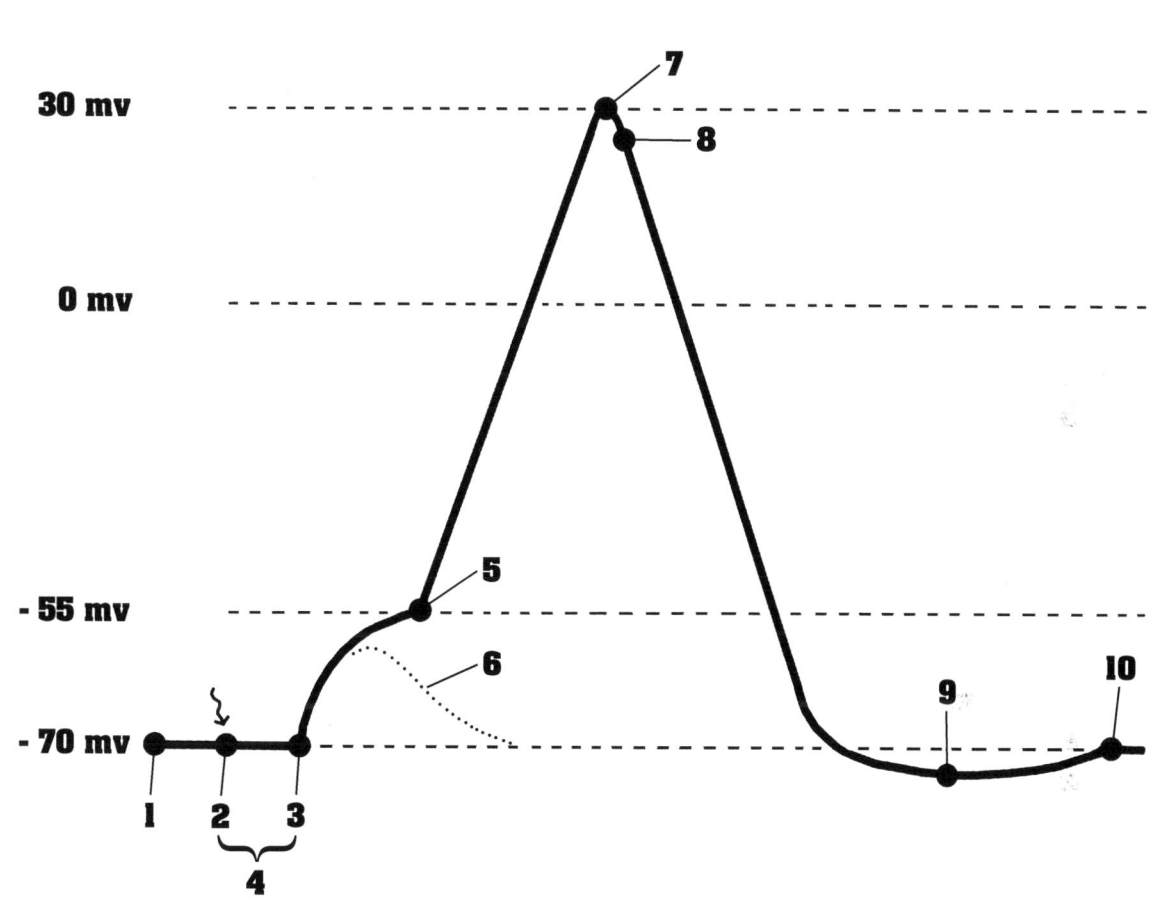

Aborted Depolarization ☐
___ Hyperpolarization ☐
___ Lag Phase ☐
___ Post AP Resting Potential ☐
___ Potassium Gates Open ☐
___ Resting Membrane Potential ☐
___ Sodium Gates Close ☐
___ Sodium Gates Open ☐
___ Stimulus ☐
___ Threshold ☐

# Synaptic Events
## At a Neuro–Neurojunction

Below is a sequence of synaptic events:

- **Presynaptic action potentials** (7) arrive at a **synaptic end bulb** (9) and cause **calcium ion gates** (5) to open and calcium ions to be released.

- **Calcium ions** (4) act on **synaptic vesicles** (3), causing them to exocytose at the **presynaptic membrane** (10) and release **ACh (acetylcholine) molecules** (2).

- Acetylcholine molecules move across the **synaptic cleft** (11) and attach to **ACh receptor sites** (13) on **sodium ion gates** (14) embedded in the **postsynaptic membrane** (12).

- When the ACh receptor sites are occupied by ACh the sodium gates open and **sodium ions** (1) pour through the gates and depolarize the postsynaptic membrane.

- Depolarization effects the generation of **postsynaptic membrane potentials** (15) and an accumulation of membrane potentials will, in turn, generate new **postsynaptic action potentials** (16).

Energy needed for synaptic events is provided by **ATP molecules** (8) which are synthesized and secreted by numerous **mitochondria** (6) packed into the synaptic end bulbs.

*Acetylcholine is only one of more than 50 different neurotransmitter substances that have been identified. Many neurotoxins and drugs (both medicinal and recreational) act at synapses to stimulate, inhibit or block the flow of transmitter substances at synaptic clefts. Transmitter substances act as "words" or "sentences" in the language of the brain.*

# Synaptic Events
## At a Neuro-Neurojunction

147

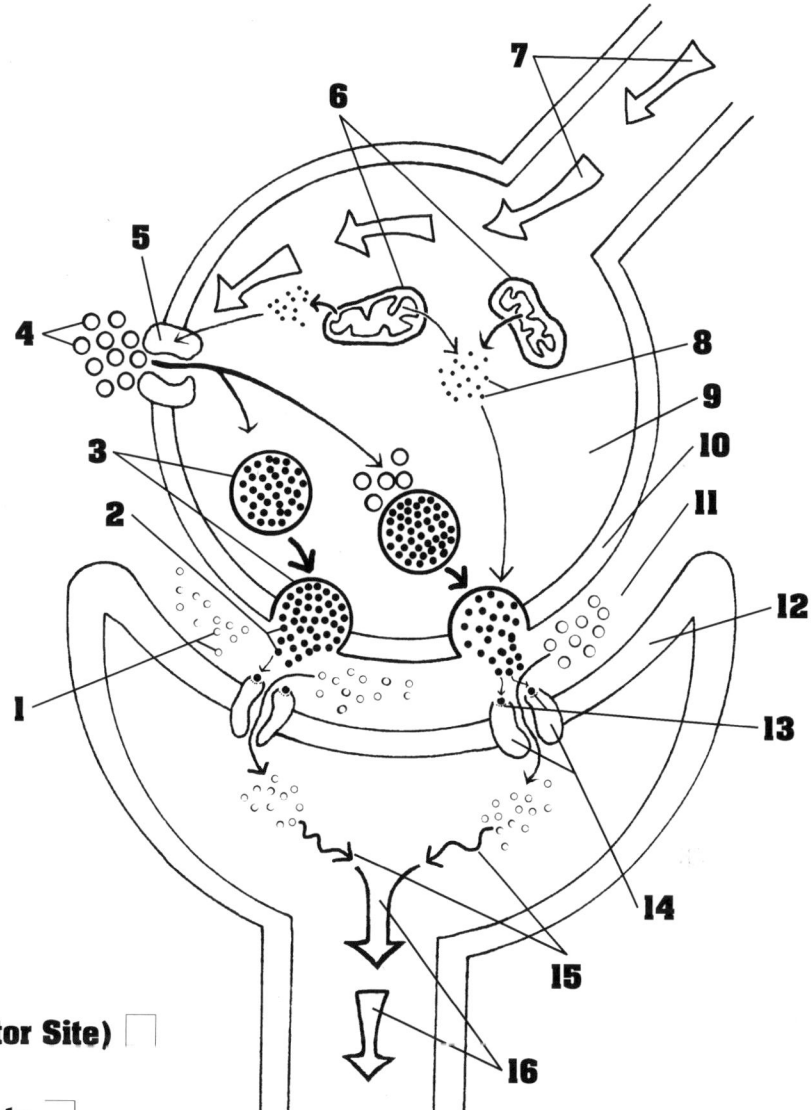

___ ACh
___ ACh (at Receptor Site)
___ ATP
___ Calcium Ion Gate
___ Calcium Ions
___ Mitochondria
___ Postsynaptic Membrane
___ Postsynaptic Action Potentials
___ Presynaptic Action Potentials
___ Presynaptic Membrane
___ Postsynaptic
    Membrane Potentials

___ Sodium Ion Gate
___ Sodium Ions
___ Synaptic Cleft
___ Synaptic End Bulb
___ Synaptic Vesicles

© Copyright 2010 Gene Johnson

# Brain
## Midsagittal Section

The brain and the spinal cord are protected by bones, meninges, and cerebrospinal fluid (CSF). The three protective meninges are the **pia mater** (2), **arachnoid membrane** (3), and **dura mater** (13). The cerebrospinal fluid, which surrounds the brain and spinal cord, circulates in the **subarachnoid space** (22). A **subdural space** (31) is located between the arachnoid membrane and the dura mater.

The cerebrospinal fluid is produced by **choroid plexuses** (18) that lie within fluid-filled chambers called ventricles. Cerebrospinal fluid from two **lateral ventricles** (16) passes through an **interventricular foramen** (14) into the **third ventricle** (11). From the third ventricle the fluid flows downward through the **cerebral aqueduct** (27) into the **fourth ventricle** (30). There are four passageways from the fourth ventricle: two **lateral apertures** (29), one **median aperture** (4), and the **central canal** (1) which continues downward through the middle of the **spinal cord** (32).

The cerebrospinal fluid eventually exits the subarachnoid space via the **arachnoid villi** (20) and enters the **superior sagittal sinus** (17), an expanded vein that runs across the top of the brain. Ciliated ependymal cells, which line the chambers and passageways, assist in keeping the fluid in circulation.

The **cerebrum** (21) consists of two cerebral hemispheres connected by the **corpus callosum** (19), **fornix** (15), **anterior commissure** (12), and **posterior commissure** (24). The cerebrum is separated from the **cerebellum** (28) by an invagination of the meninges called the **tentorium cerebelli** (25).

Six "anatomical bumps" mark the anterior aspect of the base of the brain. From inferior to superior, they are the **medulla** (5), **pons** (6), **cerebral peduncle** (7), **mammillary body** (8), **pituitary** (9), and **optic chiasma** (10).

Three "anatomical bumps" mark the posterior aspect of the brain stem. From inferior to superior, they are the **quadrigemina** (26), posterior commissure, and **pineal gland** (23).

# Brain
## Midsagittal Section

**Aperture**
___ Lateral
___ Median
___ Arachnoid Membrane
___ Arachnoid Villi
___ Central Canal
___ Cerebellum
___ Cerebral Aqueduct
___ Cerebral Peduncle
___ Cerebrum
___ Choroid Plexus
**Commissure**
___ Anterior
___ Posterior
___ Corpus Callosum
___ Dura Mater
___ Fornix
___ Interventricular Foramen
___ Mammillary Body
___ Medulla
___ Optic Chiasma
___ Pia Mater
___ Pineal Gland
___ Pituitary
___ Pons
___ Quadrigemina
___ Spinal Cord
___ Subarachnoid Space
___ Subdural Space
___ Superior Sagittal Sinus
___ Tentorium Cerebelli
**Ventricle**
___ Fourth
___ Lateral
___ Third

# Brain Stem & Ventricles

Details of the brain stem, also included on the previous exercise on the brain, are here enlarged and reviewed.

A **lateral ventricle** (11) with its **choroid plexus** (9) is seen just below the **corpus callosum** (8), and from here we can follow the cerebrospinal fluid through the **interventricular foramen** (12) into the **third ventricle** (13) and on through the **cerebral aqueduct** (4) into the **fourth ventricle** (3), and finally down through the **central canal** (2) which runs through the **spinal cord** (1).

The six "anatomical bumps" on the ventral aspect of the brain stem are, from inferior to superior:

- **Medulla** (19)
- **Pons** (18)
- **Cerebral peduncle** (17)
- **Mammillary body** 16)
- **Pituitary** (15)
- **Optic chiasma** (14)

The three "anatomical bumps" on the dorsal aspect of the brain stem are, from inferior to superior:

- **Inferior colliculus** (5)
- **Superior colliculus** (6)
- **Pineal gland** (7)

The partition between the lateral ventricles and the third ventricle is the **fornix** (10).

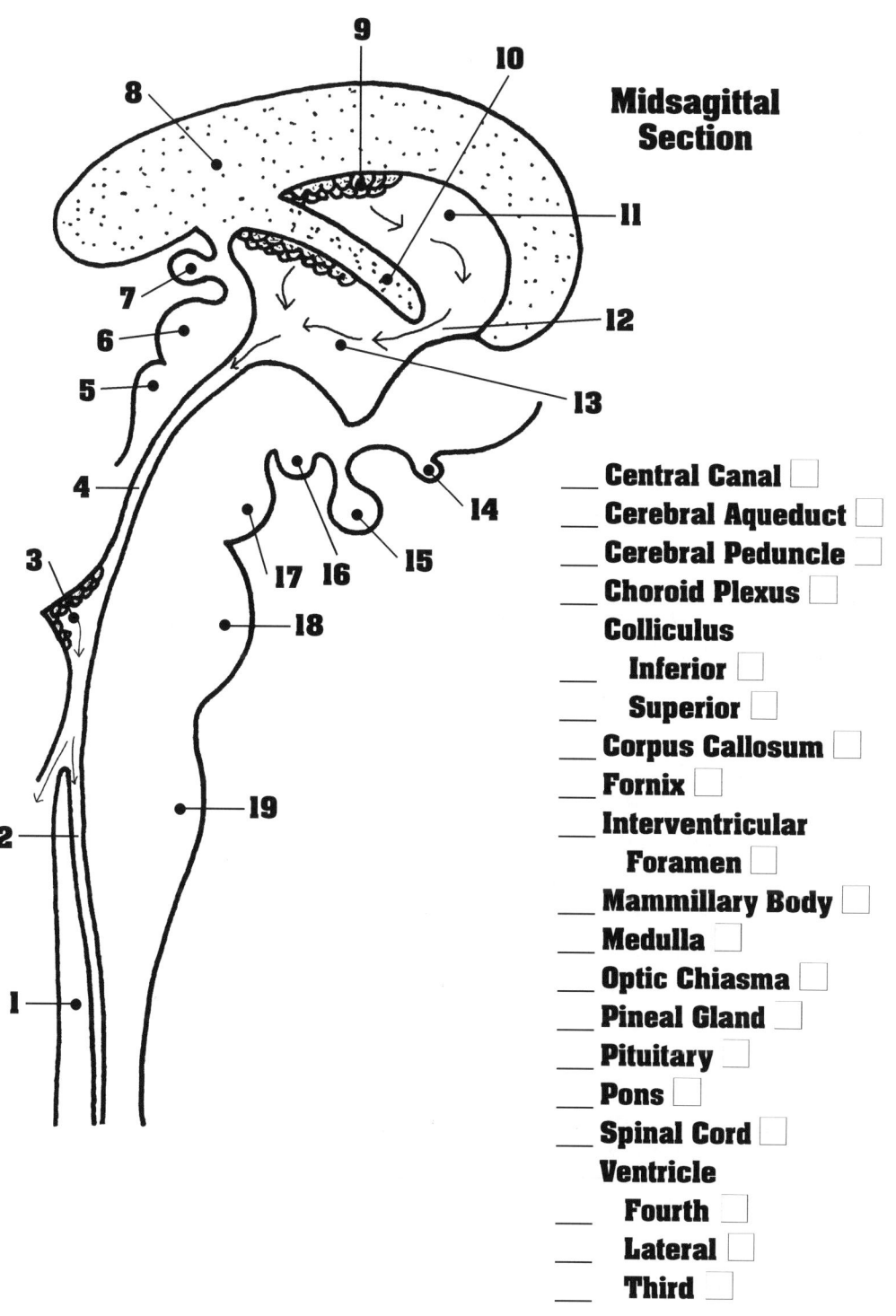

# Brain
## Frontal Section

A regional cluster of nerve cell bodies outside the central nervous system is called a ganglion. A regional cluster of nerve cell bodies within the central nervous system (the brain and spinal cord) is called a nucleus.

In this frontal section of the cerebrum we see some of the major nerve cell clusters (nuclei) that lie within the cerebrum. The entire cerebrum is surrounded with a layer thickly populated with nerve cell bodies. This layer is, of course, the **cerebral cortex** (12). The cerebral cortex, which is greatly exaggerated in most drawings — including this one! — is only about 2 mm thick, although seven or eight orders of neurons are packed into that small gray matter space!

The human brain, as with all mammalian brains, is highly convoluted to provide increased surface area for more cell bodies, more interneuronal connections, and hopefully more capacity for reason, memory, etc! The convolutions are comprised of alternating "valleys and hills" called **sulci** (2) and **gyri** (1). Additional surface area is provided by an involution in each temporal lobe called an **insula** (14). Some anatomists are now referring to the insula as the fifth lobe of the brain.

Located deep within the **white matter** (11) are the following subcortical basal nuclei:

- The **globus pallidus** (4) and the **putamen** (3), together, comprise the **lenticular (lentiform) nucleus** (5).

- The lentiform nucleus, together with the **caudate nucleus** (6), constitute the **corpus striatum** (7). The caudate nucleus is the most superior of the nuclei and is located more medially, and immediately adjacent to the **lateral ventricles** (8).

- Just below the lateral ventricles there are two large **thalamic lobes** (13), and on either side of the **third ventricle** (16) there are **hypothalamic nuclei** (15). Below the hypothalamic nuclei are two **optic tracts** (18), and most inferiorly, the **pituitary** (17).

Just inferior to the **longitudinal fissure** (9), which separates the two cerebral hemispheres superiorly, we see the **corpus callosum** (10) which is the only connection between the two cerebral hemispheres.

# Brain

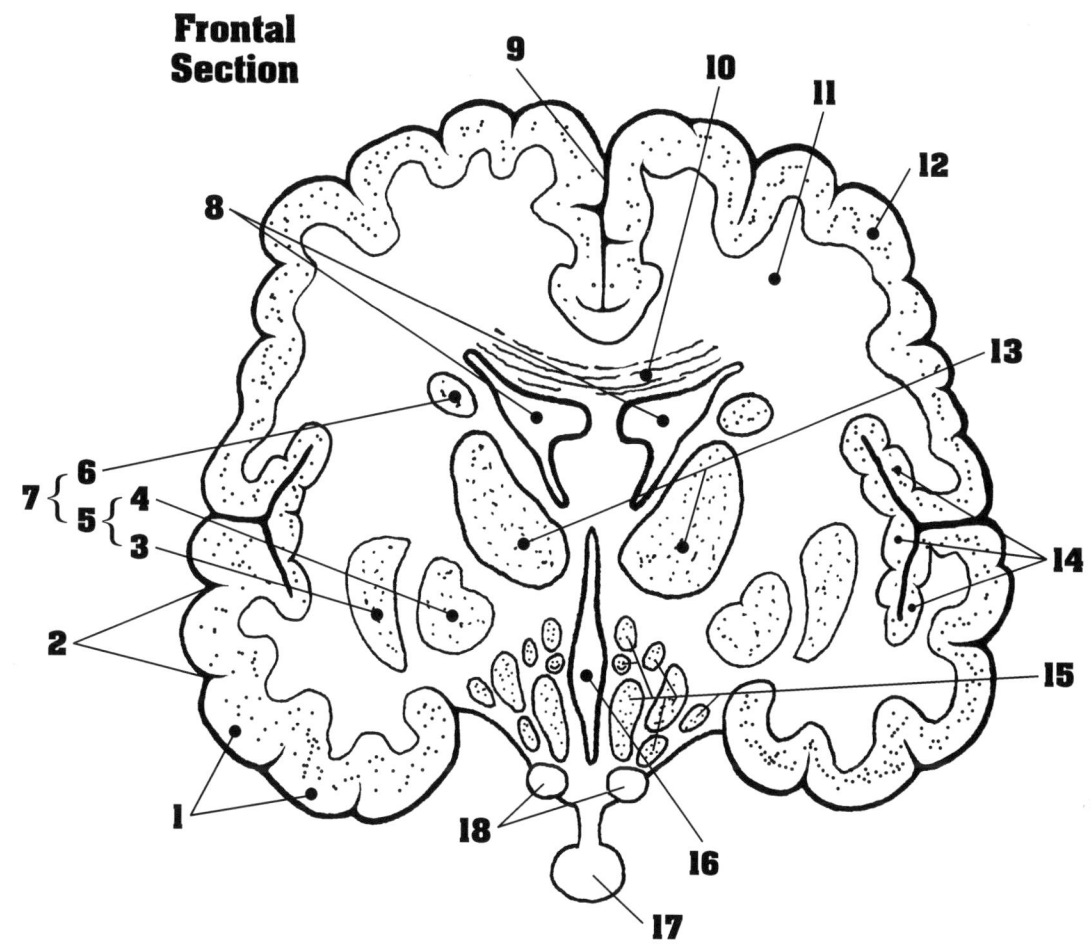

**Frontal Section**

___ Caudate Nucleus
___ Cerebral Cortex
___ Corpus Callosum
___ Corpus Striatum
___ Globus Pallidus
___ Gyri
___ Hypothalamic Nuclei
___ Insula
___ Lateral Ventricles
___ Lenticular Nucleus
___ Longitudinal Fissure
___ Optic Tracts
___ Pituitary
___ Putamen
___ Sulci
___ Thalamic Lobes
___ Third Ventricle
___ White Matter

# Ventricles

The ventricles of the brain, shown in-situ in previous exercises, are here isolated and enlarged.

Each of the two **lateral ventricles** (8) has an **anterior** (7), **posterior** (9), and **inferior** (4) **horn.**

CSF (cerebrospinal fluid) produced in the lateral ventricle flows through the **interventricular foramen** (6) into the **third ventricle** (5). Additional CSF is produced by the third ventricle, and the fluid is passed on through the **cerebral aqueduct** (3) to the **fourth ventricle** (10).

The fourth ventricle, in turn, secretes still more CSF, and from the fourth ventricle the CSF has four avenues of export:

The **central canal** (1) is a CSF-filled canal that runs through the middle of the spinal cord for its entire length.

Two **lateral apertures** (2) and one **median aperture** (11) provide avenues to the subarachnoid space in the meningeal membranes which surround both the brain and the spinal cord.

Thus, both the brain and spinal cord are surrounded and bathed in cerebrospinal fluid, and both have a fluid-core of cerebrospinal fluid. About 500 ml of CSF are formed daily. The movement of the fluid is facilitated by the ciliary-like action of long microvilli which extend from ependymal cells lining the ventricles. CSF returns to the blood via the arachnoid villi in the dural sinuses (shown on page 149).

*If, for any reason, the flow of cerebrospinal fluid is impeded or blocked, pressure will build up in the ventricles and cause discomfort, or even severe damage. If a blockage happens in a young child, before the cranial bones are fully ossified and hardened, the entire cranium will enlarge — a condition referred to as hydrocephalic.*

*Some of the benefits of CSF are quite clearly understood, while others are putative (conjectured) or even numinous (mysterious). By forming a liquid cushion the CSF essentially "floats" the brain and effectively reduces its weight by more than 90%. Without the buoyancy provided by the ventricles the heavy, but delicate, jelly-like brain mass would crush itself. The CSF does, of course, also go a long way in protecting the brain and spinal cord from accidental blows and impacts.*

*It is generally thought that the CSF helps to nourish and cleanse the brain, and evidence is accumulating that CSF also may function in the transport of signal molecules and thereby play an important regulatory role.*

# Ventricles

__ Central Canal ☐
__ Cerebral Aqueduct ☐
Horn
__ Anterior ☐
__ Inferior ☐
__ Posterior ☐
__ Interventricular Foramen ☐
__ Lateral Aperture ☐
__ Median Aperture ☐
Ventricles
__ Fourth ☐
__ Lateral ☐
__ Third ☐

**Lateral View**

# Cranial Nerves

The first two pairs of cranial nerves attach to the forebrain — the rest originate from the brain stem (which includes the midbrain, pons, and medulla oblongata). With the exception of the vagus nerve, cranial nerves serve only head and neck structures.

In most cases the names of the cranial nerves reveal the most important structures they serve, or their primary functions. The primary functions should be reviewed in your textbook. The cranial nerves are assigned Roman numerals I to XII, from the most rostral to the most caudal:

- I. **Olfactory** (1)
- II. **Optic** (2)
- III. **Oculomotor** (3)
- IV. **Trochlear** (4)
- V. **Trigeminal** (5)
- VI. **Abducens** (6)
- VII. **Facial** (7)
- VIII. **Vestibulocochlear** (8)
- IX. **Glossopharyngeal** (9)
- X. **Vagus** (10)
- XI. **Accessory** (11)
- XII. **Hypoglossal** (12)

*On Old Olympus's Towering Top A Fierce Very Gluttonous Vulture Ate Hikersbones ... (or if you don't like that, compose an acronymous sentence of your own)!*

The **first** (13) and **second** (14) **cervical nerves** appear inferior to the hypoglossal nerve.

The **spinal cord** (15), **medulla** (17), **pons** (18), **mammillary body** (20), and **pituitary** (21) are shown along the midline. The **cerebellum** (16) and the **frontal** (22) and **temporal** (19) **lobes of the cerebrum** are also shown.

# Cranial Nerves

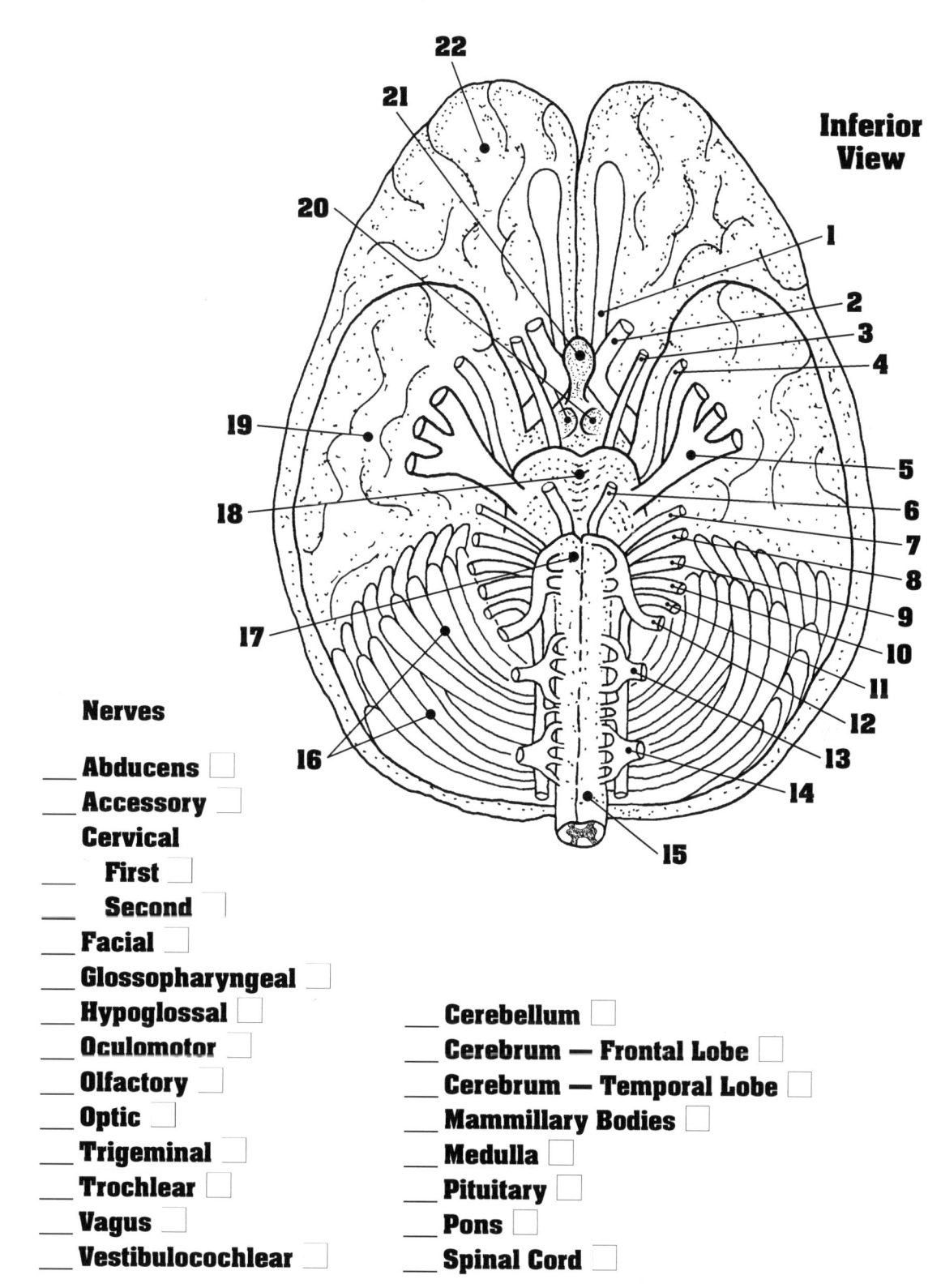

Inferior View

**Nerves**

___ Abducens
___ Accessory
   Cervical
___   First
___   Second
___ Facial
___ Glossopharyngeal
___ Hypoglossal
___ Oculomotor
___ Olfactory
___ Optic
___ Trigeminal
___ Trochlear
___ Vagus
___ Vestibulocochlear

___ Cerebellum
___ Cerebrum — Frontal Lobe
___ Cerebrum — Temporal Lobe
___ Mammillary Bodies
___ Medulla
___ Pituitary
___ Pons
___ Spinal Cord

# Spinal Nerves

Thirty pairs of spinal nerves and one unpaired nerve emerge from the spinal cord:

- C1 to C8 — eight pairs of **cervical nerves** (1)
- T1 to T12 — twelve pairs of **thoracic nerves** (2)
- L1 to L5 — five pairs of **lumbar nerves** (3)
- S1 to S5 — five pairs **sacral nerves** (4)
- And one **coccygeal** (5)

The spinal cord is enlarged at the base of the neck — the **cervical enlargement** (19), and in the lumbar region — the **lumbar enlargement** (17). Although the spinal cord ends at the **conus medullaris** (16), numerous spinal nerves continue downward forming a large tract of nerves called the **cauda (tail) equina (horse)** (11).

Four nerve plexuses arise from the spinal cord:

- **Cervical plexus** (20)
- **Brachial plexus** (18)
- **Lumbar plexus** (15)
- **Sacral plexus** (10)

Two major nerves of the lumbar plexus are the **femoral** (14) and **obturator** (13). Two major nerves of the sacral plexus are the **common fibular** (8) and **tibial** (7), which together constitute the **sciatic** (9).

The **sacrum** (6) and the **iliac crest** (12) are also shown.

# Spinal Nerves

**Posterior View**

___ Brachial Plexus
___ Cauda Equina
___ Cervical Enlargement
___ Cervical Nerves
___ Cervical Plexus
___ Coccygeal
___ Common Fibular
___ Conus Medullaris
___ Femoral
___ Iliac Crest
___ Lumbar Enlargement
___ Lumbar Nerves
___ Lumbar Plexus
___ Obturator
___ Sacral Nerves
___ Sacral Plexus
___ Sacrum
___ Sciatic
___ Thoracic Nerves
___ Tibial

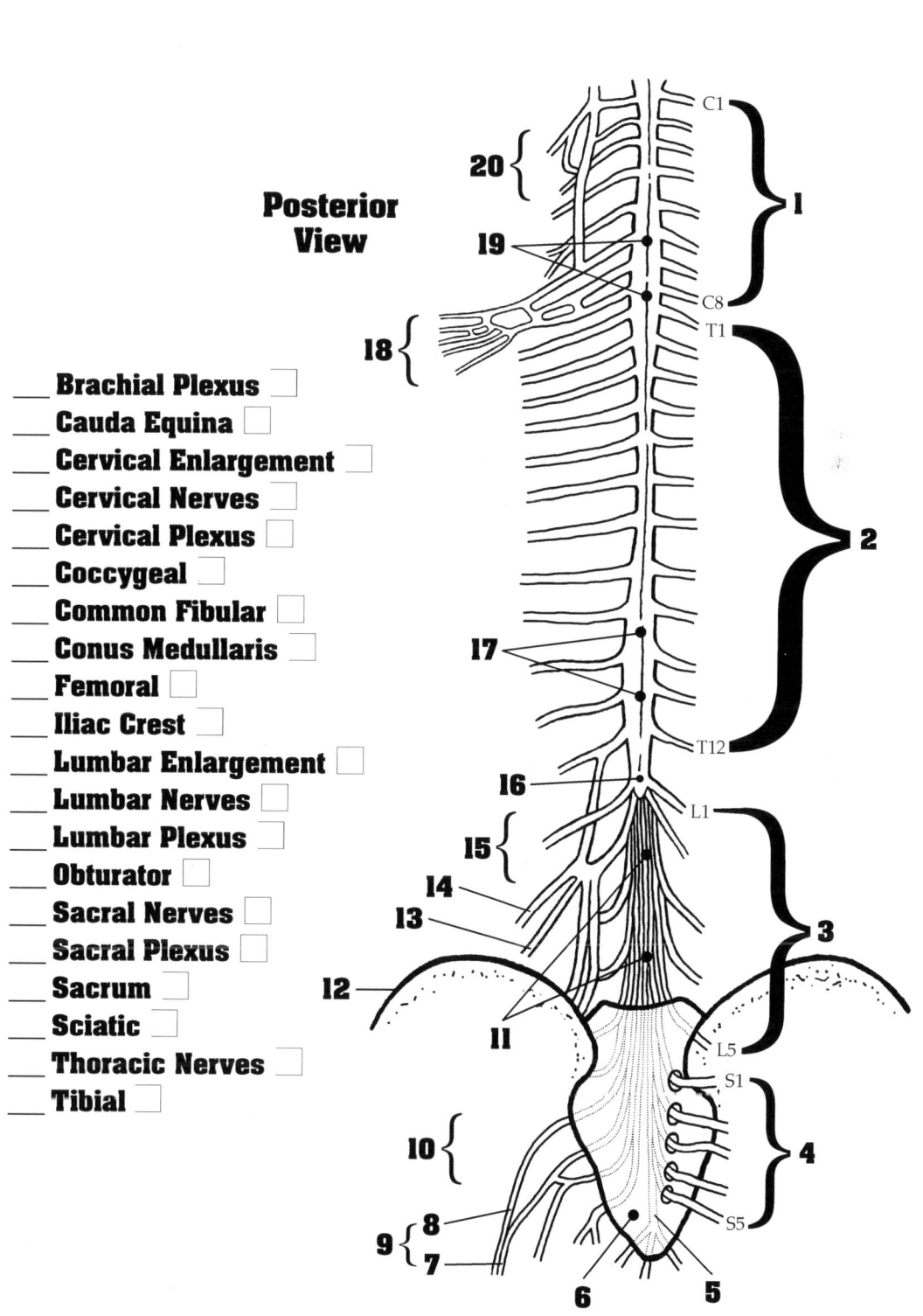

# Spinal Nerve Origin

The **spinal cord** (1) is shown here within the vertebral foramen of a vertebra. The **spinous process** (20) is posterior and the **body of the vertebra** (13) is anterior.

The spinal cord is held in position by **denticulate ligaments** (10) and protected by three meninges:

- **Pia mater** (14)
- **Arachnoid membrane** (17)
- **Dura mater** (18)

Above the dura mater the **epidural space** (12) is filled with protective **fat** (11) tissue. Beneath the dura mater is a **subdural space** (19). And beneath the arachnoid membrane, the **subarachnoid space** (15) is filled with cerebrospinal fluid.

A **dorsal root** (2) and a **ventral root** (16) merge to form a **spinal nerve** (5). Sensory (afferent) impulses come to the spinal cord via dorsal roots and synapse at a **dorsal root ganglion** (6). Motor (efferent) impulses leave the spinal cord via ventral roots.

Motor information leaving the spinal cord has a connection with a **sympathetic chain ganglion** (9) via the **gray ramus communicans** (7) and the **white ramus communicans** (8). The sympathetic chain ganglia are key components of the autonomic nervous system (see page 167).

Shortly after emerging from the spinal cord, the spinal nerve divides into a relatively large **ventral ramus** (4) and a smaller **dorsal ramus** (3).

# Spinal Nerve Origin

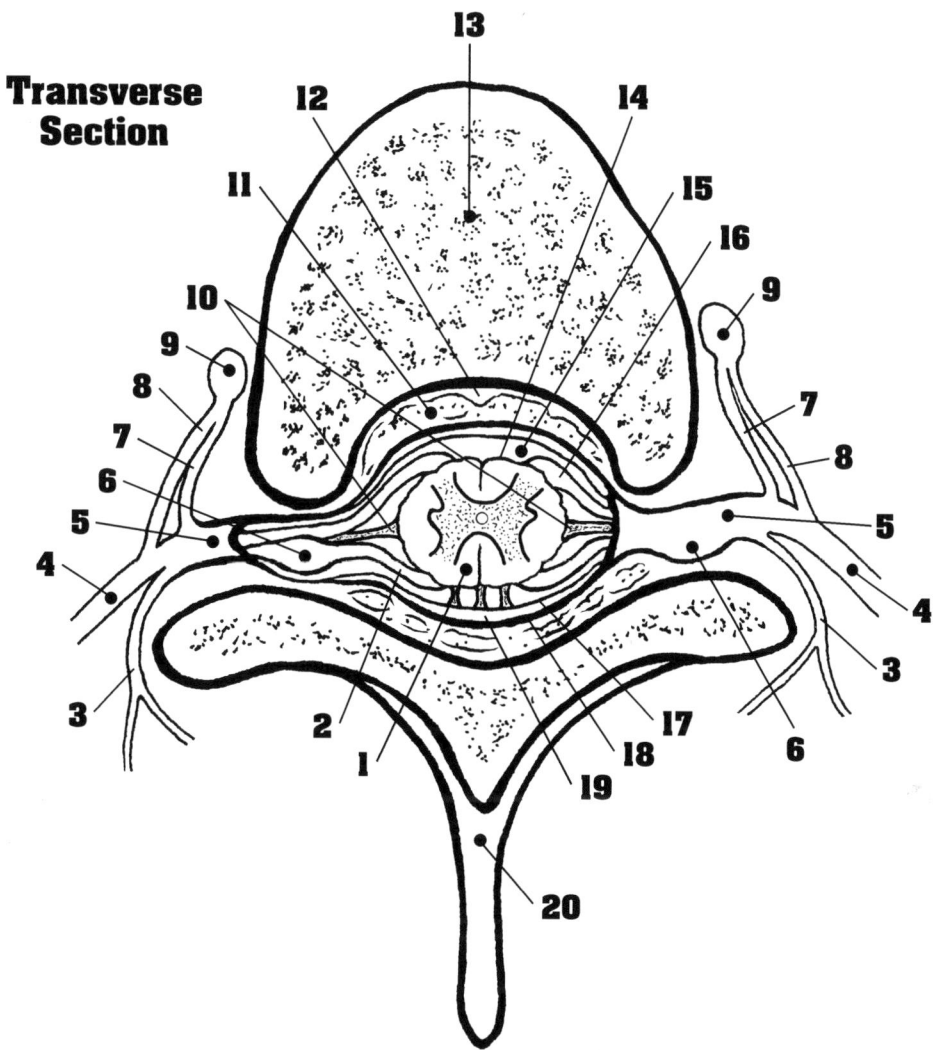

Transverse Section

___ Arachnoid Membrane
___ Body of the Vertebra
___ Denticulate Ligaments
___ Dorsal Ramus
___ Dorsal Root
___ Dorsal Root Ganglion
___ Dura Mater
___ Epidural Space
___ Fat (in Epidural Space)
___ Gray Ramus Communicans
___ Pia Mater
___ Spinal Cord
___ Spinal Nerve
___ Spinous Process
___ Subarachnoid Space
___ Subdural Space
___ Sympathetic Chain Ganglion
___ Ventral Ramus
___ Ventral Root
___ White Ramus Communicans

# Spinal Nerve

A **dorsal root** (7) and a **ventral root** (8) emerge from their origins in the **spinal cord** (9) and converge to form a **spinal nerve** (6). The dorsal root is distinguished from the ventral root by the presence of a **dorsal root ganglion** (10).

The nerve is subdivided into **fascicles** (3) and each fascicle contains many **nerve fibers** (1).

Three connective tissues support a spinal nerve:

- **Epineurium** (5) — surrounds the entire nerve.
- **Perineurium** (4) — surrounds the fascicles.
- **Endoneurium** (2) — surrounds the nerve fibers.

# Spinal Nerve

___ Dorsal Root
___ Dorsal Root Ganglion
___ Endoneurium
___ Epineurium
___ Fascicle
___ Nerve Fiber
___ Perineurium
___ Spinal Cord
___ Spinal Nerve
___ Ventral Root

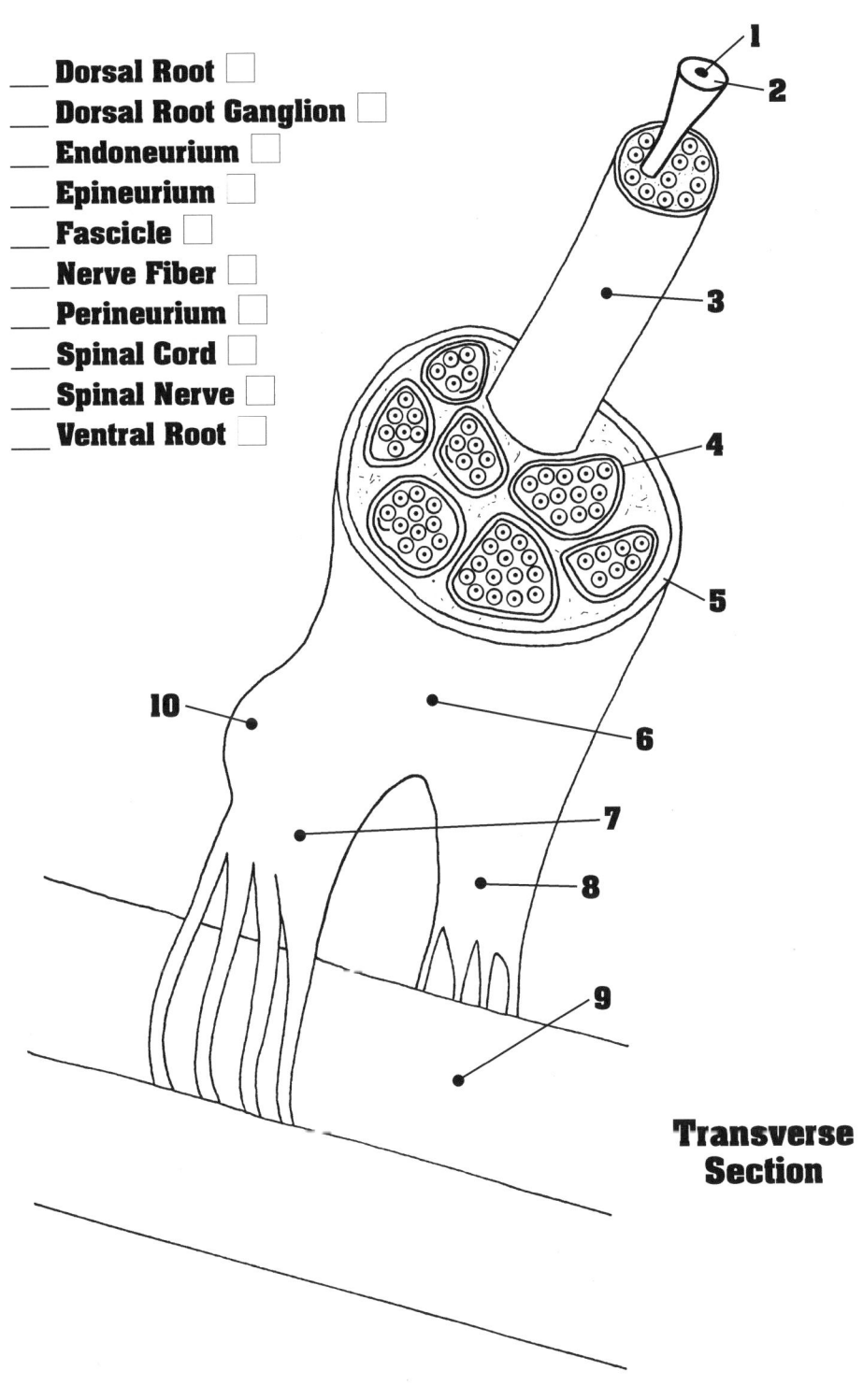

**Transverse Section**

# Spinal Cord

On each side of the spinal cord there are three regions of gray matter (where cell bodies are numerous):

- **Dorsal gray horn** (26)
- **Lateral gray horn** (29)
- **Ventral gray horn** (30)

The gray matter from one side communicates with the gray matter of the opposite side via the **gray commissure** (19) which bridges around the **central canal** (4).

The **dorsal root** (5) with its **dorsal root ganglion** (6) arise from the posterior gray horn, whereas the **ventral root** (8) derives from the region of the ventral gray horn. The two roots merge into a **spinal nerve** (7).

On each side of the spinal cord there are three regions of white matter:

- **Posterior white column** (3)
- **Lateral white column** (16)
- **Anterior white column** (18)

The columns contain:

descending motor nerve tracts & ascending sensory nerve tracts

| descending motor nerve tracts | ascending sensory nerve tracts |
|---|---|
| **lateral corticospinal** (13) | **fasciculus gracilis** (28) |
| **rubrospinal** (14) | **fasciculus cuneatus** (27) |
| **anterior corticospinal** (12) | **posterior spinocerebellar** (25) |
| **tectospinal** (10) | **anterior spinocerebellar** (23) |
| **vestibulospinal** (11) | **lateral spinothalamic** (24) |
| **anterior reticulospinal** (9) | **anterior spinothalamic** (20) |
| **lateral reticulospinal** (15) | **spinotectal** (21) |
| **septomarginal fasciculus** (2) | **spinoolivary** (22) |

The **posterior median sulcus** (1) and the **anterior median fissure** (17) are also shown.

# Spinal Cord

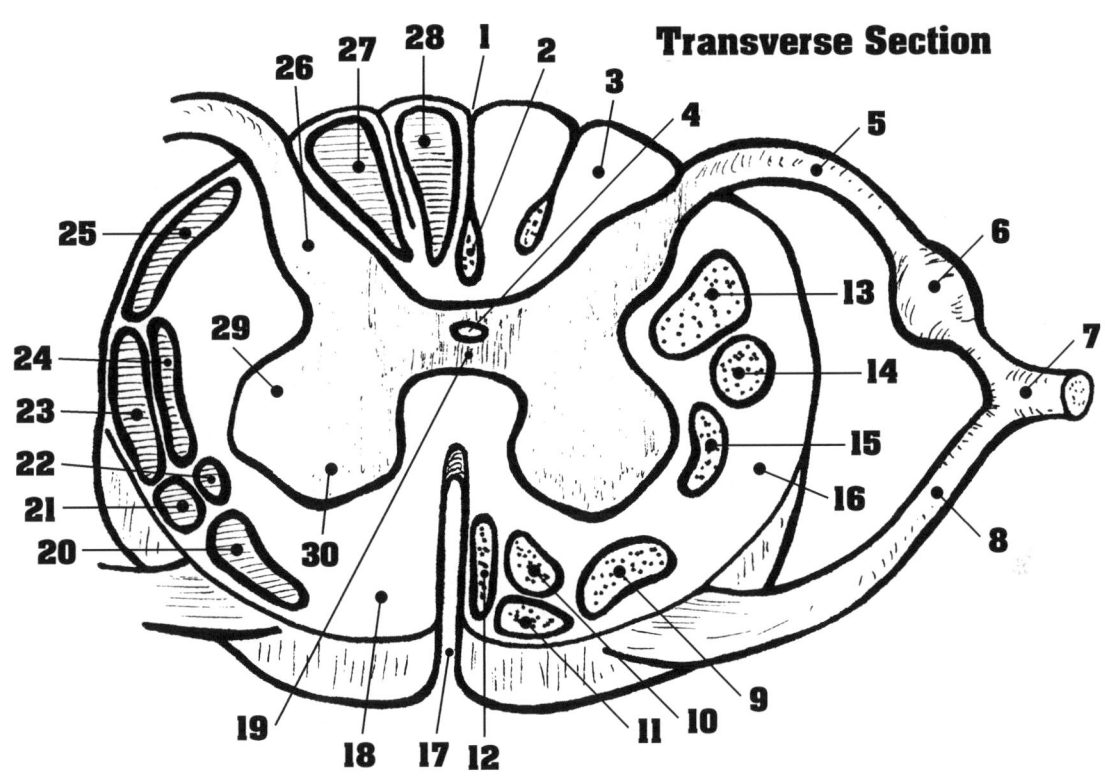

**Tracts:**
**Ascending (A) • Descending (D)**

___ Corticospinal — Anterior (D) ☐
___ Corticospinal — Lateral (D) ☐
___ Fasciculus Cuneatus (A) ☐
___ Fasciculus Gracilis (A) ☐
___ Reticulospinal — Anterior (D) ☐
___ Reticulospinal — Lateral (D) ☐
___ Rubrospinal (D) ☐
___ Septomarginal Fasciculus (D) ☐
___ Spinocerebellar — Anterior (A) ☐
___ Spinocerebellar — Posterior (A) ☐
___ Spinoolivary (A) ☐
___ Spinotectal (A) ☐
___ Spinothalamic — Anterior (A) ☐
___ Spinothalamic — Lateral (A) ☐
___ Tectospinal (D) ☐
___ Vestibulospinal (D) ☐

___ Anterior Median Fissure ☐
___ Central Canal ☐
___ Dorsal Root ☐
___ Dorsal Root Ganglion ☐
___ Gray Commissure ☐
**Gray Horn**
___ Dorsal ☐
___ Lateral ☐
___ Ventral ☐
___ Posterior Median Sulcus ☐
___ Spinal Nerve ☐
___ Ventral Root ☐
**White Column**
___ Anterior ☐
___ Lateral ☐
___ Posterior ☐

# Autonomic Nervous System
## Components of Sympathetic Division

A nerve impulse from a sensory receptor travels along a **sensory dendrite** (10) to a **dorsal root ganglion** (5) where it synapses and continues on through the **dorsal root** (3) via a **sensory axon** (4) to reach the **spinal cord** (2).

After synapsing in the spinal cord, action potentials exit the spinal cord through the **ventral root** (19) via a **motor axon** (20). Action potentials follow the motor axon, also known as a **preganglionic neuron** (13) through a **white ramus communicans** (11) to one of the many **sympathetic chain ganglia** (6) found along the **sympathetic chain** (7).

Synapses then occur within the sympathetic chain ganglion and action potentials are sent, via a **postganglionic neuron** (14), out through the **gray ramus communicans** (12) to a peripheral effector or through the **splanchnic nerve** (15) to a **prevertebral ganglion** (16) and on through another postganglionic neuron **to a deep visceral effector** (17).

Note that the dorsal and ventral roots merge to form a **spinal nerve** (18), and that the spinal nerve shortly thereafter divides into a **ventral branch** (9) and a **dorsal branch** (8).

The **central canal** (1) of the spinal cord is also noted.

# Autonomic Nervous System
## Components of Sympathetic Division

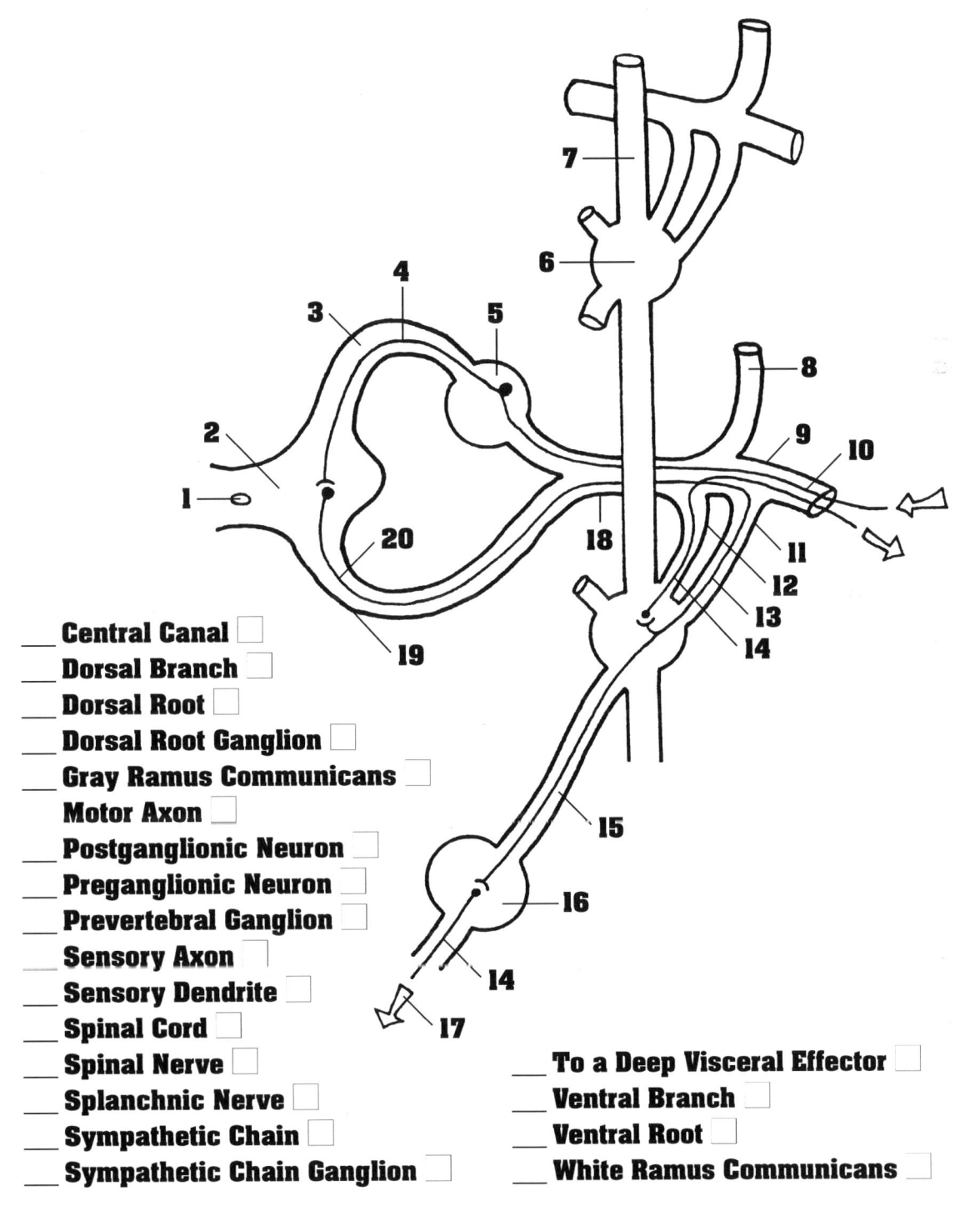

___ Central Canal
___ Dorsal Branch
___ Dorsal Root
___ Dorsal Root Ganglion
___ Gray Ramus Communicans
___ Motor Axon
___ Postganglionic Neuron
___ Preganglionic Neuron
___ Prevertebral Ganglion
___ Sensory Axon
___ Sensory Dendrite
___ Spinal Cord
___ Spinal Nerve
___ Splanchnic Nerve
___ Sympathetic Chain
___ Sympathetic Chain Ganglion
___ To a Deep Visceral Effector
___ Ventral Branch
___ Ventral Root
___ White Ramus Communicans

# Gustatory

Different kinds of taste buds in different regions of the tongue are collectively involved in the gustatory response to food molecules.

The major gustatory receptor areas are:

- **Sweet** (8)
- **Salty** (7)
- **Sour** (6)
- **Bitter** (5)

A complex **circumvallate papilla** (9) is magnified to show numerous **taste buds** (11) embedded around the outer edge. A single taste bud is further magnified to show **gustatory cells** (12) and **gustatory hairs** (13). Action potentials from gustatory cells are transmitted via the **gustatory nerve** (10) to the gustatory center in the cerebrum.

The general theory of taste suggests that food molecules of different shapes are detected by receptor sites of different shapes embedded in sensory hairs. The precise shape of a molecule often determines its precise function. A great deal of physiology is ultimately understood with reference to the architecture of molecules.

And so long as we are sticking out our tongues we thought we may as well note a few other features:

- **Lingual tonsils** (4)
- **Palatine tonsils** (3)
- **Epiglottis** (2)
- **Vocal cords** (1)

*A third pair of tonsils, the pharyngeal tonsils, are found further back in the throat. The lymphoid tissues of the tonsils not only contain numerous diffusely scattered lymphocytes and macrophages that play important roles in immunity and body defense, but also harbor colonies of symbiotic bacteria which assist in maintaining a strong immune system ... don't give up your tonsils without a fight!*

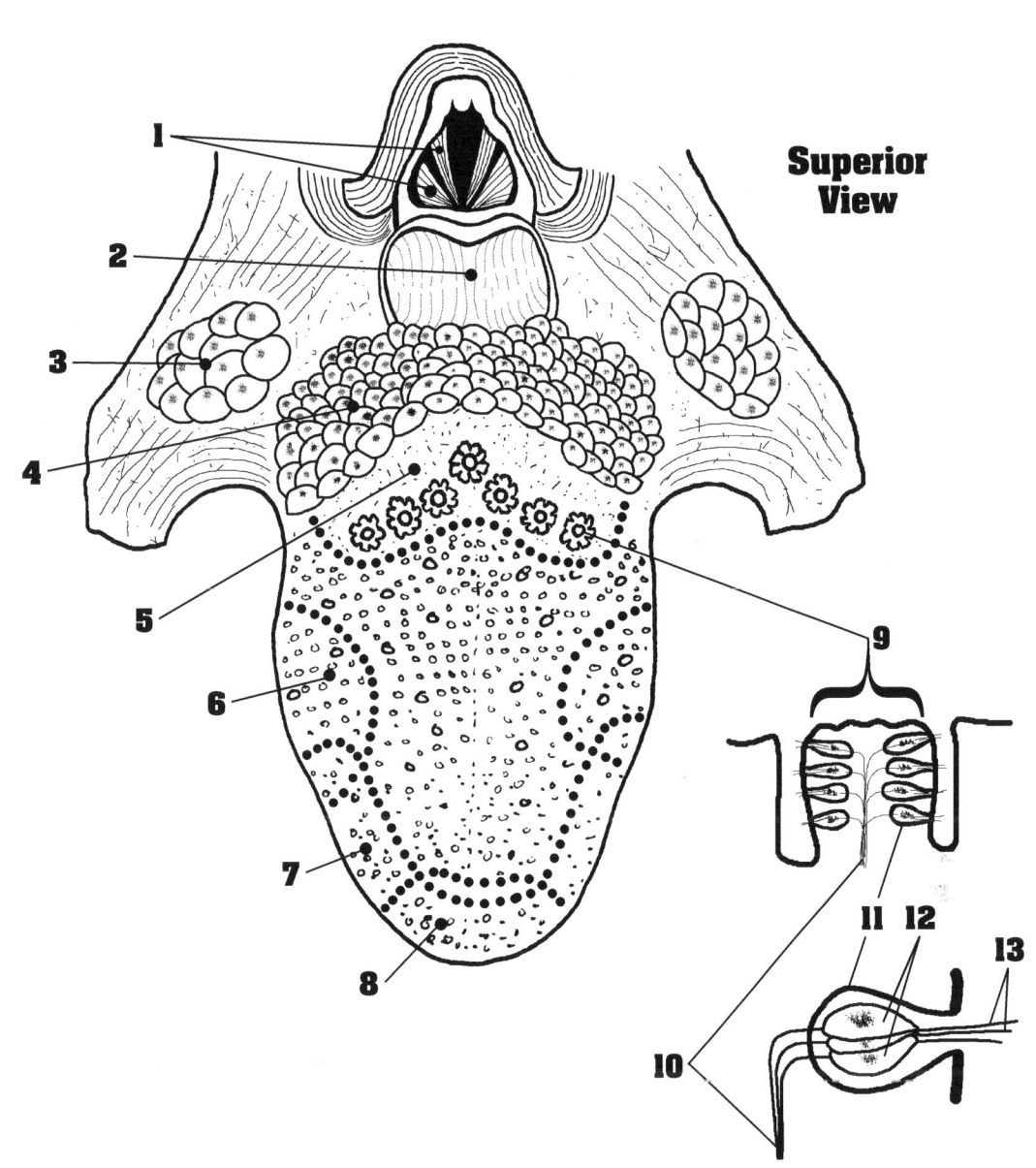

# Olfactory Receptors

**Molecular air flow** (10) moves through the **nasal cavity** (6) and makes contact with receptors on **olfactory cilia** (7) which extend from **olfactory vesicles** (8) embedded in a **mucus layer** (5).

Olfactory receptors respond by generating action potentials which travel via an **olfactory dendrite** (9) to an **olfactory receptor cell body** (12). From the cell body, action potentials proceed upward, via an **olfactory axon** (14), through the **lamina propria** (3), and on through a **foramen** (15) in the **cribiform plate** (2) of the ethmoid bone.

Above the cribiform plate bipolar neurons synapse in the **olfactory bulb** (1) and the nerve impulses continue toward the cerebrum via the **olfactory tract** (16). The olfactory tract then carries action potentials to the olfaction center in the cerebrum.

**Supporting cells** (11) stabilize the receptor cells in the **epithelial layer** (4), and **mucus glands** (13), embedded in the lamina propria, secrete mucus and covey it via mucus ducts through the epithelium to the mucus layer.

# Olfactory Receptors

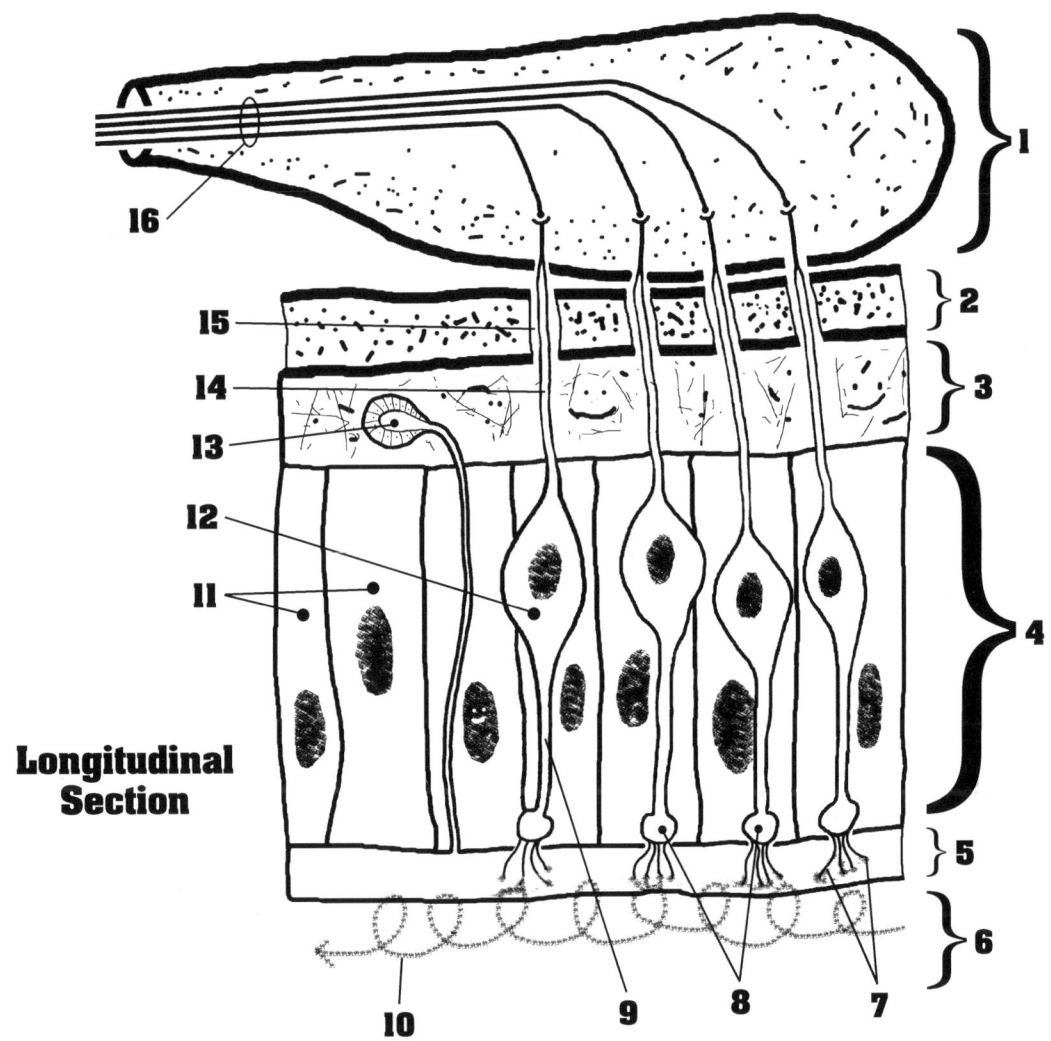

**Longitudinal Section**

___ Cribiform Plate
___ Epithelial Layer
___ Foramen of Cribiform Plate
___ Lamina Propria
___ Molecular Air Flow
___ Mucous Gland
___ Mucous Layer
___ Nasal Cavity
___ Olfactory Axon
___ Olfactory Bulb
___ Olfactory Cilia
___ Olfactory Dendrite
___ Olfactory Receptor Cell Body
___ Olfactory Tract
___ Olfactory Vesicles
___ Supporting Cells

# Lacrimal Apparatus

When "the window to the soul" needs cleaning, tears from **lacrimal glands** (7) come down — like falling rain — via **lacrimal excretory ducts** (6).

Tears wash across the eye and pass through a **lacrimal punctum** (3) into a **lacrimal canal** (8), then into the **lacrimal sac** (9), and on downward through the **nasolacrimal duct** (2) … and then, sure enough, your nose starts dripping!

So our eyes have cleaning fluid but where are the wipers? The wipers are, of course, the eyelids, which are automatically set on every individual to intermittently "blink."

The **iris** (5) surrounds the **pupil** (4).

*Grief and anxiety promote toxic build-ups in the body and there is evidence that some accumulated toxins may be eliminated in tears. Thus, and perhaps, one ought not to hold back "the **toxic tears of grief**" (1). A good cleansing cry, after a really hard unfair A&P test really does make you feel better … sometimes … doesn't it?*

# Lacrimal Apparatus

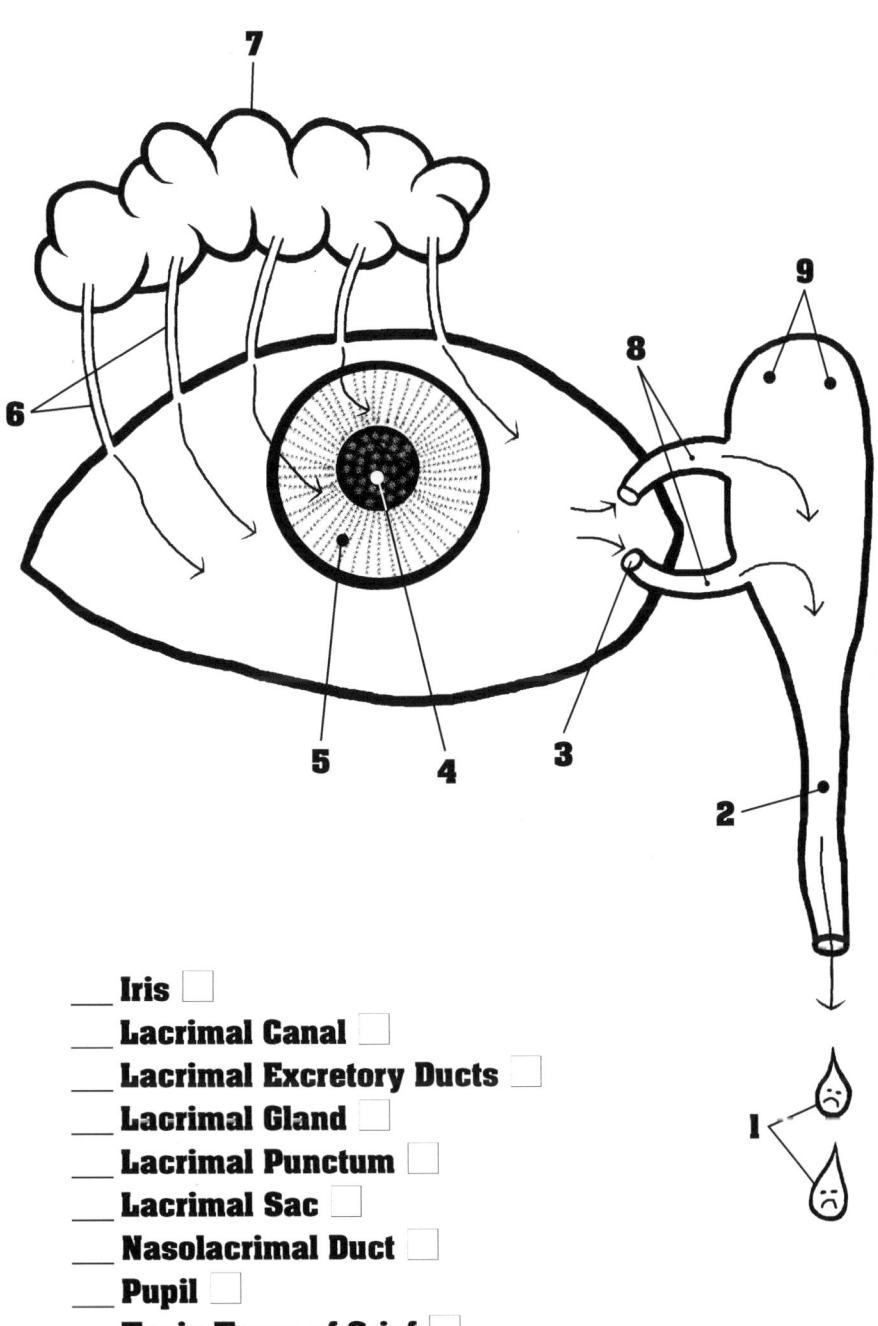

___ Iris
___ Lacrimal Canal
___ Lacrimal Excretory Ducts
___ Lacrimal Gland
___ Lacrimal Punctum
___ Lacrimal Sac
___ Nasolacrimal Duct
___ Pupil
___ Toxic Tears of Grief

# Eyeball

The **tendon of the lateral rectus** (1) and the **tendon of the medial rectus** (12) attach to the **sclera** (19). At the region where it evaginates and merges into the **cornea** (6), the sclera loses its pigmentation. Whereas the sclera (scleroid) is a tough protective outercoat, the **choroid** (18) is a softer vascular coat. The innermost coat, the **retina** (17) is a sensitive coat which contains rods and cones (sensory receptors).

Although sensory receptors are scattered throughout the retina there is only one small spot — an indentation called the **fovea centralis** (20) — where the sensory receptors are arranged densely enough to facilitate a sharp image, hence, it is the function of the cornea and **lens** (14) to refract light so as to focus precisely on the fovea.

The lens is held in position by **suspensory ligaments** (15) anchored in **ciliary processes** (3). Just beneath the ciliary processes is the **ciliary muscle** (2) which acts via suspensory ligaments to change the shape of the lens to make focusing adjustments. The ciliary processes, together with the ciliary muscle, comprise the **ciliary body** (4).

The retina is a thin, delicate, cream-colored coat that, in the normal eyeball, is always "ironed" in place by a jelly-like **vitreous body** (11) that fills the posterior cavity of the eyeball. A serrated border called the **ora serrata** (16) is found toward the anterior of the eyeball where the retina ends.

Nerve impulses that arise from the retinal response to light are transmitted to the brain via the **optic nerve** (23). A **central artery and vein** (22) pass through the center of the optic nerve. The optic nerve contains approximately one million sensory fibers. The juncture where the optic nerve enters the eyeball is called the **optic disk** (21). Because there are no sensory receptors in the region of the optic disk, it is called the blind spot.

A thin muscular diaphragm, the **iris** (13) is located immediately in front of the lens. The iris diaphragm controls the amount of light that enters the eye by changing the size of the **pupil** (7).

The thin space between the lens and the iris is the **posterior chamber** (9), while the larger space between the iris and the cornea is the **anterior chamber** (8). These two chambers together comprise the **anterior cavity** (10) which is filled with aqueous humor.

The aqueous humor is continuously circulated through the anterior cavity, being produced at the site of the ciliary processes and exiting through the **scleroid sinus** (5), the details of which are shown on the next page.

# Eyeball

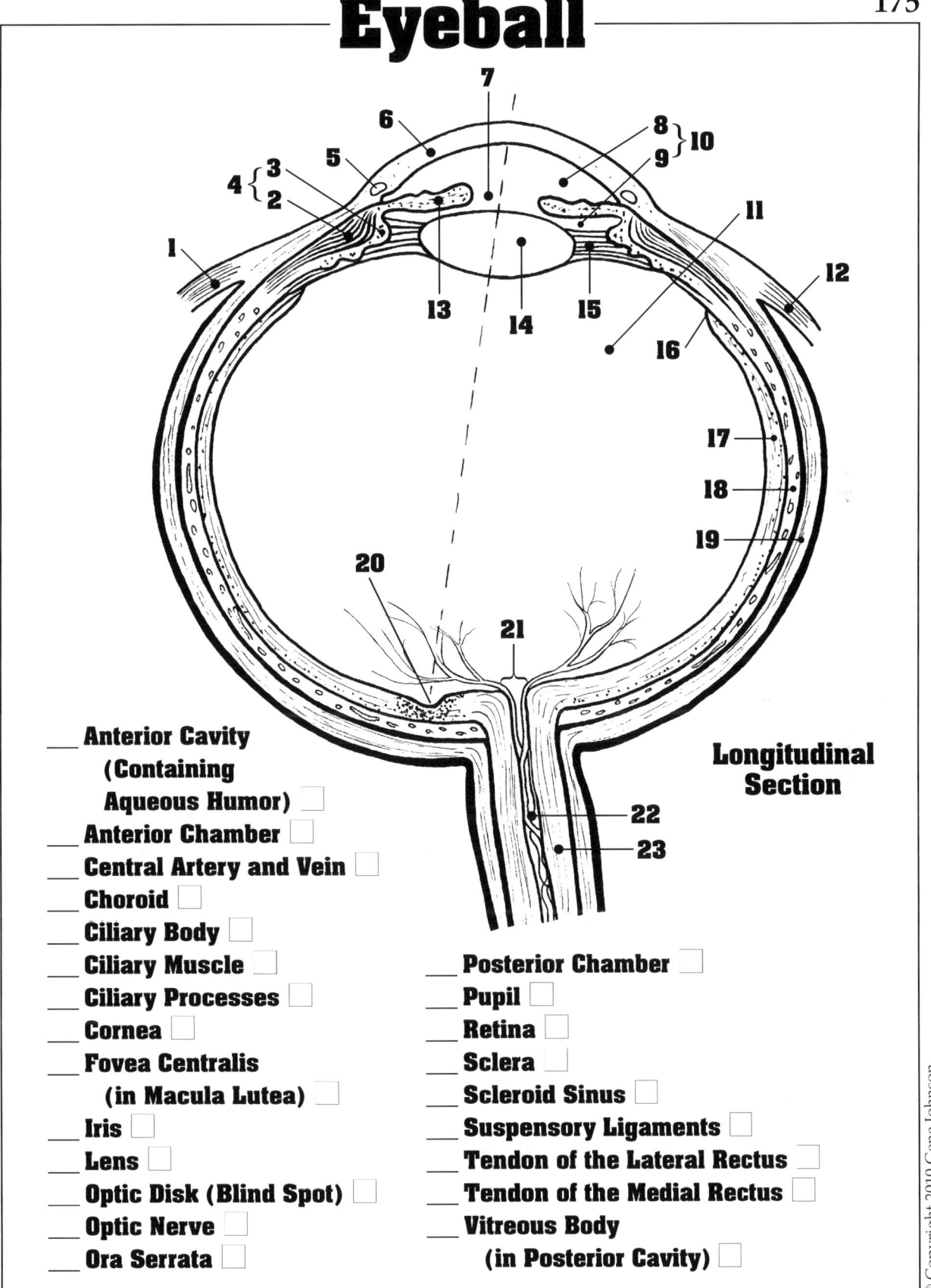

**Longitudinal Section**

___ Anterior Cavity (Containing Aqueous Humor) ☐
___ Anterior Chamber ☐
___ Central Artery and Vein ☐
___ Choroid ☐
___ Ciliary Body ☐
___ Ciliary Muscle ☐
___ Ciliary Processes ☐
___ Cornea ☐
___ Fovea Centralis (in Macula Lutea) ☐
___ Iris ☐
___ Lens ☐
___ Optic Disk (Blind Spot) ☐
___ Optic Nerve ☐
___ Ora Serrata ☐
___ Posterior Chamber ☐
___ Pupil ☐
___ Retina ☐
___ Sclera ☐
___ Scleroid Sinus ☐
___ Suspensory Ligaments ☐
___ Tendon of the Lateral Rectus ☐
___ Tendon of the Medial Rectus ☐
___ Vitreous Body (in Posterior Cavity) ☐

© Copyright 2010 Gene Johnson

# Aqueous Humor Circulation

Focusing on the **anterior cavity (segment)** (9), which is composed of a **posterior chamber** (7) and an **anterior chamber** (8), we track the flow of **aqueous humor** (10) from its origin in **ciliary processes** (2) through the posterior chamber, then through the **pupil** (11) into the anterior chamber, and finally to its exit point at a **scleral venous sinus** (6).

The aqueous humor, which is normally produced and drained at the same rate, maintains a constant intraocular pressure which aids internal support for the eyeball. However, if the flow of aqueous humor is blocked, or partially blocked, the intraocular pressure is elevated above the normal 16 mm Hg. Increased pressure in the anterior cavity reflects back through the vitreous humor into the posterior cavity, and impinges on the retina and optic nerve. Sustained, abnormally high, intraocular pressure eventually causes glaucoma, one of the most common forms of blindness, or partial blindness.

We encourage you to review the anatomy of the anterior aspect of the eye:

Note how the **lens** (5), is surrounded and supported by **suspensory ligaments** (4) anchored in ciliary processes. The ciliary processes, together with the **ciliary muscle** (1) comprise the **ciliary body** (3). An adjustable, muscular, **iris** (13) regulates the size of the pupil, hence controlling the amount of light that passes through the lens. Because the **cornea** (12), at the anterior surface of the eye, is essentially free of protein it is a "privileged tissue" and can be transplanted from one individual to another without the usual antigen/antibody complications that challenge most transplants. The cornea is protected by a **corneal epithelium** (14).

*The remarkable capacity of the lens to change shape, and thus variably focus on objects at different distances — a process called accommodation — is based upon the ability of the ciliary muscle to pull on the anterior aspect of the eyeball and thereby ease the tension on the suspensory ligaments. As an object is brought closer to the eye the ciliary muscle contracts more forcibly, thus releasing the tension on the suspensory ligaments and thereby allowing the lens to become more rounded. As a person grows older the lens becomes less elastic, hence less able to accommodate, and at some point reading glasses are usually required. With age the lens may also become "cloudy" and less able to transmit light, a condition called a cataract. And at any age the lens, or the cornea, may develop irregularities on their surfaces, called astigmatisms, which can cause "blurry spots" in the vision.*

# Aqueous Humor Circulation

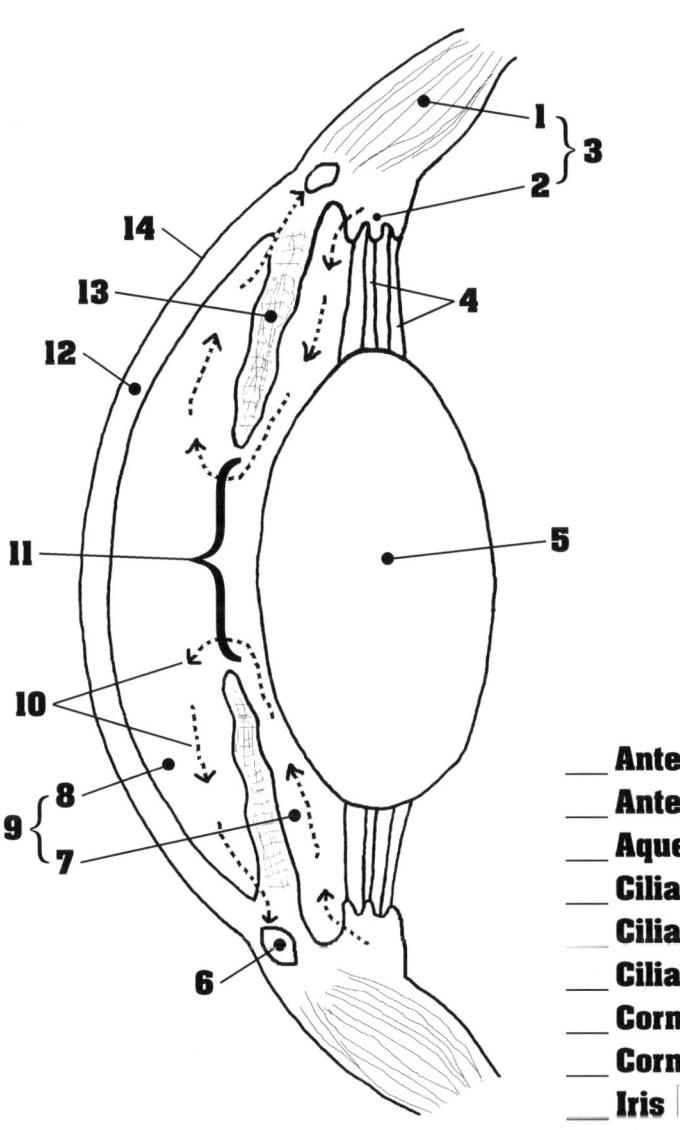

___ Anterior Cavity (Segment)
___ Anterior Chamber
___ Aqueous Humor
___ Ciliary Body
___ Ciliary Muscle
___ Ciliary Processes
___ Cornea
___ Corneal Epithelium
___ Iris
___ Lens
___ Posterior Chamber
___ Pupil
___ Scleral Venous Sinus
___ Suspensory Ligaments

# Photoreceptor

A single rod is shown here, extracted from the retina of the eye where many millions more occur, along with many millions of cones!

The **inner segment** (2) is packed with **mitochondria** (6) ... Ah! ATP for energy!

The **outer segment** (1) is composed of membranous **discs** (11), stacked like pancakes.

**Visual pigments** (7) are embedded in the **phospholipid bi-layers** (10) of the discs.

Each visual pigment has an **opsin** (8) protein component and a light-absorbing molecule called **retinal** (9) — derived from vitamin A. When the elbow-shaped retinal body is struck by light, and absorbs photons, it performs an isomeric conformational shift and straightens out (technically changing from a *cis* isomer form to a *trans* isomer form). If enough retinal isomeric "kicks" are performed nerve action potentials will be generated. The action potentials will then pass through the **cell body** (5) of the receptor cell and on to the **synaptic endings** (3).

Beyond the synaptic endings the action potentials pass through two additional orders of neurons in the retinal layer, and then, on to the brain, via the optic nerve.

... And did we almost forget to tell you that the **nucleus** (4) of the bipolar cell body is also shown?

# Photoreceptor

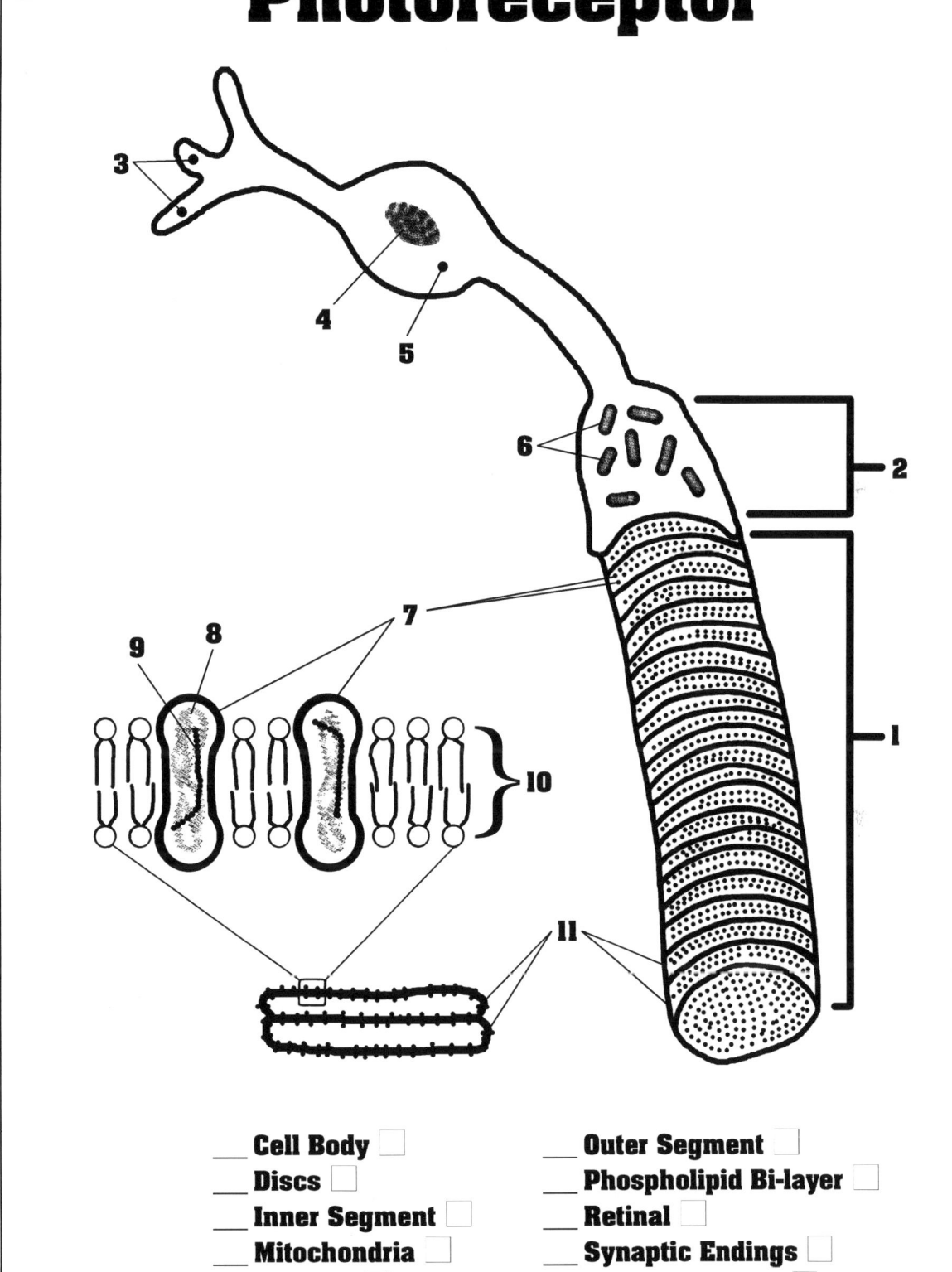

___ Cell Body  
___ Discs  
___ Inner Segment  
___ Mitochondria  
___ Nucleus  
___ Opsin  
___ Outer Segment  
___ Phospholipid Bi-layer  
___ Retinal  
___ Synaptic Endings  
___ Visual Pigments

# Outer Ear
## (Pinna/Auricle)

The various ridges, valleys, bumps, lobes and depressions that comprise the outer ear are as follows:

An outer (posterior) ridge, the **helix** (2) expands at its inferior border to become the **lobule** (9). Running parallel to the helix is the **antihelix** (1) which branches out superiorly to become the **crura of the antihelix** (3). The fossa formed between the branches of the crura is the **triangular fossa** (4).

The **tragus** (7) is located on the anterior border of the **external auditory meatus** (6), whereas the **antitragus** (8) is located on the superior border of the lobule.

The central hollow of the auricle is called the **concha** (5).

# Outer Ear
## (Pinna/Auricle)

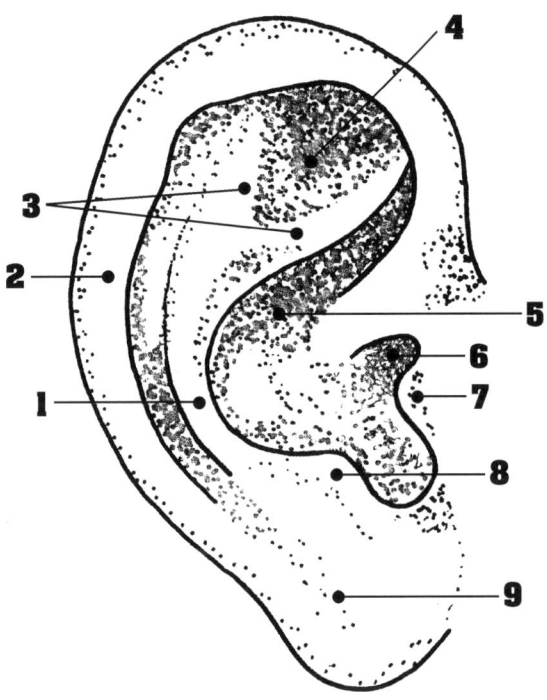

___ **Antihelix** ☐
___ **Antitragus** ☐
___ **Concha (Hollow of the Auricle)** ☐
___ **Crura of the Antihelix** ☐
___ **External Auditory Meatus** ☐
___ **Helix** ☐
___ **Lobule** ☐
___ **Tragus** ☐
___ **Triangular Fossa** ☐

# External & Middle Ear

A **cartilaginous meatus** (8) extends inward from the outer ear, or **pinna** (6). The cartilaginous meatus gives way to the **osseous meatus** (7) as it proceeds inward through the **petrous portion of the temporal bone** (5).

At the end of the osseous meatus the **tympanic membrane** (10) marks the lateral boundary of the **tympanic cavity** (11). The three auditory **ossicles (ear bones)** (2) within the middle ear will be shown in more detail in the next exercise.

The **auditory tube** (1) connects the tympanic cavity with the pharynx and facilitates the regulation of air pressure within the tympanic cavity.

The **cochlea** (3) and the **semicircular canals** (4) which comprise the inner ear, are also shown superiorly adjacent to the middle ear.

Note also the **styloid process** (9) which protrudes downward from the region of the osseous meatus.

# External & Middle Ear

**Frontal Section**

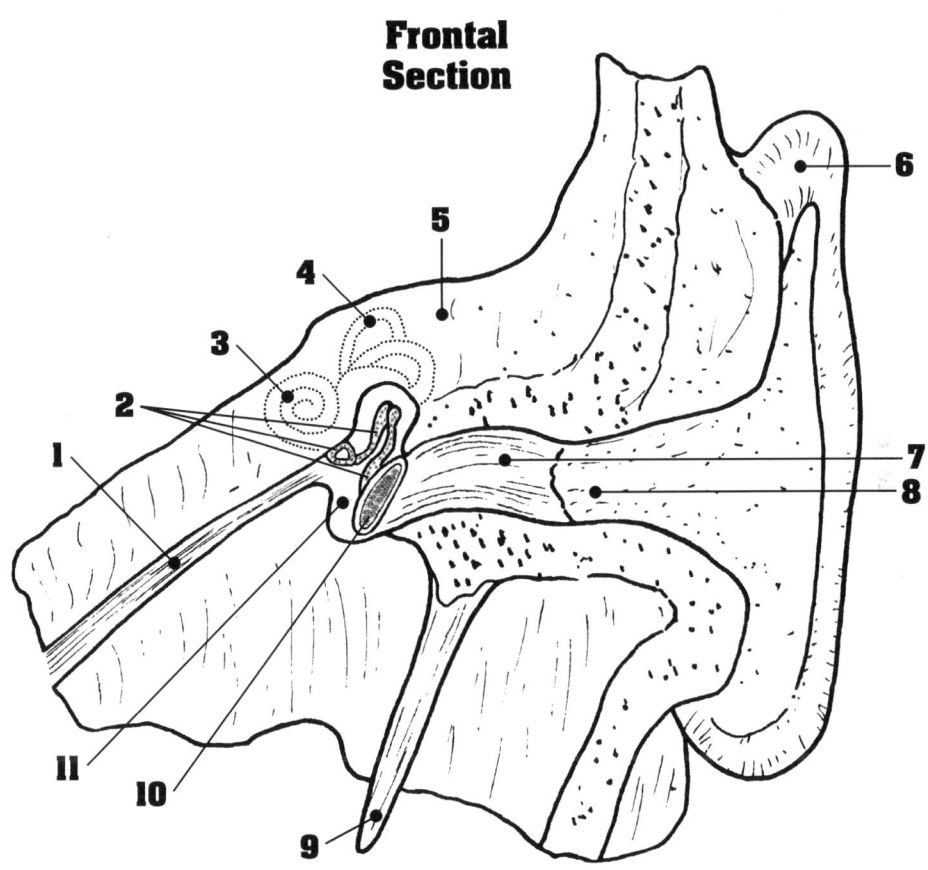

___ Auditory Tube
___ Cartilaginous Meatus
___ Cochlea
___ Osseous Meatus
___ Ossicles
___ Petrous Portion of the Temporal Bone
___ Pinna
___ Semicircular Canals
___ Styloid Process
___ Tympanic Cavity
___ Tympanic Membrane

# Middle Ear

Here we see the middle ear embedded in the **petrous portion of the temporal bone** (6). The three auditory ossicles, the **malleus** (4), **incus** (7) and **stapes** (11) are held in place by the **anterior** (3), **superior** (5) and **posterior** (8) ligaments.

Sound waves arriving via the **external auditory meatus** (1) cause the **tympanic membrane** (2) to vibrate. The vibrations are then transmitted through the auditory ossicles where they are amplified up to 100X.

The vibrating stapes then vibrates the oval window. The vibration of the oval window is then transmitted to fluid in the cochlea. Cochlear details are covered in the next exercise.

The middle ear contains not only some of the smallest bones but also some of the smallest muscles. The **tensor tympanic muscle** (10) and the **stapedius muscle** (9) both act to dampen excessive vibrations that might otherwise damage the ear.

The **auditory tube** (12), which connects the **tympanic cavity** (13) with the pharynx, is shown beneath the tensor tympanic muscle.

*The auditory tube facilitates the regulation of air pressure within the tympanic membrane. An increase in air pressure makes it more difficult for the ossicles and tympanic membrane to vibrate freely and the ears become "plugged." When the excessive air pressure is suddenly released via the auditory tube, the tympanic membrane will sometimes snap quickly back into its normal position, causing the ear to "pop."*

# Middle Ear

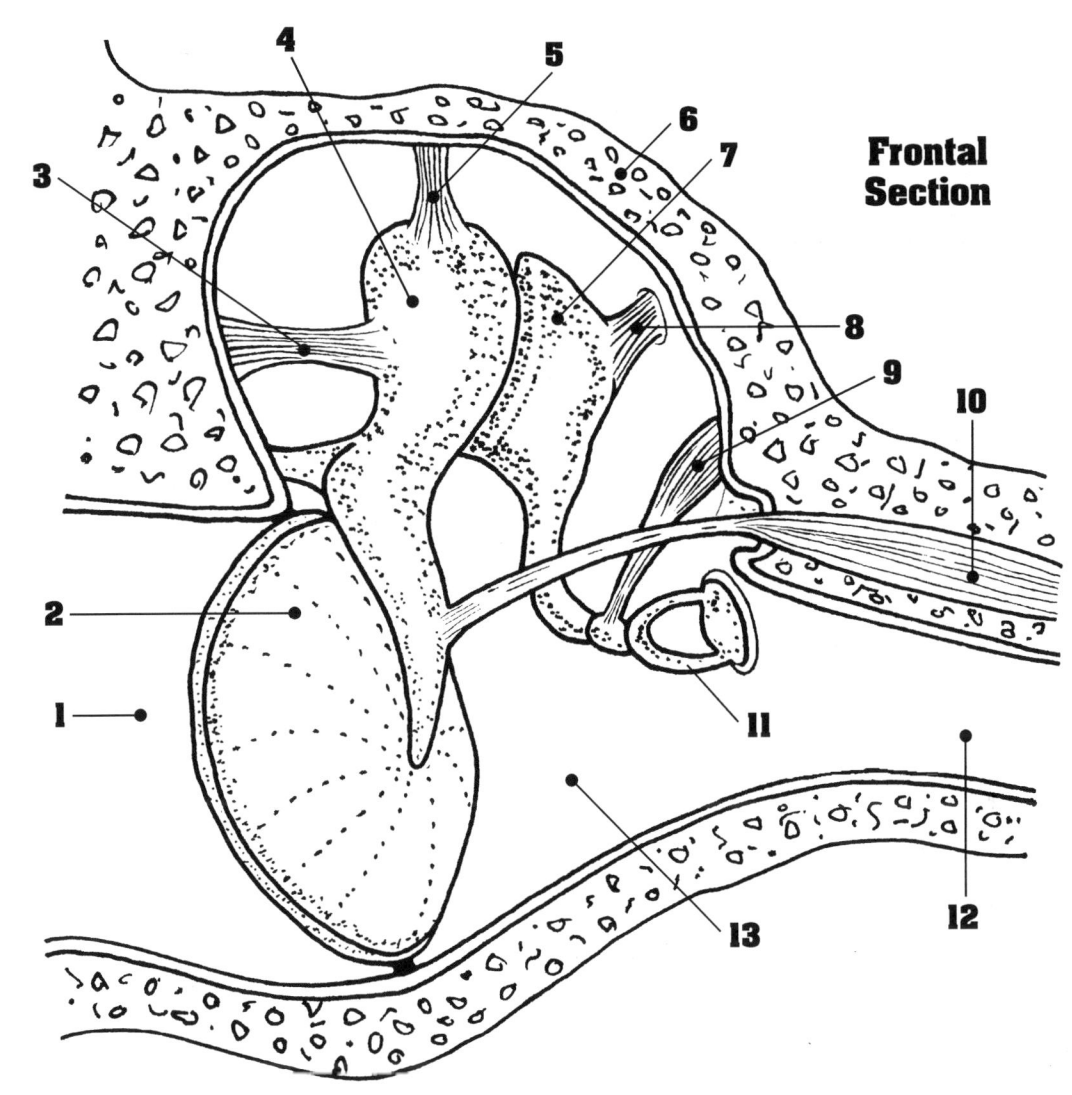

Frontal Section

___ Auditory Tube
___ External Auditory Meatus
___ Incus
    Ligament
    ___ Anterior
    ___ Posterior
    ___ Superior
___ Malleus
___ Petrous Portion of the Temporal Bone
___ Stapedius Muscle
___ Stapes
___ Tensor Tympanic Muscle
___ Tympanic Cavity
___ Tympanic Membrane

# Cochlea
## A Transverse Section Through One Turn

Three fluid-filled chambers of the cochlea, the **scala vestibuli** (5), **scala media (cochlear duct)** (8) and **scala tympani** (14) are located within the **osseous cochlea** (4). The scala vestibuli is separated from the scala media by the **vestibular membrane** (7).

The thick base of the osseous cochlea is called the **modiolus** (3) and the bony shelf which supports the **organ of Corti** (10) is called the **spiral lamina** (15).

The organ of Corti rests on the **basilar membrane** (12). When the basilar membrane vibrates it pushes the **inner** (13) and **outer hair cells** (11) up against the **tectorial membrane** (9), which causes the hair cells to transmit nerve impulses through the **spiral ganglion** (2) to the **cochlear nerve** (1) and on to the auditory region of the cerebrum.

Hair cells within the organ of Corti are supported by **pillar cells (rods of Corti)** (6). The basilar membrane varies in its structure and width from one end of the cochlea to the other. Because of these variances the basilar membrane responds to different vibrations at different structural regions. The basilar membrane is more narrow at the beginning and wider toward the end of the cochlea. High pitched, high frequency sound waves activate hair cells near the beginning of the cochlea whereas low pitched, low frequency sound waves activate hair cells near the end of the cochlea.

The frequencies of the sound waves that affect the organ of Corti range between 20,000/second (affecting the basilar membrane at the beginning of the cochlea) to 16/second (affecting the basilar membrane near the end of the cochlea).

# Cochlea
## A Transverse Section Through One Turn

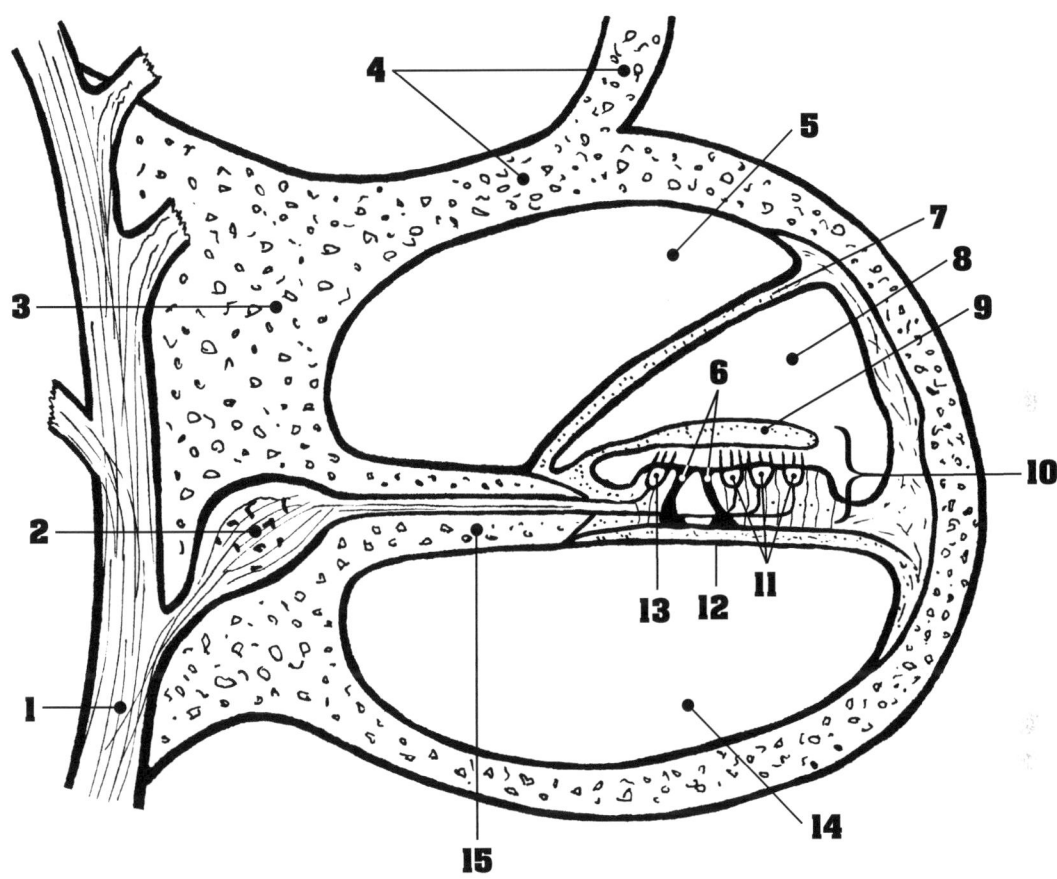

___ Basilar Membrane
___ Cochlear Nerve
    Hair Cells
___   Inner
___   Outer
___ Modiolus
___ Organ of Corti
___ Osseous Cochlea
___ Pillar Cells (Rods of Corti)
___ Scala Media (Cochlear Duct)
___ Scala Tympani
___ Scala Vestibuli
___ Spiral Ganglion
___ Spiral Lamina
___ Tectorial Membrane
___ Vestibular Membrane

# Osseous Labyrinth

The osseous labyrinth is an intricate cavity within the petrous portion of the temporal bone, thus the picture for this exercise, and most pictures shown in textbooks, are more properly "casts" of the cavity. The membranous labyrinth shown in the next exercise is an intricate system of fluid-filled tubes that is found inside the osseous labyrinth.

Amplified vibrations from the auditory ossicles of the middle ear are transmitted to the **cochlea** (7) at the **oval window** (8). (Recall that the stapes was connected with the oval window.)

Within the cochlea, sensory cells respond to the vibrations and send nerve impulses via the **cochlear branch of VIII** (6) to the auditory region of the brain.

The semicircular canals, together with the **ampullae** (3), function in balance and equilibrium. Sensory cells in the ampullae respond to moving fluid and send nerve impulses via the **vestibular branch of VIII** (5) to the cerebellum.

The **anterior** (2), **posterior** (4) and **lateral** (1) **semicircular canals** are at right angles to each other.

# Osseous Labyrinth

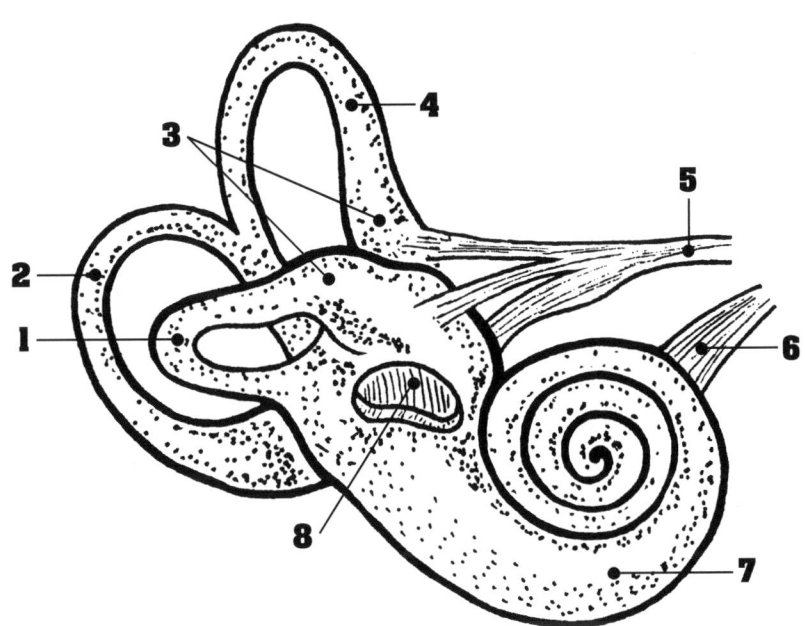

___ Ampullae ☐
___ Cochlea ☐
___ Cochlear Branch of VIII ☐
___ Oval Window ☐
   Semicircular Canal
___    Anterior ☐
___    Lateral ☐
___    Posterior ☐
___ Vestibular Branch of VIII ☐

# Membranous Labyrinth

Housed within the osseous labyrinth is a delicate system of fluid filled membranous ducts. The **cochlear duct (scala media)** (8) contains sensory cells that respond to the vibration of the cochlear endolymph fluid and send impulses to the cerebrum where they are perceived as sound.

**Ampullae** (2) contain sensory cells (see next exercise) that respond to the directional flow of the semicircular endolymph fluid and send impulses to the cerebellum where they are perceived as movements through space (dynamic balance).

The **utricle** (1) and the **saccule** (6) respond to the directional pressure of gelatinous masses embedded with otoliths (ear stones) (see next exercise) by sending impulses to the cerebellum where they are perceived as positions in space (static balance).

The **anterior** (5), **posterior** (3) and **lateral** (4) **semicircular labyrinths** are at right angles to each other.

The entire labyrinth is filled with endolymph and the **ductus reuniens** (7) provides a pathway for endolymph flow from the saccule to the cochlear duct.

# Membranous Labyrinth

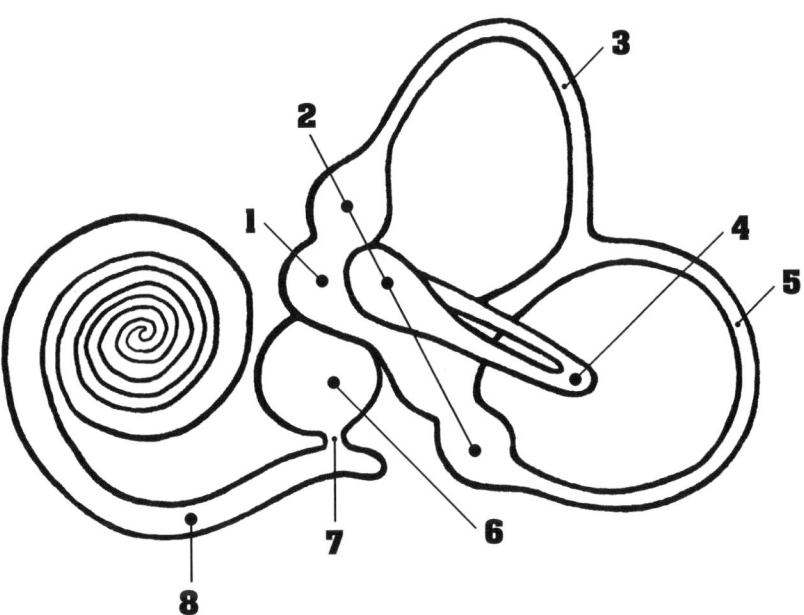

___ Ampullae □
___ Cochlear Duct (Scala Media) □
___ Ductus Reuniens □
___ Saccule □
    Semicircular Labyrinth
___   Anterior □
___   Lateral □
___   Posterior □
___ Utricle □

# Macula & Crista

Recall that the vestibular apparatus is comprised of semicircular canals and a vestibule, and that the vestibule is comprised of the utricle and saccule. The vestibule is involved with what is called static equilibrium (our position in space) while the semicircular canals are involved with what is called dynamic equilibrium (our movement through space).

## Static Equilibrium

In each wall of the utricle and saccule there is a sensory organ called a macula. Each macula contains **sensory cells** (3) embedded in a layer of **supporting cells** (4). The entire apparatus rests on a **basement membrane** (2).

Two types of sensory hairs extend upward from each of the sensory cells: numerous long microvilli called **stereocilia** (7) and a single, sturdy, true cilia, called a **kinocilium** (6).

The sensory hairs are embedded in a gelatinous **otolithic membrane** (5) studded with numerous small calcium carbonate crystals called **otoliths (ear stones)** (8). The otoliths stabilize the membrane by making it heavier and more dense.

The macular hair cells bend in different directions with different body positions, and signals are then sent to the cerebellum via a branch of the **vestibular nerve** (1).

## Dynamic Equilibrium

Each of the three semicircular canals has an ampulla at its base. Inside each ampulla there is a small sensory organ called a crista.

The micro-anatomy of the crista is essentially the same as for the macula except that the gelatinous membrane is called a **cupula** (9) and there are no calcium carbonate crystals embedded in it.

When we move, fluid in the semicircular canals is set in motion. The moving fluid bends the cupula, along with the sensory hairs, in the direction of the motion. Signals are then sent to the cerebellum via a branch of the vestibular nerve.

*Hypersensitivity to signals coming from the cristae sometimes causes "motion sickness," and some people have suffered so severely from motion sickness that they have had their vestibular nerves severed to stop the signals. Hypersensitivity has been observed for most of the senses. There are super-tasters, super-smellers, and super-feelers (super-sensitive to touch). One basis for this is that, for a given sense, some people have more sensory receptors per square centimeter, or millimeter, than others. Do you have a super-sense?*

# Macula & Crista

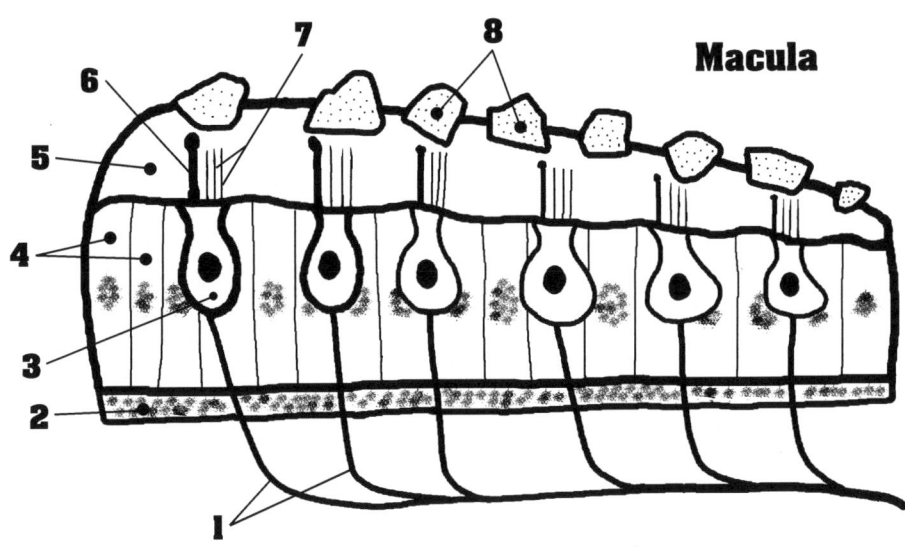

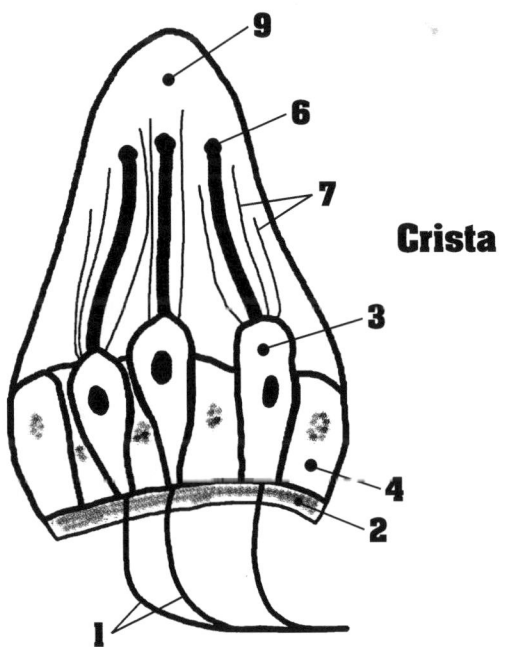

___ Basement Membrane
___ Cupula
___ Kinocilium
___ Otoliths (Ear Stones)
___ Otolithic Membrane
___ Sensory Cell
___ Stereocilia
___ Supporting Cells
___ Vestibular Nerve Fibers

# Endocrine System

As previously mentioned the endocrine system intimately interacts with the nervous system to coordinate and integrate metabolic activities. The endocrine system produces, and sends out, chemical messages called endocrines (hormones) which travel via blood and bind to receptors on target cells. Major endocrine glands include the **pineal** (1), **pituitary** (2), **parathyroids** (3), **thyroid** (4), **thymus** (5), **pancreas** (6), **adrenals** (7), and **gonads** (8). A variety of other tissues and organs also produce hormones. Some cells produce "local hormones" called autocrines or paracrines. Whereas autocrines affect the same cells that produce them, paracrines affect clusters of nearby cells. The regulatory, chemical messages which mediate our metabolisms are proving to be more diverse and abundant than we could have previously imagined. Discovering "local hormones" is presently a burgeoning business in the field of endocrinology.

Whereas most hormones — "local hormones" excluded — are comprised of amino acids in either short chains (peptides), or longer chains (proteins), gonadal and adrenocortical hormones are lipid-based steroids. Because the lipid-soluble (steroid) hormones can penetrate lipid-based cell membranes they act directly on intra-cellular receptors. In contrast, the water-soluble (protein) hormones, because they cannot penetrate cell membranes, bind to cell-surface (extra-cellular) receptors. In response to binding, intra-cellular second messengers (such as cyclic AMP) are activated and subsequently carry signals to intra-cellular targets.

Whereas the nervous system sends electrochemical signals (impulses) directly to target effectors via precisely located nerve fiber pathways, the endocrine system delivers its chemical messengers into the blood stream where they are carried to all regions of the body. But although endocrines are carried to all regions of the body they affect only the target cells and tissues which have the specific receptors to which the specific endocrines can bind. Specificity is a hallmark of the endocrine system as well as of the nervous system.

The pineal gland, located close to the center of the brain, secretes melatonin (an amine hormone derived from serotonin). The highest levels of melatonin occur around midnight and the lowest levels around noon. Although the regulation of sleep and diurnal rhythms are implicated, mysteries abound as to the precise mechanisms and modes of pineal function.

# Endocrine System

The thymus gland, which produces a family of peptide hormones including thymosins, thymopoietins, and thymic factor, is large and well developed in infants and children, greatly diminished in adults, and almost non-existent in old-age. Thymic hormones play vital roles in the development of the immune system.

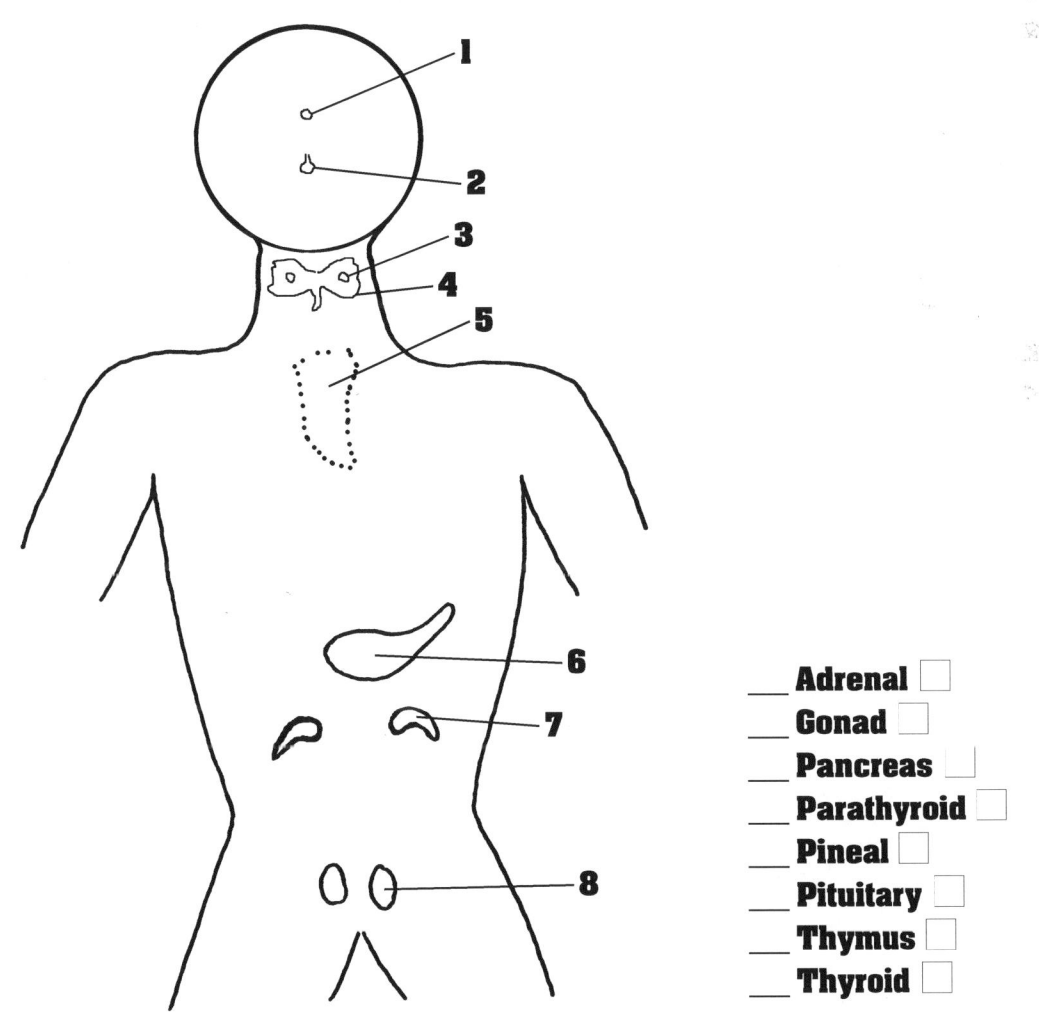

___ Adrenal ☐
___ Gonad ☐
___ Pancreas ☐
___ Parathyroid ☐
___ Pineal ☐
___ Pituitary ☐
___ Thymus ☐
___ Thyroid ☐

# Pituitary & Its Hormones

The pituitary gland (hypophysis) is comprised of two lobes which are attached to the **hypothalamus** (4) by a stalk called the **infundibulum** (3). The anterior lobe, the **adenohypophysis** (2), is the larger of the two lobes and secretes seven hormones, whereas the smaller posterior lobe, the **neurohypophysis** (1), secretes two hormones (which are actually produced by **neurosecretory cells** (5) in a **hypothalamic nucleus** (6) and then delivered via a nerve to the posterior hypophysis where they are later released into the blood stream).

Although hormones are chemical messengers in the blood, there is an intimate connection between the nervous and endocrine systems. The secretion of the seven hormones of the anterior hypophysis is also under the control of a variety of releasing and inhibiting factors associated with several of the hypothalamic nuclei.

## Hormones of the Neurohypophysis

- **Oxytocin (OT)** (14) — stimulates the contraction of the uterine muscles thus facilitating childbirth, and also assists in the initiation of milk ejection from the breasts.
- **Antidiuretic Hormone (ADH)** (15) — opens water gates in the collecting ducts of the kidneys to facilitate the reabsorption of water into the blood and thereby prevent dehydration. Because a loss of water will be reflected as a loss of blood volume, ADH also has a vasoconstricting effect on the arterioles, so that a normal blood pressure is maintained even when the body is somewhat dehydrated.

*A vicious cycle: Alcohol inhibits the pituitary secretion of ADH, hence the more alcohol one has in the blood the less ability one has to retain water, hence the more often one drinks alcoholic beverages the more often one urinates and loses water, hence the more thirsty one becomes … and drinks more alcoholic beverages?*

## Hormones of the Adenohypophysis

- **Melanocyte Stimulating Hormone (MSH)** (7) — effects the production and release of melanin in the integument.
- **Prolactin (PRL)** (8) — stimulates milk production.
- **Growth Hormone (GH)** (9) — has a wide range of effects on body growth.
- **Thyroid Stimulating Hormone (TSH)** (10) — stimulates the thyroid gland to produce the thyroid hormones.
- **Adrenocorticotropic Hormone (ACTH)** (11) — stimulates the release of the hormones from the cortex of the adrenal gland.
- **Follicle Stimulating Hormone (FSH)** (13) — stimulates follicle maturation in the female and sperm production in the male.
- **Luteinizing Hormone (LH)** (12) — promotes the production of estrogen and progesterone in females and testosterone in males.

# Pituitary & Its Hormones

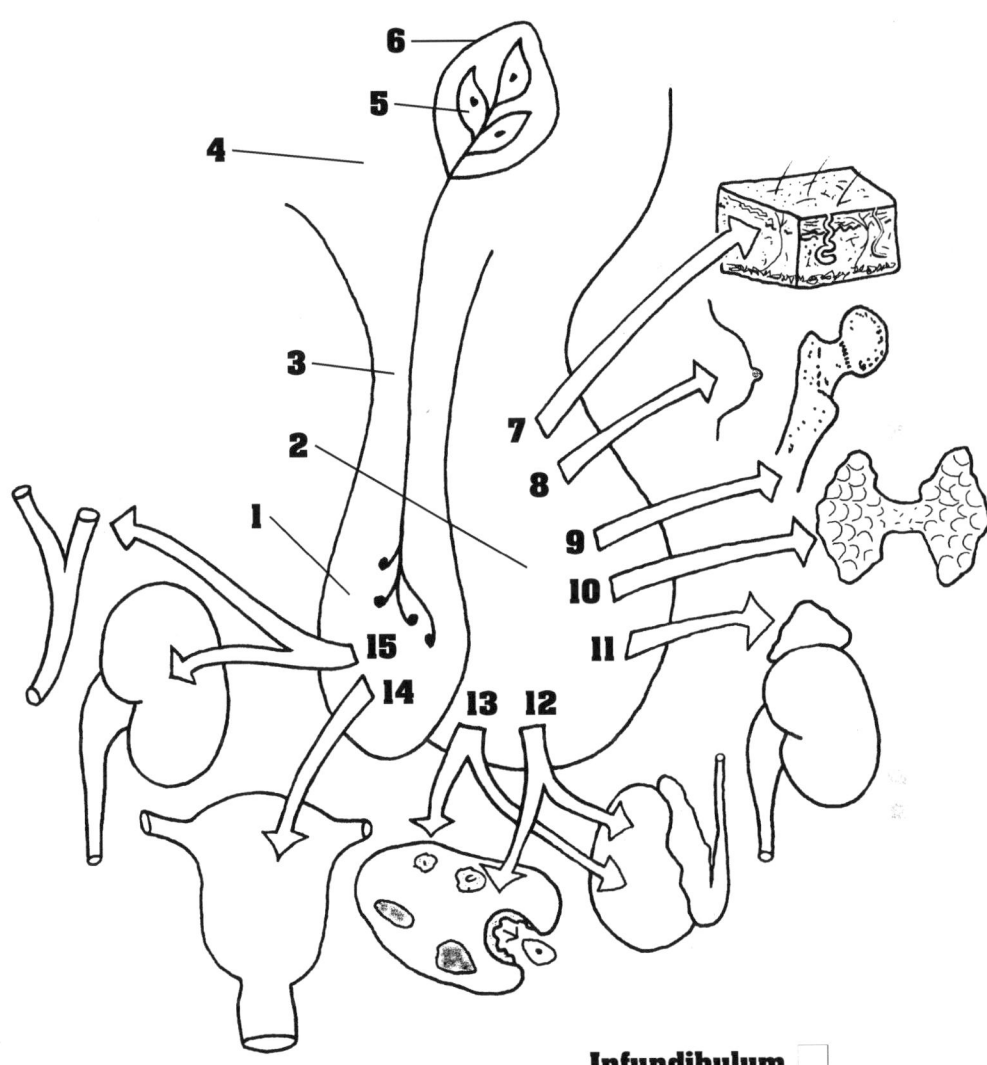

___ Adenohypophysis ☐
___ Adrenocorticotropic
   Hormone (ACTH) ☐
___ Antidiuretic Hormone (ADH) ☐
___ Follicle Stimulating
   Hormone (FSH) ☐
___ Growth Hormone (GH) ☐
___ Hypothalamus ☐
___ Hypothalamic Nucleus ☐
___ Infundibulum ☐
___ Luteinizing Hormone (LH) ☐
___ Melanocyte Stimulating
   Hormone (MSH) ☐
___ Neurohypophysis ☐
___ Neurosecretory Cells ☐
___ Oxytocin (OT) ☐
___ Prolactin (PRL) ☐
___ Thyroid Stimulating
   Hormone (TSH) ☐

# Thyroid & Parathyroid

The **thyroid gland** (14) is located on the anterior-superior border of the **trachea** (17) at the base of the **thyroid cartilage** (6). An **isthmus of the thyroid** (12) connects the two lateral lobes of the thyroid and a smaller **pyramidal lobe of the thyroid** (11) extends upward from the isthmus. Each of the lobes is comprised of **thyroid follicles** (13). An exploded view shows **follicular cells** (3) which produce **thyroid hormones (T3 & T4) (1)**, and **parafollicular cells** (4) which produce **calcitonin** (2). The effect of **thyroid stimulating hormone (TSH)** (5) from the pituitary is also indicated.

Thyroid hormones affect metabolism broadly. One of their many effects is to stimulate the enzymes involved in glucose metabolism, thereby increasing the metabolic rate and heat production (the calorigenic "heat producing" effect).

Calcitonin lowers the blood calcium level by stimulating the reabsorption of calcium into the bones.

As is the case with all endocrine glands, the thyroid gland is in intimate association with blood vessels. A **superior thyroid artery** (10) derives from the **external carotid** (8) branch of the **common carotid** (15). The **carotid sinus** (9) is shown at the base of the **internal carotid** (7), and an **inferior thyroid artery** (16) is shown servicing the gland from below.

The **parathyroid glands** (20) are embedded in the posterior aspects of the two lateral lobes of the thyroid, which, in the posterior view at the bottom of the page, can be seen wrapped around the **pharynx** (18), **esophagus** (21), and trachea.

**Parathyroid hormone (PTH)** (19), also called parathormone, raises the blood calcium level by stimulating the bones to release calcium into the blood. The fact that there are two endocrine regulators for calcium bears witness to the importance of calcium's central role in metabolism. We will later see that the sugar level in the blood is also "doubly regulated" by two different endocrine regulators, insulin and glucagon.

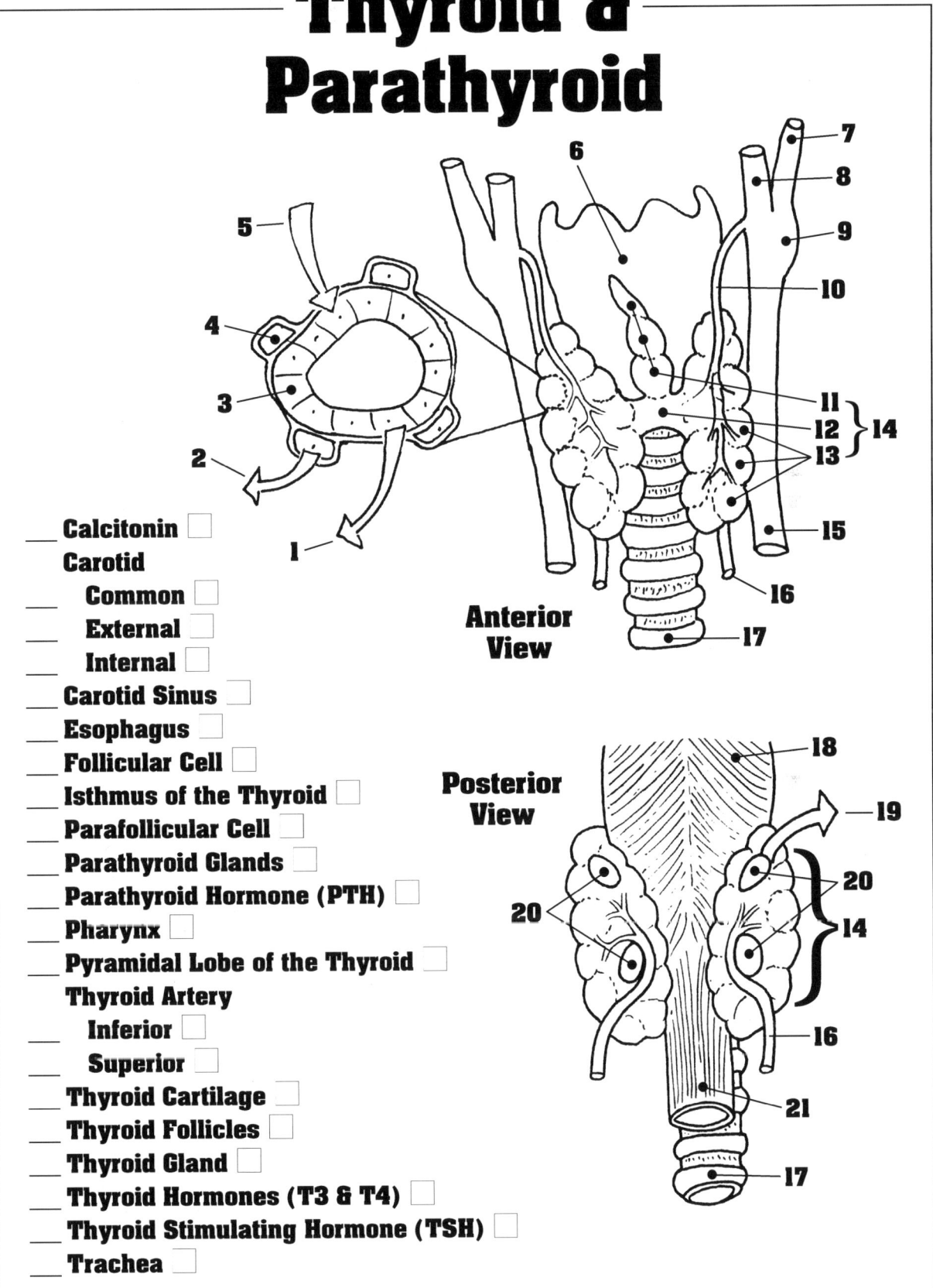

# Pancreas
## With Stomach Reflected Cephalad

The **stomach** (10) is reflected upward to show the **pancreas** (13) in conjunction with the **spleen** (12) and **duodenum** (3).

The pancreas produces three hormones:

- **Insulin** (14) — facilitates the conversion of glucose to glycogen and is produced by beta cells.

- **Glucagon** (15) — facilitates the conversion of glycogen to glucose and is produced by alpha cells.

- **GHIH (growth hormone-inhibiting hormone)** (16) — is produced by delta cells.

Whereas pancreatic enzymes are delivered to the duodenum via the **pancreatic duct** (4), the pancreatic hormones are delivered via the blood.

Along its upper margin the pancreas is serviced by arterial branches from the **splenic** (11). The splenic, in turn, derives from the **celiac** (7).

The **left gastric** (9) derives from the celiac trunk. The **right gastric** (6) derives from the **hepatic** (5).

The **superior mesenteric** (2) arises from the **abdominal aorta** (1) at the inferior border of the pancreas.

The **aortic hiatus** (8) is seen beneath the reflected stomach.

# Pancreas
## With Stomach Reflected Cephalad

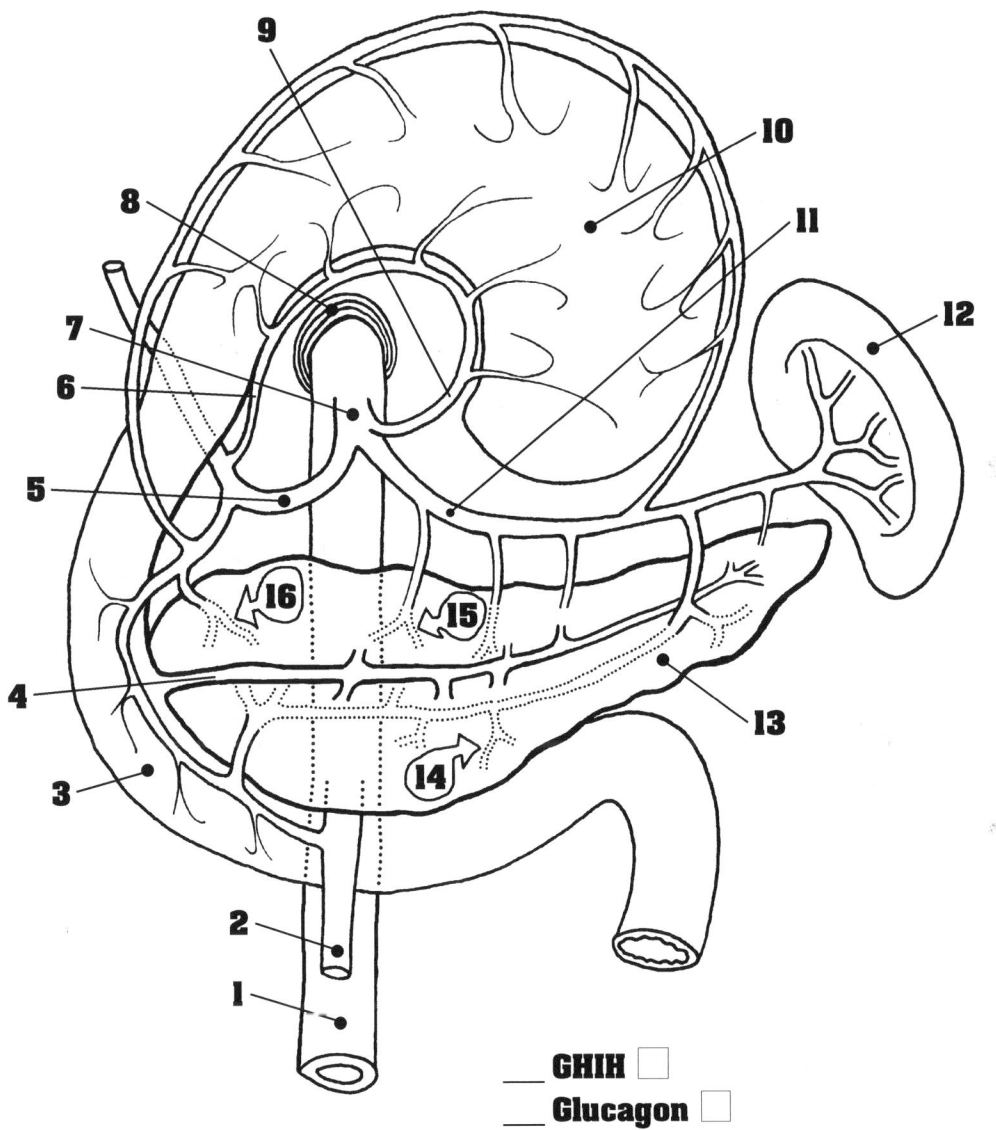

___ GHIH
___ Glucagon
___ Hepatic
___ Insulin
**Abdominal Aorta**
**Aortic Hiatus**
___ Pancreas
**Celiac**
___ Pancreatic Duct
**Duodenum**
___ Spleen
**Gastric**
___ Splenic
___ **Left**
___ Stomach
___ **Right**
___ Superior Mesenteric

# Adrenals

The **adrenal glands** (14) are adjacent to the renals, or **kidneys** (3). Each adrenal gland has an **adrenal cortex** (12) and an **adrenal medulla** (13).

The adrenal cortex produces three categories of hormones:

- **Mineralocorticoids** (4) — chiefly aldosterone. Aldosterone stimulates the reabsorption of sodium into the blood and because water follows salt this increases the blood volume, which in turn, increases the blood pressure.

- **Glucocorticoids** (5) — chiefly cortisol. Cortisol broadly effects cell metabolism, especially energy metabolism, and also depresses inflammation and the immune response.

- **Gonadocorticoids** (6) — chiefly testosterone. The role of the adrenal sex hormones is still in question. (Ahh! many things are still in question.) The gonadocorticoids from the adrenals are thought to have an insignificant effect on adults.

The adrenal medulla produces the "fight or flight" hormones, **epinephrine** (7) and **norepinephrine** (8), which mobilize the body's emergency response systems.

The vascular connections and associations with the kidneys and adrenal glands are shown and will serve as a nice primer for upcoming exercises on the circulatory blood vessel routes.

Blood is brought to the adrenals via three arteries: the **superior suprarenal artery** (11) which derives from the **inferior phrenic artery** (10), the **inferior suprarenal artery** (16) which derives from the **renal artery** (18), and the **middle suprarenal artery** (15) which derives directly from the aorta.

The **celiac** (9) and **superior mesenteric** (17) arteries are seen near the top of the **abdominal aorta** (19).

Blood returns from the kidneys to the **inferior vena cava** (20) via **renal veins** (2). The **ureters** (1) transport urine from the kidneys to the urinary bladder.

# Adrenals

## Anterior View

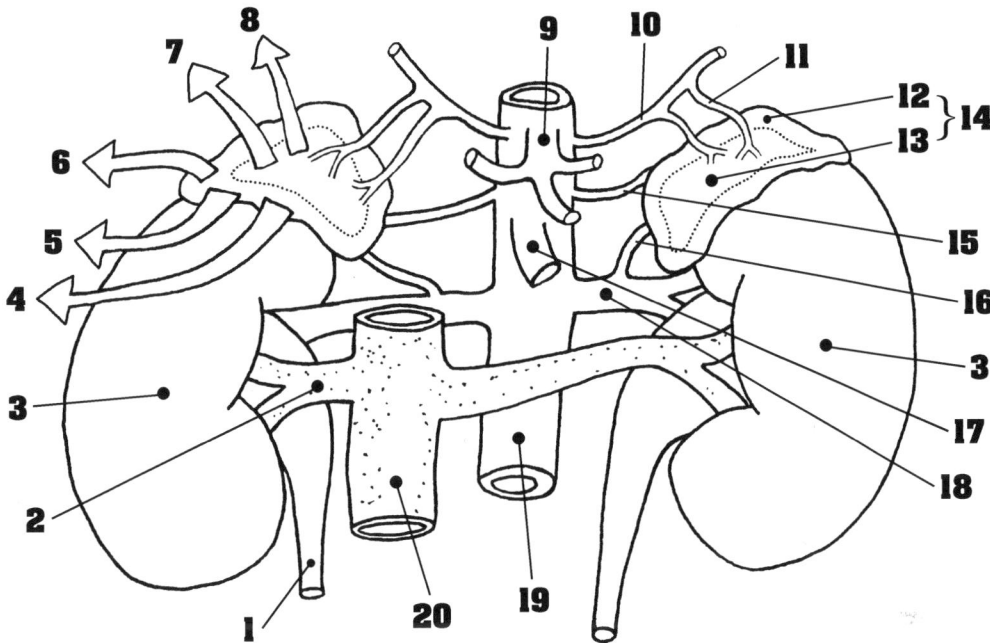

___ Abdominal Aorta
___ Adrenal Cortex
___ Adrenal Gland
___ Adrenal Medulla
___ Celiac Artery
___ Epinephrine
___ Glucocorticoids
___ Gonadocorticoids
___ Inferior Phrenic Artery
___ Inferior Vena Cava
___ Kidneys
___ Mineralocorticoids
___ Norepinephrine
___ Renal Artery
___ Renal Vein
___ Superior Mesenteric Artery
___ Suprarenal Artery
   ___ Inferior
   ___ Middle
   ___ Superior
___ Ureter

# Circulatory System

Pumps (heart), pipes (vessels), and fluid (blood) = a circulatory system. But no ordinary circulatory system to be sure. Flexible, distensible, variable, durable pumps force multi-functional fluid through flexible, distensible, variable, durable pipes. Consider the functions of the blood:

- **Delivers oxygen and nutrients to all cells.**
- **Transports wastes from cells to elimination sites.**
- **Transports chemical messages (hormones) to target tissues.**
- **Transports antibodies, and leukocytes to infection sites.**
- **Transports platelets and clotting proteins to damaged areas.**
- **Assists temperature regulation by absorbing and distributing heat.**
- **Transports pH buffers which assist acid/base balance.**
- **Provides hydrostatic pressure for some pressure depended activities (pssh! … such as a penile erection).**

The delivery of oxygen-deficient blood to the lungs, and the subsequent return of freshly oxygenated blood to the heart, is referred to as **pulmonary circulation**. The right ventricle serves as the pump for the pulmonary circulation.

The delivery of freshly oxygenated blood to all regions of the body, and the return of oxygen-deficient blood to the heart, is called **systemic circulation**. The left ventricle serves as the pump for systemic circulation.

The two ventricular pumps beat synchronously, delivering the same volume of blood, at the same time, with each beat. Although the two pumps beat as one, the left ventricular pump has considerable more muscle and pushes the blood out with a great deal more pressure. This is because the right ventricle needs only to push blood to the nearby lungs, while the left ventricle must push blood to all regions of the body.

Whereas arteries carry blood away from the heart, veins carry blood back to the heart. Intermediate, between the arteries and the veins, are the capillaries. Because the thin-walled capillaries directly deliver and receive materials from the body cells they are often referred to as the "functional units" of the circulatory system. Because, in the case of humans, the blood is always contained within blood vessels it is referred to as a **closed circulatory system**.

A good way to review the pathway of blood circulation, as it occurs within the heart, is to pretend you are a red blood cell returning to the heart via the superior vena cava (see page 215 for the anatomical correlates for this story).

# Circulatory System

After entering the right atrium you fall through the right atrioventricular valve into the right ventricle. From the right ventricle you are pumped upward through the pulmonary semilunar valve into the pulmonary artery which delivers you to the lungs where iron atoms on your hemoglobin molecules bind with oxygen. After being oxygenated you are carried, via a pulmonary vein to the left atrium of the heart. From the left atrium you drop down through the left atrioventricular valve into the left ventricle. From the left ventricle you are pumped upward through the aortic semilunar valve into the aorta.

Given the vital functions of the blood, as listed above, it is easy to see how our good health would be jeopardized if our circulatory pumps and pipes were not kept in good condition. The bad news is: circulatory disease is, in fact, presently the number one killer in America. The good news is: it is by and large preventable. Hoping it might encourage you to make good lifestyle choices we present the following sequence of drawings, illustrating what happens to the arteries over time if you do not make good lifestyle choices.

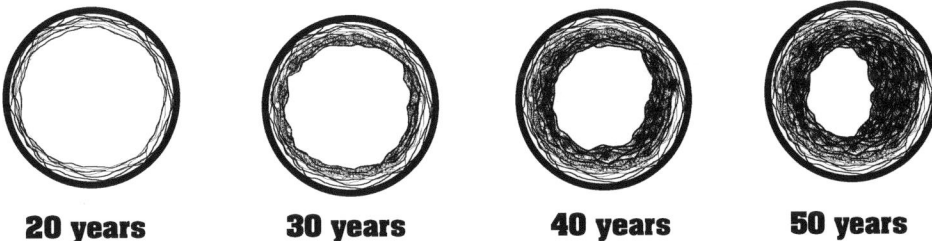

**20 years**  **30 years**  **40 years**  **50 years**

As the lumens of the arteries become more occluded and narrowed, they also become less elastic and more rigid, and dangers now lurk:

- The heart must now work harder to push the same volume of blood through the arteries, and over-worked hearts are dangerous hearts!

- Pushing blood through partially occluded arteries creates high blood pressure, and high blood pressure is risky business!

- As the lumens of the arteries become narrower the danger of a traveling blood clot causing a complete blockage of an artery increases, and as you probably know, a blood clot occluding a coronary artery can cause a heart attack. And as you probably already know a blood clot occluding a cerebral artery can cause a "stroke!" And as you probably already know a lot of people die from heart attacks or strokes.

# Blood Composition

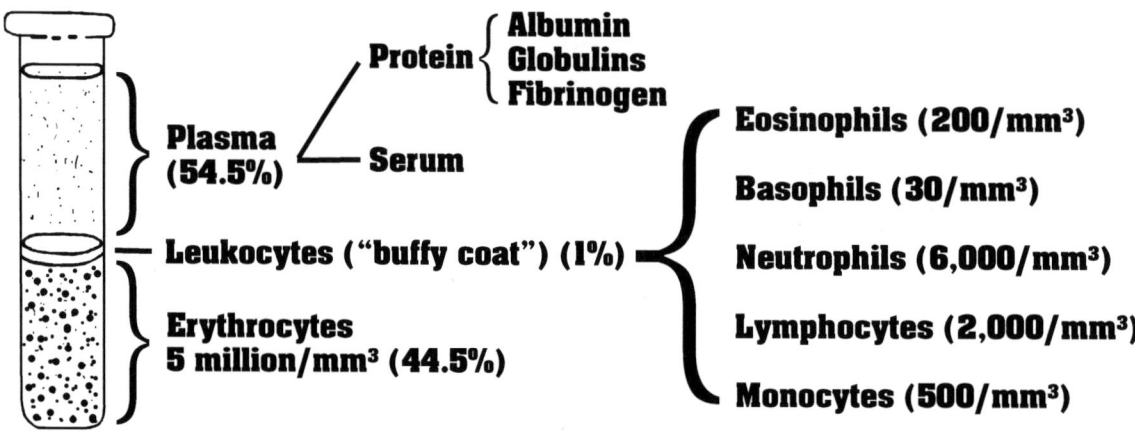

When a mixture of substances (such as blood cells) is centrifuged substances are sorted out by weight, with the lightest at the top and the heaviest at the bottom. The percentage of erythrocytes in whole blood (normally about 45%) is known as the **hematocrit**. Leukocytes and platelets contribute less than 1% to the blood volume. (The percentages and numbers in the above table are all approximations.)

## Plasma Proteins

Plasma proteins comprise about 8% of the blood plasma volume. Although all plasma proteins contribute to maintaining osmotic pressure, **albumin** plays a major osmotic role. Three major types of protein globulins occur in plasma. Alpha and beta globulins play important roles in protein transport. **Gamma globulins are antibodies** released by lymphocyte plasma cells which play important roles in the immune response.

## Erythrocytes

Because red blood cells are almost totally devoted to transporting oxygen molecules, they are, as mature cells, little more than "bags" of **hemoglobin molecules**. (As RBCs mature they cast out their nuclei to make room for more hemoglobin!) A single RBC contains approximately 250 million hemoglobin molecules, and each hemoglobin molecule has the capacity to bind with one oxygen molecule, thus, a fully "oxygen-saturated" RBC could potentially carry one billion oxygen atoms! — and there are 5 million RBCs in 1 cubic millimeter of blood!

# Blood Composition

## Leukocytes
There are three types of granular leukocytes and two types of agranular leukocytes.

### Agranular Leukocytes

#### Lymphocytes
Lymphocytes develop into plasma cells which produce and secrete antibodies which contribute to the immune response by combating alien proteins such as antigens.

#### Monocytes
Certain monocytes develop into **macrophages** which contribute to the immune response by patrolling body tissues and physically engaging alien proteins such as bacteria.

### Granular Leukocytes

#### Neutrophils
Neutrophils secrete a potent brew of antimicrobial proteins and are particularly proficient at fighting and destroying bacteria. Neutrophils multiply rapidly during acute bacterial infections. An elevated neutrophil count almost always indicates an infection.

#### Eosinophils
Eosinophils, as compared with neutrophils, produce and secrete different classes of enzymes, which are thought to be more proficient at fighting larger "aliens" such as parasitic worms.

#### Basophils
Basophils, the rarest of the leukocytes, assist the immune response by releasing histamine which acts as a vasodilator and also attracts other white blood cells to the site of an infection.

## Platelets
Platelets, which are fragmentary cytoplasmic "pieces" produced from the disintegration of extraordinarily large cells called megakaryocytes, play essential roles in hemostasis (blood clotting). Platelets interact with other clotting factors (over 30 different substances are involved!) to coagulate (clot) the blood.

# Hemopoiesis

All formed elements (blood cell types) arise from pleuripotent hematopoietic stem cells, called **hemocytoblasts** (1). Pleuripotency implies a capacity for becoming many different things, and in this case it implies the capacity of hemocytoblasts to give rise to:

- **Erythroblasts** (2) — when **erythrocytes (red blood cells)** (22) are needed.
- **Lymphoid stem cells** (4) — when **lymphocytes** (27) are needed.
- **Myeloid stem cells** (3) — when **basophils** (24), **eosinophils** (23), **neutrophils** (25), or **monocytes** (26) are needed.
- **Megakaryoblasts** (5) — when **blood platelets** (28) are needed.

In the case of red blood cells, erythroblasts become **normoblasts** (6) which then **eject their nuclei** (11) and become immature RBCs called **reticulocytes** (17). Reticulocytes then mature into erythrocytes.

In the case of blood platelets, megakaryoblasts transform into **megakaryocytes** (21), which later fragment into platelets.

In the case of eosinophils, **type 1 myeloblasts** (7) give rise to **type 1 promyelocytes** (12) which give rise to **eosinophilic band cells** (18), which then become eosinophiles.

In the case basophils, **type 2 myeloblasts** (8) give rise to **type 2 promyelocytes** (13) which give rise to **basophilic band cells** (19), which then become basophiles.

In the case of neutrophils, **type 3 myeloblasts** (9) give rise to **type 3 promyelocytes** (14) which give rise to **neutrophilic band cells** (20), which then become neutrophils.

In the case of monocytes, myeloid stem cells give rise to **type 3 myeloblasts** (9), which in turn give rise to **promonocytes** (15). Promonocytes then become monocytes.

In the case of lymphocytes, **lymphoblasts** (10) give rise to **prolymphocytes** (16), which later become mature lymphocytes.

*As we consider the remarkable pleuripotency of the hemocytoblast "mother" cells, which can give rise to seven different blood cell types, we are led to ponder the even more remarkable pleuripotency of the early embryonic cells which can give rise to all the cell types which occur in the human body.*

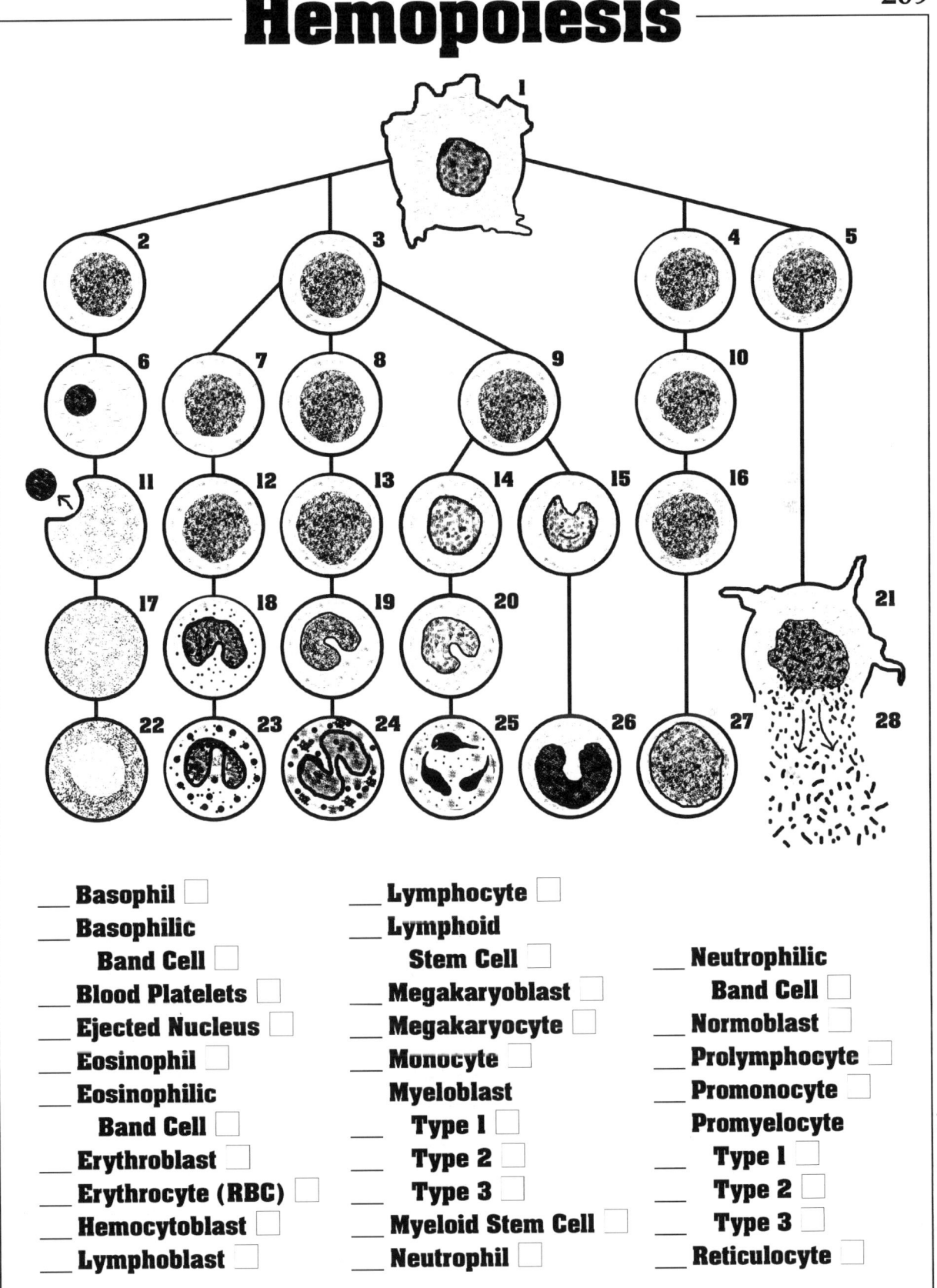

# Capillary Net & Microcirculation

Approximately 60,000 miles of capillaries service the tissues. Because thin-walled capillaries facilitate the exchange of nutrients, gases, hormones, wastes, etc., between blood and cells, they are referred to as the "functional units" of the circulatory system. No living body cell can be far removed from a capillary.

After arriving at the **capillary net** (3) via a **metarteriole** (6) branch from a **pre-capillary arteriole** (1), blood may either pass into a capillary or by-pass the capillary via a **vascular shunt** (7). The vascular shunt, which runs through the capillary net, is comprised of a metarteriole, and a continuation of the metarteriole called a **thoroughfare channel** (5). **Postcapillary venules** (4) drain blood away from the capillary net.

Whether or not blood will flow into the capillaries, or pass through the vascular shunt is determined by whether or not the **pre-capillary sphincters** (8) are open or closed. Because a variety of subtle, complex, signaling mechanisms, including local hormones, play roles in regulating the opening and closing of the pre-capillary sphincters, the microcirculation of blood is precisely controlled. Arterioles also have smooth **muscle rings** (2) which by constricting or dilating control the flow of blood.

*The benefits of exercise and conditioning are clearly manifest at the microcirculation level where, over time, hundreds, or even thousands, of miles of capillaries can be added or deleted by exercise or the lack thereof. Because a domestic chicken does not use its wings for flight, the breast muscles, which operate the wings, have relatively few capillaries, and hence are offered to the consumer as "white meat." The legs and thighs, however, which are constantly used, and have more dense populations of capillaries, are offered to the consumer as "dark meat." Thus, to some degree or another, depending on exercise or, the lack thereof, the muscles throughout our bodies are "capillary-rich" or "capillary-poor." Best, we suggest, to exercise regularly and be "rich."*

# Capillary Net & Microcirculation

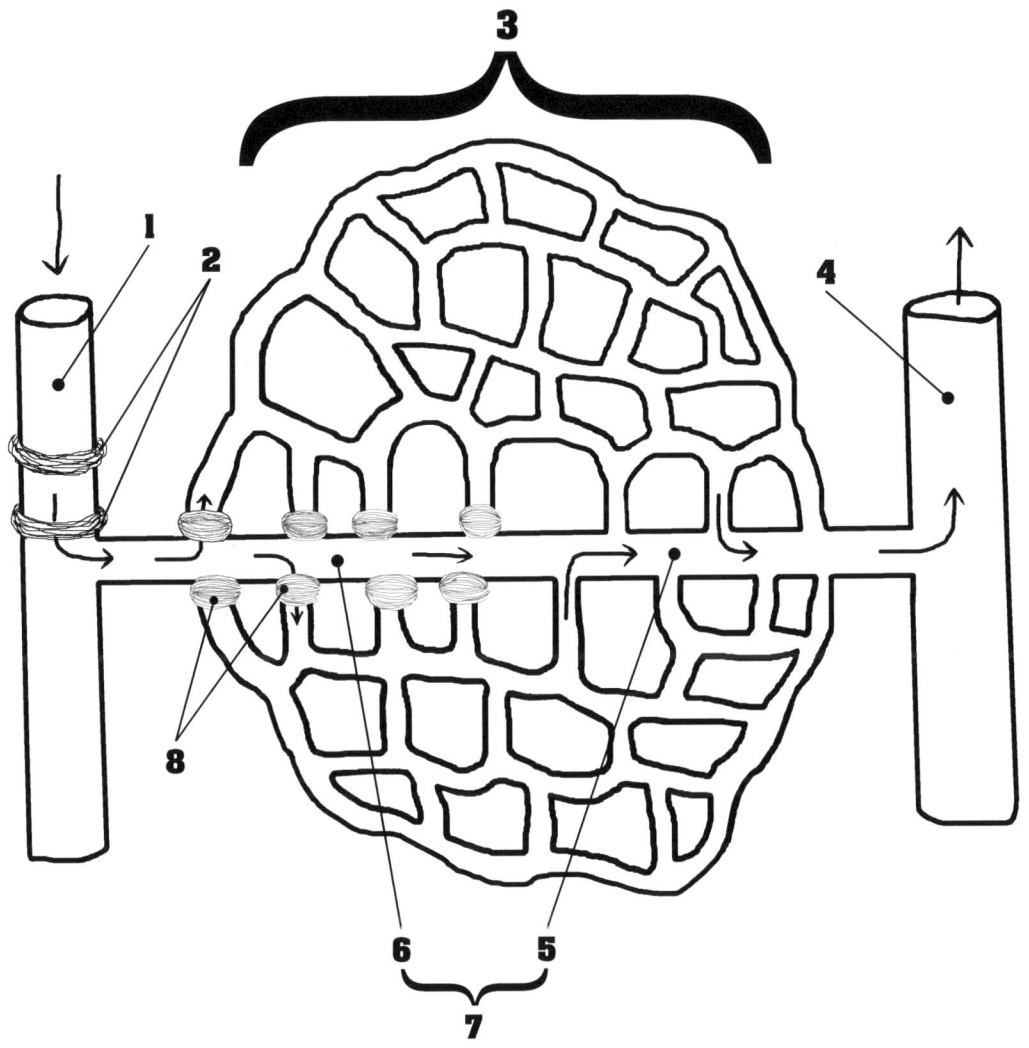

___ Capillary Net ☐
___ Metarteriole ☐
___ Muscle Rings ☐
___ Pre-Capillary Arteriole ☐
___ Pre-Capillary Sphincters ☐
___ Postcapillary Venules ☐
___ Thoroughfare Channel ☐
___ Vascular Shunt ☐

# Heart
## Anterior View with By-pass

Although coronary circulation is the main emphasis of this drawing, it also provides an excellent opportunity for observing the major vessels which enter and exit the heart.

The **right** (6) and **left** (18) **coronary arteries** arise from the extreme base of the **aortic trunk** (8). The left coronary artery divides into an **anterior interventricular branch** (20) which serves the muscles of the anterior aspect of the heart and a **circumflex artery** (21) that curves around to the posterior aspect of the heart.

The booming business of by-pass surgery is based upon transplanting blood vessels from other places in the body, sewing them into the base of the aorta, and by-passing cholesterized and incompetent coronary arteries. In this drawing we see a **transplanted vessel** (4) sewn in to the aorta, by-passing a **right coronary occlusion** (3).

The **left coronary vein** (19), along with other coronary veins, returns blood to the **coronary sinus** (22), which is located on the upper backside of the heart. The venous blood then enters the **right atrium** (7) via the **orifice of the coronary sinus** (5). (When a vein is expanded to serve as a blood reservoir as well as a passageway, it is called a sinus. Another good example is the superior sagittal sinus at the very top of the head where the venous blood returning from cerebral circulation is collected (see page 233). Actually, the entire venous system acts, in part, as a blood reservoir.)

Besides the coronary sinus, other veins which bring blood to the heart are the **superior** (9) and **inferior** (1) **vena cavas**, which return blood to the right atrium, and the **pulmonary veins** (16), which return blood to the **left atrium** (15).

The two great arteries that carry blood away from the heart are the aorta and the **pulmonary** (17). Three major branches arise at the top of the **aortic arch** (13): the **brachiocephalic** (10), **left common carotid** (11), and **left subclavian** (12). The pulmonary trunk divides into the right and **left pulmonary artery** (14). The regions of the **right** (2) and **left** (23) **ventricles** are also indicated.

# Heart
## With By-pass

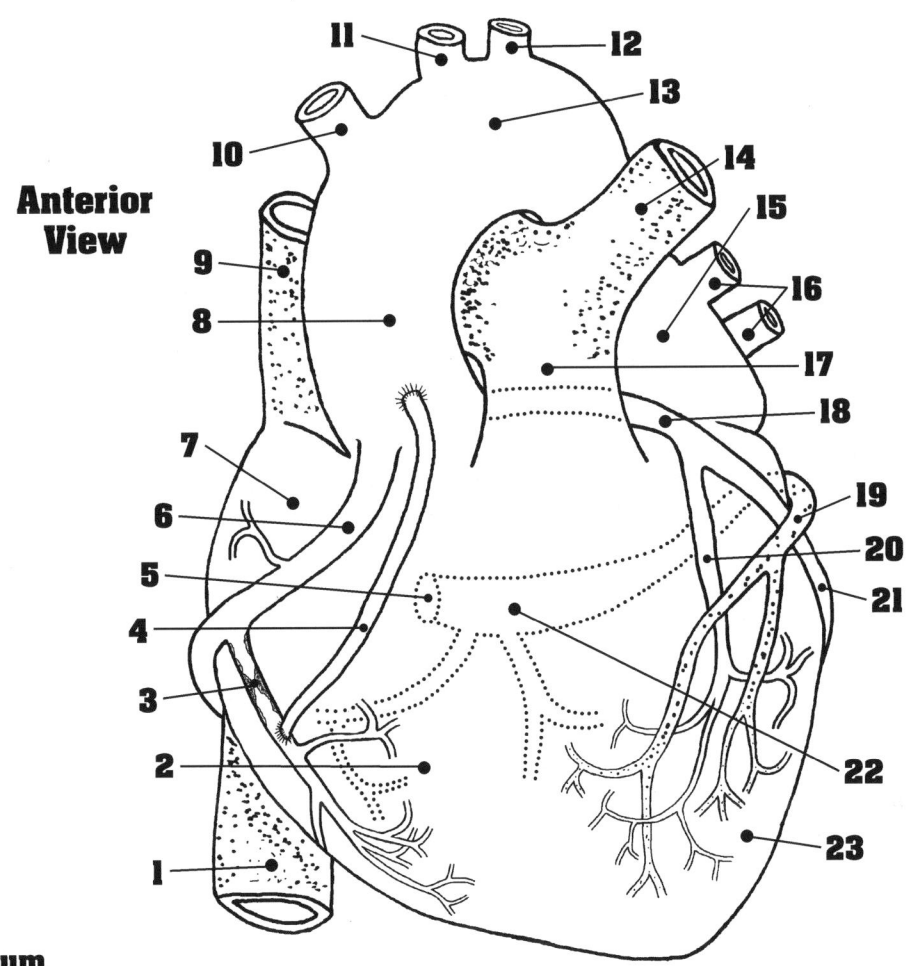

Anterior View

**Atrium**
___ Left
___ Right
___ **Anterior Interventricular Artery**
___ **Aortic Arch**
___ **Aortic Trunk**
___ **Brachiocephalic Artery**
___ **Circumflex Artery**
**Coronary Artery**
___ Left
___ Right
___ **Coronary Sinus**
___ **Left Common Carotid**
___ **Left Coronary Vein**

___ **Left Pulmonary Artery**
___ **Left Subclavian Artery**
___ **Orifice of the Coronary Sinus**
___ **Pulmonary**
___ **Pulmonary Veins**
___ **Right Coronary Occlusion**
___ **Transplanted Vessel**
**Vena Cava**
___ Inferior
___ Superior
**Ventricle**
___ Left
___ Right

# Heart
## Frontal Section

Unoxygenated blood, returning from the body, enters the **right atrium** (9) through three portals of entry. The **superior vena cava** (12), via its **cardiac entry** (11), returns blood from the head, neck, shoulders and arms. The **inferior vena cava** (1), via its **cardiac entry** (6) returns blood from body regions below the heart. The coronary sinus (see page 213), via its **cardiac entry** (7) returns blood from the muscle tissues of the heart itself.

The returning blood then passes through the **tricuspid (right atrioventricular) valve** (5) into the **right ventricle** (3). As ventricular blood pressure increases the cuspid valve flaps are pushed upward and close, but are prevented from introverting (prolapsing) by the contraction of the **papillary muscles** (2) which put tension on the **chordae tendinae** (4). The chordae tendinae ("heart strings") are attached to the undersides of the cuspid valves.

The right and **left ventricle** (27) contract at the same time, and each pumps out approximately 75 ml of blood. The blood from the right ventricle is pumped up through the **pulmonic semilunar valve** (18) into the **trunk of the pulmonary artery** (19). The pulmonary trunk divides into the **right** (10) and **left** (20) **pulmonary arteries**.

After gas exchange in the lungs (between air and blood), the refreshed blood returns to the **left atrium** (21) via the **right** (8) and **left** (22) **pulmonary veins**. There are two **cardiac entries** (24) for the right pulmonary veins, and two **cardiac entries** (23) for the left pulmonary veins.

From the left atrium the blood passes downward through the **bicuspid (left atrioventricular) valve** (25) into the left ventricle, which, in turn, pumps the blood upward through the **aortic semilunar valve** (26) into the **aortic arch** (13), and then downward through the **thoracic aorta** (29), and on to all parts of the body.

From the top of the aortic arch, three major arterial branches arise upward: the **brachiocephalic** (14), **left common carotid** (16), and **left subclavian** (17).

The two ventricular pumps are divided by an **interventricular septum** (28). The significantly thicker muscular wall of the left ventricle bears evidence to the fact that whereas the right ventricle pumps blood only to the lungs, the left ventricle pumps blood to all body parts.

We see also the **scar of the ductus arteriosus** (15) which, during the fetal circulation, allowed blood to by-pass the inoperative fetal lungs and go directly from the trunk of the pulmonary artery to the aorta.

# Heart

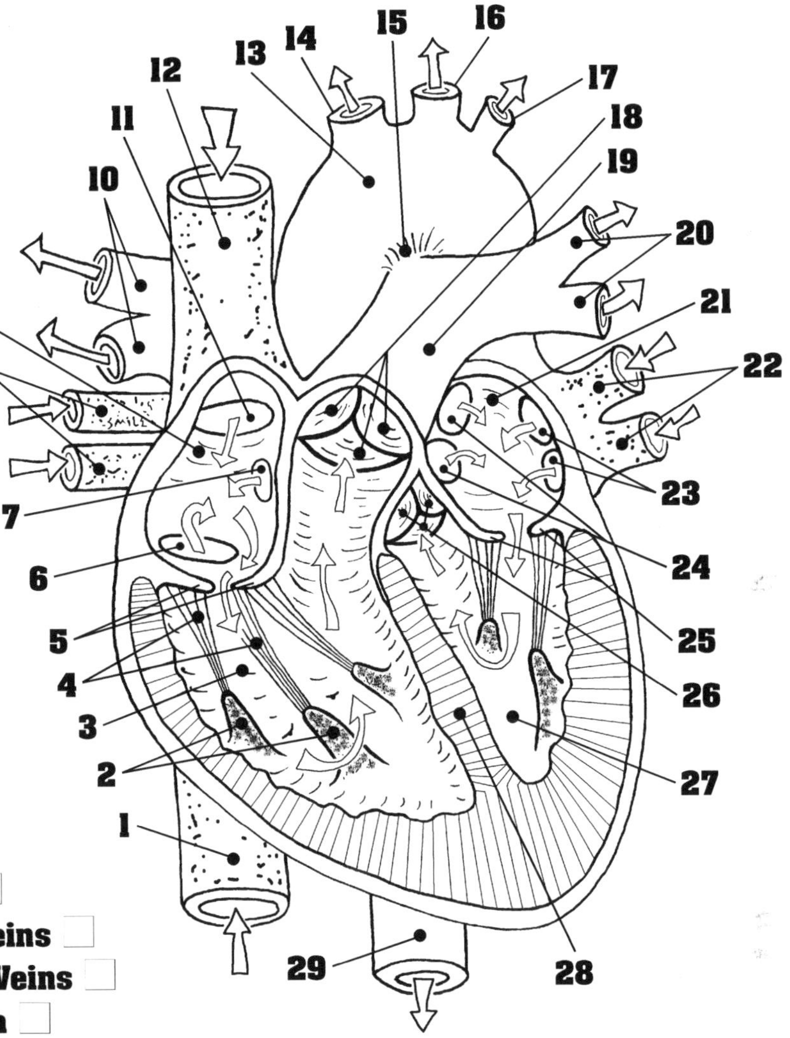

**Frontal Section**

___ Aortic Arch
___ Aortic Semilunar Valve
___ Atrium
   ___ Left
   ___ Right
___ Bicuspid (Left Atrioventricular) Valve
___ Brachiocephalic Artery
___ Cardiac Entry of
   ___ Coronary Sinus
   ___ Left Pulmonary Veins
   ___ Right Pulmonary Veins
   ___ Inferior Vena Cava
   ___ Superior Vena Cava
___ Chordae Tendinae
___ Interventricular Septum
___ Left Common Carotid Artery
___ Left Subclavian Artery
___ Papillary Muscles
___ Pulmonary Arteries
   ___ Left
   ___ Right
___ Pulmonary Veins
   ___ Left
   ___ Right
___ Pulmonic Semilunar Valve
___ Scar of the Ductus Arteriosus
___ Thoracic Aorta
___ Tricuspid (Right Atrioventricular) Valve
___ Trunk of the Pulmonary Artery
___ Vena Cava
   ___ Inferior
   ___ Superior
___ Ventricle
   ___ Left
   ___ Right

# Cardiac Center

The heart rate, usually designated in beats per minute (BPM), is normally regulated by blood pressure. **Pressoreceptors** (14) are located in sinuses in both the **aortic arch** (15) and the **carotid artery** (13).

Nerve action potentials travel, via the **glossopharyngeal nerve** (12), from the pressoreceptors to the **cardiac center** (11) in the medulla of the brain. The cardiac center contains both a **CAC (cardiac accelerator center)** (10) and a **CIC (cardiac inhibitory center)** (9).

If the blood pressure in the aortic arch and carotid artery is too low the CAC sends impulses via the **spinal cord** (7) and **cardiac nerve** (6) to the pacemaker, also called the **SA (sino-atrial) node** (5). The SA node, which is located in the wall of the right atrium, responds by increasing its signal output, both to the atrial muscles, and to the **AV (atrioventricular) node** (4).

The AV node, in turn, sends more facilitory action potentials through the **atrioventricular bundle (bundle of His)** (3) and **bundle branches** (1) to the **Purkinje fibers** (2) which innervate the ventricular muscles.

Electrical changes (voltage changes) associated with the depolarization of heart muscle membranes prior to their contraction can be recorded, and such a recording is called an EKG (electrokardiogram).

The first wave of depolarization — the P wave — reflects the depolarization of the atrial muscles. The second, much larger, wave of depolarization — the QRS wave — reflects the depolarization of the ventricular muscles. The third wave on the EKG — the T wave — reflects the repolarization of the ventricular muscles.

If the blood pressure is too high, rather than too low, the CIC sends signals via the **vagus nerve** (8) to the SA node. And in response to these signals the SA node decreases its signal output.

The synaptic inhibitory neurotransmitter substance used by the vagus nerve is ACh (acetylcholine), while the facilitory neurotransmitter used by the cardiac nerve is epinephrine. If, in the laboratory, you drip acetylcholine on a live frog heart it will slow down, and conversely, if you drip epinephrine on it, it will speed up. The drug digitalis, derived from the foxglove plant, mimics the effect of epinephrine, and has long been used as a heart stimulant.

*If the brain is so severely damaged that the cardiac center cannot operate, the SA node will take over and send out signals on its own. Hence, if one removes the heart from a frog, puts it in a petri plate, and keeps it moist, it will continue to beat (and entertain A&P students) for several hours without any connection to the central nervous system.*

# Cardiac Center

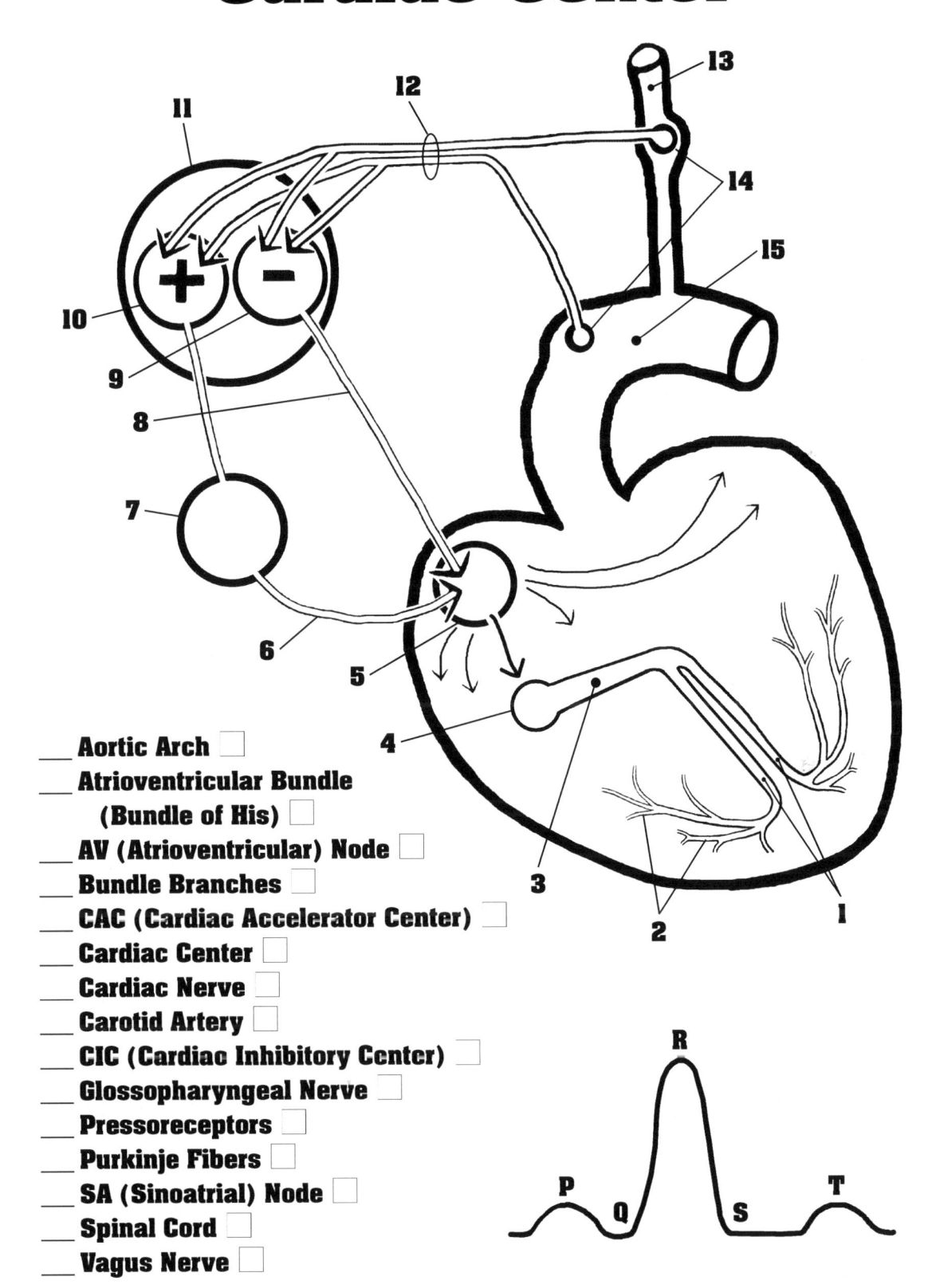

___ Aortic Arch ☐
___ Atrioventricular Bundle
    (Bundle of His) ☐
___ AV (Atrioventricular) Node ☐
___ Bundle Branches ☐
___ CAC (Cardiac Accelerator Center) ☐
___ Cardiac Center ☐
___ Cardiac Nerve ☐
___ Carotid Artery ☐
___ CIC (Cardiac Inhibitory Center) ☐
___ Glossopharyngeal Nerve ☐
___ Pressoreceptors ☐
___ Purkinje Fibers ☐
___ SA (Sinoatrial) Node ☐
___ Spinal Cord ☐
___ Vagus Nerve ☐

# Abdominopelvic Blood Vessels

The two great vessels of the abdomen, the **abdominal aorta** (6) and the **inferior vena cava** (7), seen together in this drawing, offer an excellent opportunity to observe the major vessels — both arteries and veins — of the abdominal region.

## Abdominal Aorta

The **phrenic arteries** (18) arise first, then the **celiac artery** (19) with its tributaries: the **gastric artery** (20), **splenic artery** (21), and **hepatic artery** (17).

A **superior mesenteric artery** (22) arises just prior to where the **renal arteries** (13) branch off toward the **kidneys** (11). **Adrenal arteries** (14) ascend up from the renal arteries to the **adrenal glands** (16).

The **gonadal arteries** (8) arise next, and further down the **inferior mesenteric artery** (23) arises. At its termination the abdominal aorta splits into two **common iliac arteries** (24), which, in turn, branch into the **external** (3) and **internal** (1) **iliac arteries**. A **sacral artery** (26) descends from the juncture of the common iliacs.

The **ureters** (10) are shown descending from the kidneys.

## Inferior Vena Cava

**Adrenal veins** (15) arise from the **renal veins** (12). Note the asymmetry of the **gonadal veins** (9), with one arising directly from the vena cava, while the other arises from the left renal vein.

Most inferiorly, the **external** (4) and **internal** (2) **iliac veins** converge to form the **common iliac veins** (5) and a **sacral vein** (25) also merges at the juncture of the common iliac veins.

*We are sure you are asking why there are so many more arteries carrying blood away from the abdominal aorta than veins returning blood to the inferior vena cava? That is because there is an important alternative venous return system — the hepatic portal system — shown on page 243.*

# Abdominopelvic Blood Vessels

- ___ Abdominal Aorta
- Adrenal
  - ___ Arteries
  - ___ Glands
  - ___ Veins
- ___ Celiac Artery
- ___ Gastric Artery
- Gonadal
  - ___ Arteries
  - ___ Veins
- ___ Hepatic Artery
- Iliac Arteries
  - ___ Common
  - ___ External
  - ___ Internal
- Iliac Veins
  - ___ Common
  - ___ External
  - ___ Internal
- ___ Inferior Vena Cava
- ___ Kidneys
- Mesenteric Artery
  - ___ Inferior
  - ___ Superior
- ___ Phrenic Arteries
- Renal
  - ___ Arteries
  - ___ Veins
- Sacral
  - ___ Artery
  - ___ Vein
- ___ Splenic Artery
- ___ Ureters

**Anterior View**

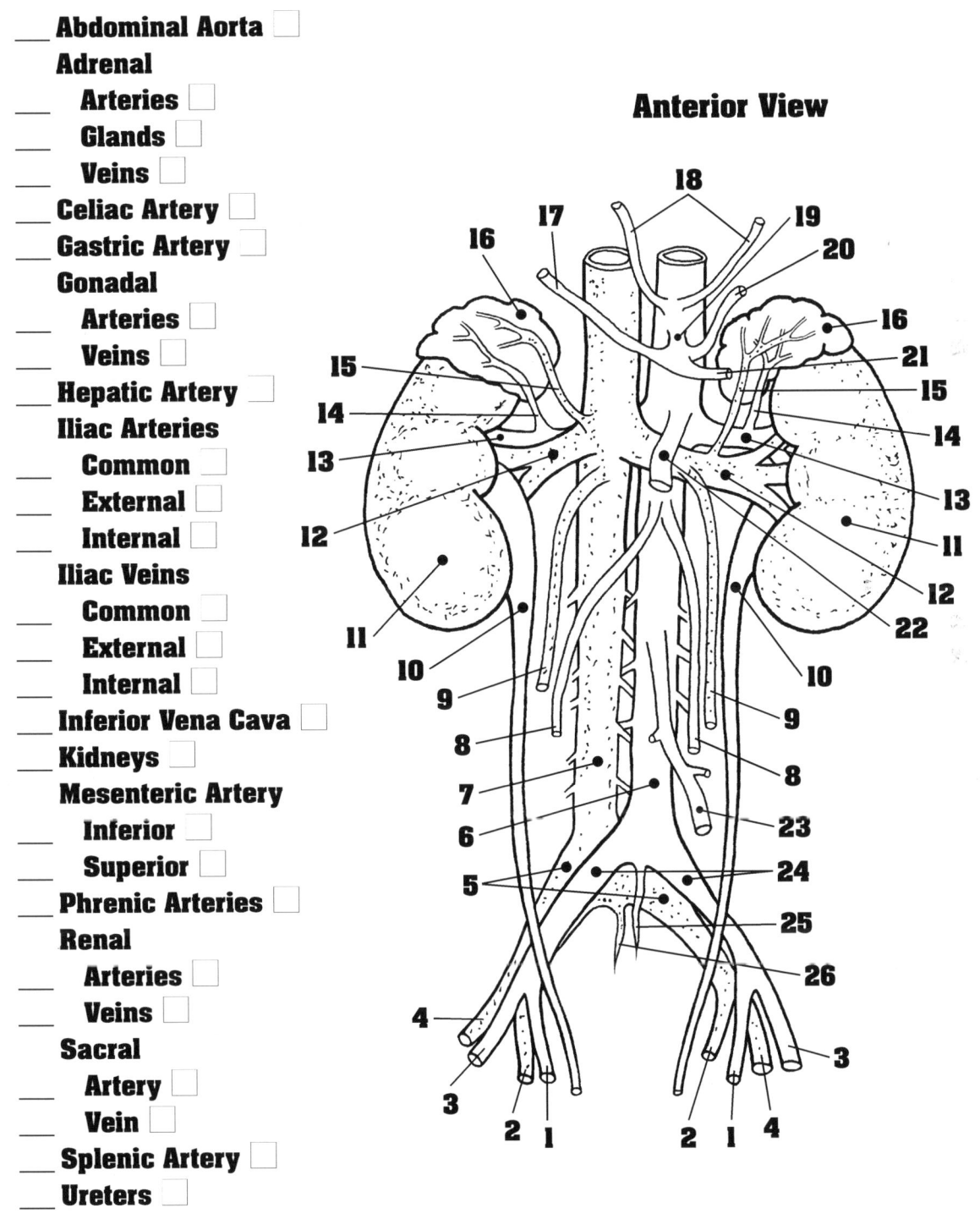

# Aorta — In Situ

In this introductory exercise to the arterial maps of the body the lungs and heart are removed in order to more clearly see the major pipes and vessels of the thoracic cavity, as well as the muscular, circular hiatuses the pipes and vessels use to pass through the **diaphragm** (27).

The **inferior vena cava** (7) carries venous blood upward through the **vena cava hiatus** (6). The **esophagus** (8) carries food downward through the **esophageal hiatus** (26) to the **stomach** (28). The **thoracic aorta** (25) carries arterial blood downward through the **aortic hiatus** (4).

The **trachea** (11), with its two **bronchi** (10), is seen medially between the **aortic arch** (22) and the esophagus. The **ribs** (9) radiate out from the vertebral column (not shown) posterior to the esophagus.

Other than the inferior vena cava, the only vessels of the venous system shown in this drawing are the **hepatic veins** (5).

Three major arteries arise from the top of the aortic arch: the **brachiocephalic** (12), **left common carotid** (21), and **left subclavian** (20).

On the right side of the aortic arch three major arteries ascend upward from the region where the brachiocephalic artery merges into the **right subclavian artery** (14): the **right common carotid artery** (19), **vertebral artery** (15), and **thyrocervical artery** (16). The thyrocervical artery branches upward into the **thyroid artery** (18) and the **cervical artery** (17). Descending downward from the same region is the **internal thoracic artery** (13).

On the left side of the aortic arch a vertebral artery and a thyrocervical artery ascend upward from the left subclavian artery, and an internal thoracic artery descends downward.

As the thoracic aorta descends through the thorax it sends out **intercostal arteries** (23) to service the intercostal muscles of the rib cage, and **esophageal arteries** (24) to service the esophagus.

Below the diaphragm, the thoracic aorta becomes the **abdominal aorta** (1). The first arteries to arise from the abdominal aorta are the **inferior phrenic arteries** (3), which supply blood to the inferior surface of the diaphragm. Below that we see the **celiac artery** (30) which immediately branches into a notable trinity of arteries: the **gastric artery** (29), **splenic artery** (31), and **hepatic artery** (2).

# Aorta — In Situ

**Anterior View**

- ___ Abdominal Aorta ☐
- ___ Aortic Arch ☐
- ___ Aortic Hiatus ☐
- ___ Brachiocephalic Artery ☐
- ___ Bronchi ☐
- ___ Celiac Artery ☐
- ___ Cervical Arteries ☐
- Common Carotid
  - ___ Left ☐
  - ___ Right ☐
- ___ Diaphragm ☐
- ___ Esophageal Arteries ☐
- ___ Esophageal Hiatus ☐
- ___ Esophagus ☐
- ___ Gastric Artery ☐
- Hepatic
  - ___ Artery ☐
  - ___ Veins ☐
- ___ Inferior Phrenic Arteries ☐
- ___ Inferior Vena Cava ☐
- ___ Intercostal Arteries ☐
- ___ Internal Thoracic Artery ☐
- Subclavian Artery
  - ___ Left ☐
  - ___ Right ☐
- ___ Ribs ☐
- ___ Splenic Artery ☐
- ___ Stomach ☐
- ___ Thoracic Aorta ☐
- ___ Thyrocervical Arteries ☐
- ___ Thyroid Arteries ☐
- ___ Trachea ☐
- ___ Vena Cava Hiatus ☐
- ___ Vertebral Arteries ☐

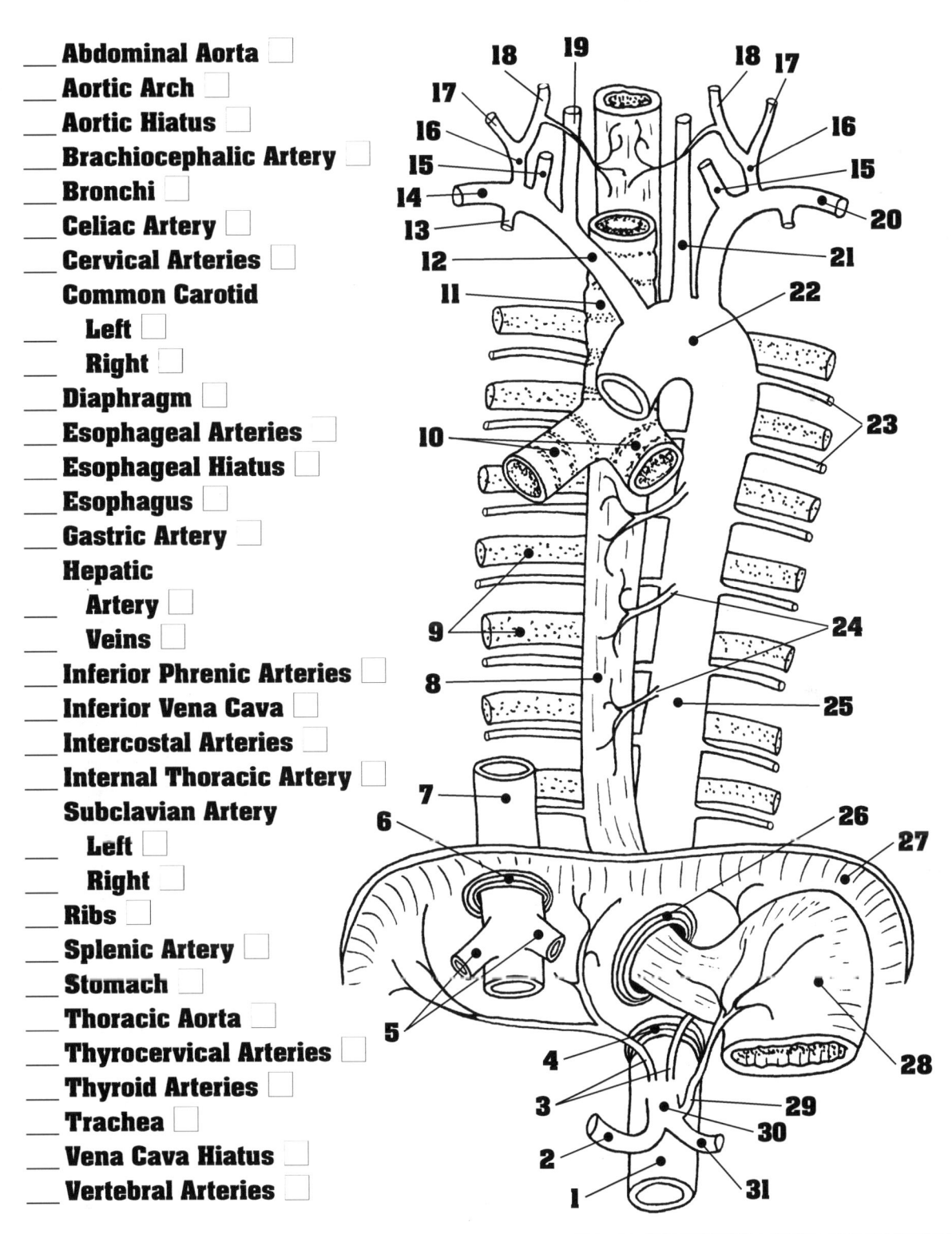

221

# Aortic Arterial Map

The aorta is regionally subdivided into the **aortic trunk** (11), **aortic arch** (20), **thoracic aorta** (23), and, below the **diaphragm** (24), the **abdominal aorta** (28). Knowing the major arterial branches of the great aorta is fundamental to good basic human anatomy.

## Aortic Trunk

**Coronaries** (10), which carry blood to the heart muscles, arise from the base of the trunk.

## Aortic Arch

The **brachiocephalic** (17), **left common carotid** (18), and **left subclavian** (19), all arise from the top of the aortic arch.

On the right side of the arch, in the region where the brachiocephalic artery gives way to the **right subclavian** (13), the **right common carotid** (16), **vertebral** (15), and **thyrocervical** (14) ascend upward. An **internal thoracic** (12) descends.

On the left side of the aortic arch the left common carotid originates directly from the aortic arch, while a vertebral and a thyrocervical derive from the left subclavian, and ascend from the same region where an internal thoracic descends.

## Thoracic Aorta

The **bronchials** (21) arise at the region where the aortic arch becomes the thoracic aorta. Downward through the thorax, **intercostals** (22) emerge from the aorta to service the intercostal muscles which move the rib cage. **Esophageals** (9) also emerge to service the esophagus, and just prior to reaching the diaphragm **superior phrenics** (8) service the superior surface of the diaphragm.

## Abdominal Aorta

From superior to inferior, we encounter the **inferior phrenics** (7) and the **celiac** (25) with its trinity of branches — the **gastric** (26), **splenic** (27), and **hepatic** (6). Below the celiac are the **superior mesenteric** (5), **renals** (29), **gonadals** (3), and **inferior mesenteric** (30).

**Lumbars** (4) arise throughout the lumbar region. At its termination the abdominal aorta divides into the **common iliacs** (31), and each of the common iliacs divides into an **external** (1) and an **internal** (2) iliac.

Finally the **sacral** (32) descends from the region where the abdominal aorta terminates.

# Aortic Arterial Map

**Anterior View**

___ Abdominal Aorta
___ Aortic Arch
___ Aortic Trunk
___ Brachiocephalic
___ Bronchials
___ Celiac
    Common Carotid
___     Left
___     Right
___ Coronaries
___ Diaphragm
___ Esophageals
___ Gastric
___ Gonadals
___ Hepatic
    Iliacs
___     Common
___     External
___     Internal
___ Intercostals
___ Internal Thoracics
___ Lumbars
    Mesenteric
___     Inferior
___     Superior
    Phrenics
___     Inferior
___     Superior
___ Renals
___ Sacral
___ Splenic
___ Subclavian — Left
___ Subclavian — Right
___ Thoracic Aorta
___ Thyrocervicals
___ Vertebrals

# Back & Thorax Veins

Although veins, which return blood to the heart, do, in general, follow similar patterns to the corresponding arteries that carry blood away from the heart, veins are not only larger than arteries, but in most regions of the body are also more numerous. Hence, veins usually carry about 65% of the body's blood. In this drawing the major veins connecting with the **superior vena cava** (11) and **inferior vena cava** (9), together with the major veins connecting with the **azygos** (10) and **hemiazygos** (19) accessory venous return system, are shown.

Blood returns from the head principally through four major pairs of veins: the **external jugulars** (15), **internal jugulars** (16), **vertebrals** (17), and most medially, the **inferior thyroids** (18). The blood from all four pairs of veins flows into the paired **brachiocephalics** (12), then into the superior vena cava, and on to the heart (the heart has been removed to show the accessory azygos venous system). Blood from the arm and shoulder regions return via the **subclavians** (14).

The **external** (1) and **internal** (2) **iliacs** return blood from the legs to the **common iliacs** (3). A **sacral** (27) also joins the common iliacs.

From the common iliacs upward, the blood has two alternate routes. Most blood is returned via the inferior vena cava, but some blood returns via the **ascending lumbars** (4), which also receive blood from the muscles of the lower back.

Blood from the gonads is picked up by the **right** (5) and **left** (26) **gonadals**. Blood from the kidneys is received by the **renals** (25). Blood from the liver is returned via the **hepatics** (7). There are no major direct venous connections between the major digestive organs (stomach, intestines, pancreas and spleen) and the inferior vena cava. Those organs return blood to the liver via the separate hepatic portal system. After monitoring the digestive blood, the liver then returns it to the inferior vena cava via the previously noted hepatics. (The hepatic portal system is shown in a separate exercise on page 243.)

Following the azygos and hemiazygos veins up through the **aortic hiatus** (6) — and, yes, this is the same hiatus through which the thoracic aorta passed — we note that the right ascending lumbar becomes the azygos and that the left ascending lumbar becomes the hemiazygos. The azygos and hemiazygos pick up blood from the intercostal muscles via **intercostals** (20) and blood from the **esophagus** (13) via **esophageals** (21).

In addition to the previously noted aortic hiatus, the **inferior vena cava hiatus** (8) and the **esophageal hiatus** (22) are shown as portages through the **diaphragm** (23). The proximal end of the **stomach** (24) is also shown.

# Back & Thorax Veins

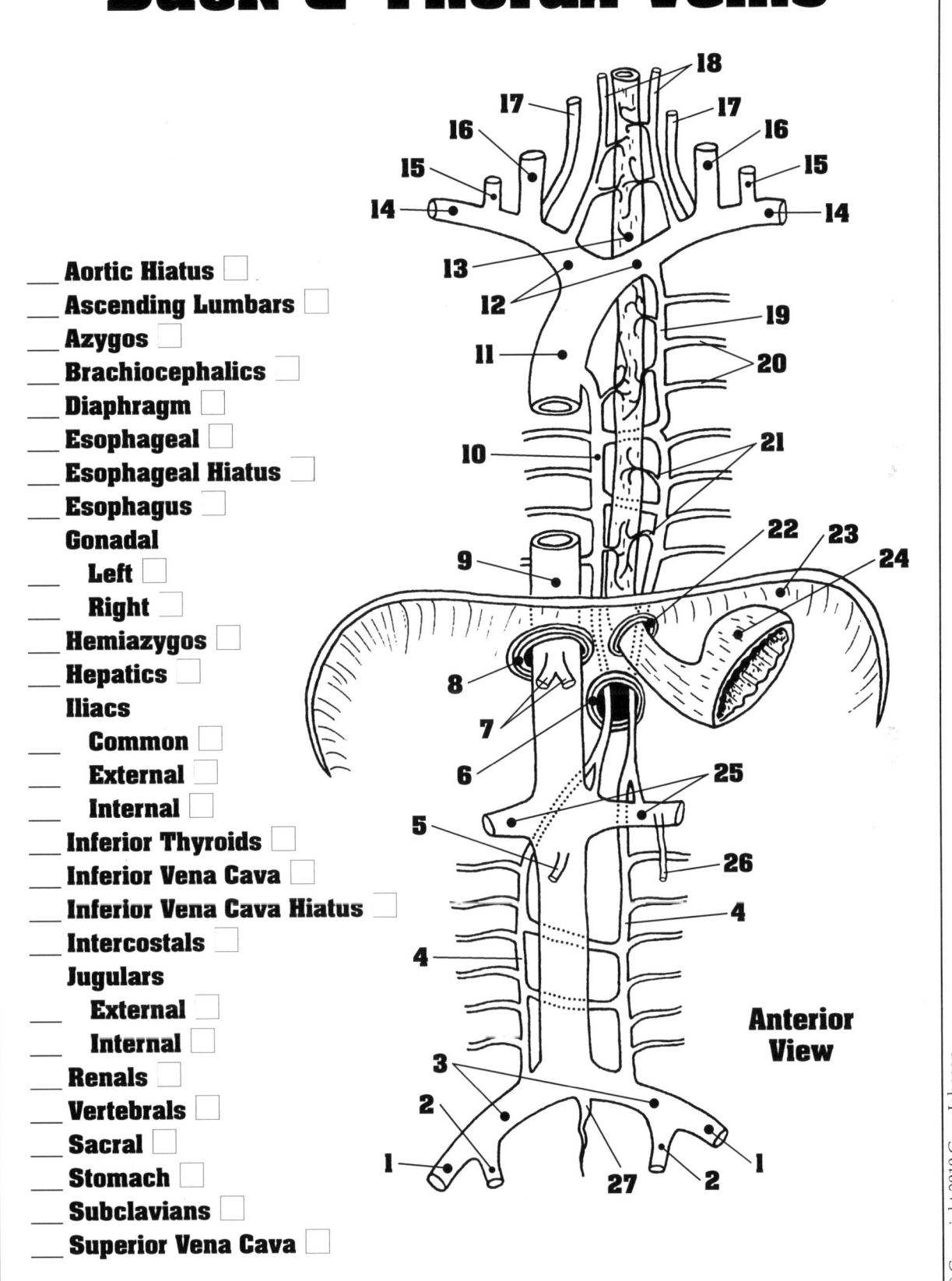

# Neck Arteries
## Posterior View

In this posterior view we see the arteries of the neck in conjunction with the pharynx, **esophagus** (19), **trachea** (20), **thyroid glands** (7), and **parathyroid glands** (8).

On the left side the **left common carotid** (9) and **left subclavian** (2) arise directly from the **aortic arch** (1), while on the right side the **right common carotid** (18) arises from the **brachiocephalic** (22), and the brachiocephalic merges into the **right subclavian** (21). Recall that this is a posterior view and that in all posterior views the right and left sides are opposite to what they would be in an anterior view.

Lateral to each common carotid, a **vertebral** (5) and **thyrocervical** (4) ascend, and an **internal thoracic** (3) descends.

Higher up in the neck each of the common carotids branches into an **external** (15) and **internal** (14) **carotid**. At the junction of the branching the internal carotid is slightly expanded into a **carotid sinus** (12) with an associated **carotid body** (13). While the carotid sinuses aid in monitoring blood pressure the carotid bodies serve as chemoreceptors and are involved in the control of the respiratory rate.

Thyroid glands, like all endocrine glands, are richly supplied with blood. The superior regions of the glands are supplied by **superior thyroids** (10) which branch downward from the bases of the internal carotid arteries, while the inferior regions are supplied by **inferior thyroids** (6) that derive from the thyrocervicals.

From superior to inferior, the major pharyngeal muscles are the **superior constrictor** (17), **middle constrictor** (16), and **inferior constrictor** (11).

# Neck Arteries

**Posterior View**

___ Aortic Arch
___ Brachiocephalic
___ Carotid Bodies
___ Carotids
  ___ External
  ___ Internal
___ Carotid Sinuses
___ Common Carotid
  ___ Left
  ___ Right
___ Constrictor Muscle
  ___ Inferior
  ___ Middle
  ___ Superior
___ Esophagus
___ Internal Thoracics
___ Parathyroid Glands
___ Subclavian
  ___ Left
  ___ Right
___ Thyrocervicals
___ Thyroid Glands
___ Thyroids
  ___ Inferior
  ___ Superior
___ Trachea
___ Vertebrals

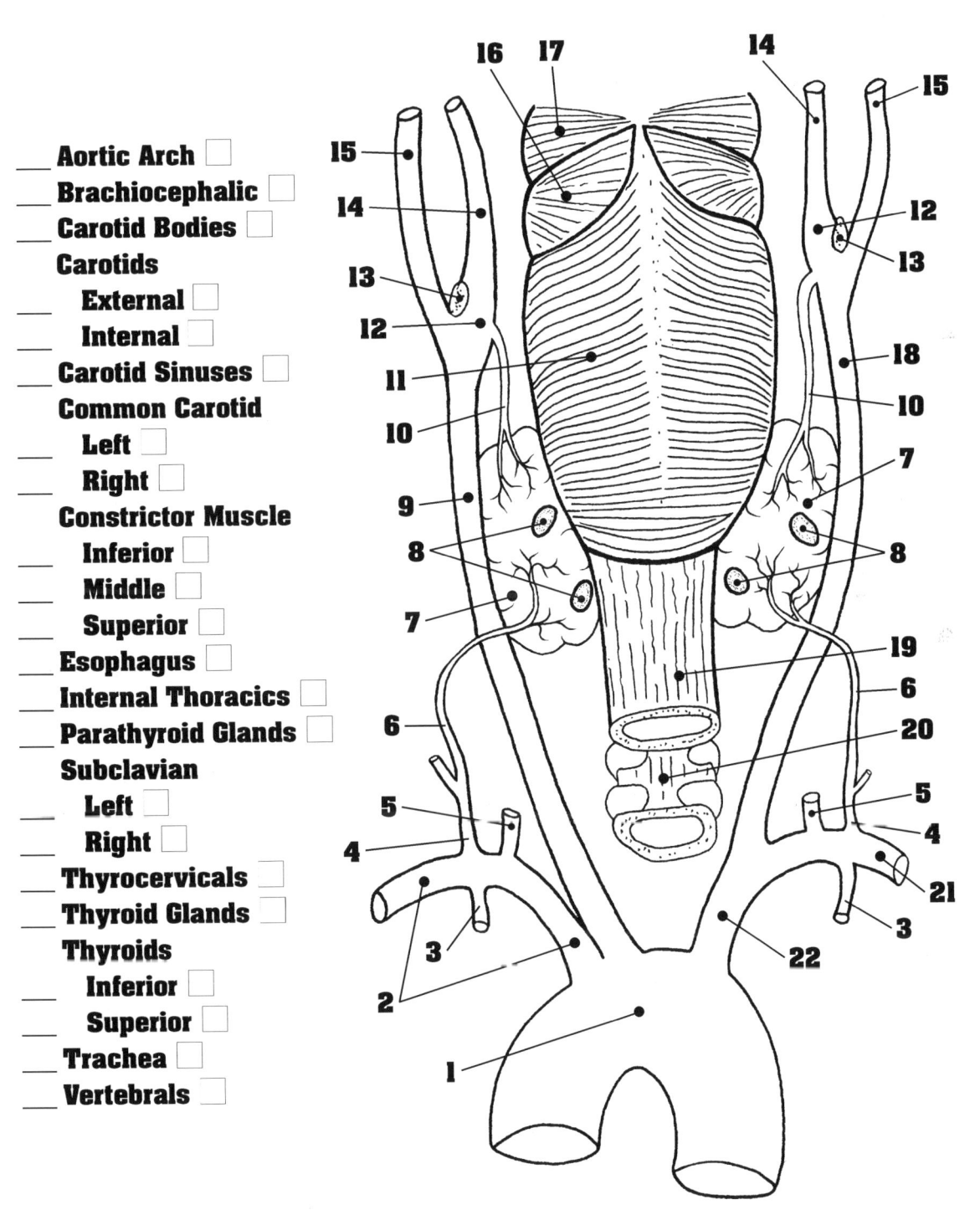

227

# Head & Neck
## Lateral View

The **brachiocephalic** (17), which is the first ascending branch from the aortic arch, divides into a **common carotid** (16) — which carries blood up through the neck and head, and a **subclavian** (2) — which carries blood out to the shoulder region.

The **vertebral** (5), **thyrocervical** (4) and **costocervical** (3) ascend from the subclavian. The **internal thoracic** (1) descends from the subclavian.

The **carotid sinus** (15) is located high in the neck, in the region where the common carotid divides into the **external** (14) and **internal** (13) **carotids**.

Transverse foramina in the **cervical vertebrae** (6) provide protected passageways for both the vertebral arteries and the vertebral veins (not shown). The vertebral arteries, along with the internal carotid arteries supply blood to the brain via the Circle of Willis.

At the base of the brain the two vertebral arteries merge into a **basilar** (7). The basilar branches laterally into the **posterior cerebrals** (8) and anteriorly into the **posterior communicatings** (9). At the juncture where the internal carotids join the Circle of Willis the posterior communicatings branch into the **middle cerebrals** (10) and **anterior cerebrals** (11). At the anterior rim of the Circle of Willis the two anterior cerebrals are connected by the **anterior communicating** (12).

The Circle of Willis is shown in more detail in the next exercise.

# Head & Neck

229

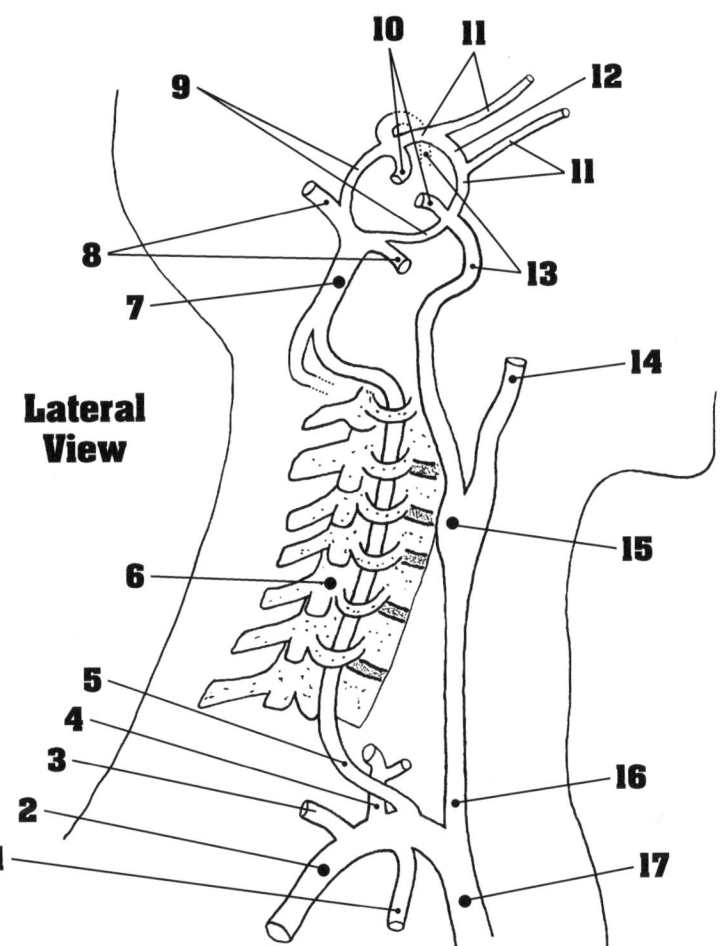

Lateral View

___ Anterior Communicating
___ Basilar
___ Brachiocephalic
Carotids
___ Common
___ External
___ Internal
___ Carotid Sinus

Cerebrals
___ Anterior
___ Middle
___ Posterior
___ Cervical Vertebra
___ Costocervical
___ Internal Thoracic
___ Posterior Communicatings
___ Subclavian
___ Thyrocervical
___ Vertebral

© Copyright 2010 Gene Johnson

# Circle of Willis

Looking up at the base of the brain we see the **pituitary gland** (11) and **optic chiasma** (10) surrounded by the Circle of Willis.

The Circle of Willis is directly supplied by blood from the **internal carotids** (7), and indirectly from the **vertebrals** (1) via the **basilar** (3).

**Posterior communicatings** (5) connect the **posterior cerebrals** (4) with the **middle cerebrals** (6). The **anterior communicating** (9) connects the two **anterior cerebrals** (8).

The **anterior inferior cerebellars** (2) arise at the region where the vertebrals merge into the basilar.

The **pons** (12), **medulla** (13) and **spinal cord** (14) are shown in conjunction with the vertebral and basilar arteries.

The Circle of Willis not only provides alternate routes for blood to reach the brain (perhaps in the case of a severe injury or occlusion), but also serves to equalize the blood pressure in the two cerebral hemispheres.

# Circle of Willis

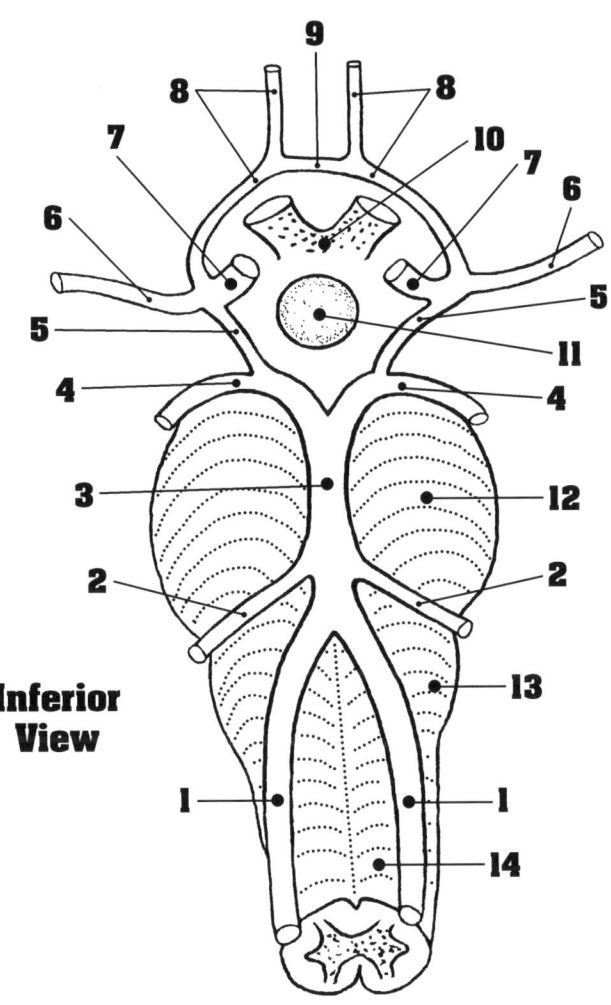

Inferior View

___ Anterior Communicating
___ Anterior Inferior Cerebellars
___ Basilar
___ Cerebrals
   ___ Anterior
   ___ Middle
   ___ Posterior
___ Internal Carotids
___ Medulla
___ Optic Chiasma
___ Pituitary
___ Pons
___ Posterior Communicatings
___ Spinal Cord
___ Vertebrals

# Head & Neck Veins

In this drawing we see three major veins that deliver blood from the brain toward the heart via the **internal jugular** (11):

- **Superior sagittal sinus** (9)
- **Inferior sagittal sinus** (8)
- **Facial** (10)

The inferior sagittal sinus merges with the superior sagittal sinus via a **straight sinus** (7). Recall that the cerebrospinal fluid is picked up by the superior sagittal sinus.

At their juncture the straight and superior sagittal sinuses diverge into two **transverse sinuses** (6). The transverse sinuses connect with the internal jugulars via **sigmoid sinuses** (5).

Veins, when designated as sinuses, are more expanded than regular veins, and can collect and hold more blood.

Also returning blood from the central head region (particularly from the base of the brain) is the **vertebral** (3) vein, which, along with the vertebral artery runs through the transverse foramina of the **cervical vertebrae** (4).

Blood from the more lateral region of the head returns via the **external jugular** (2) which merges with the **subclavian** (1). The vertebral and internal jugular merge into the **brachiocephalic** (12).

# Head & Neck Veins

233

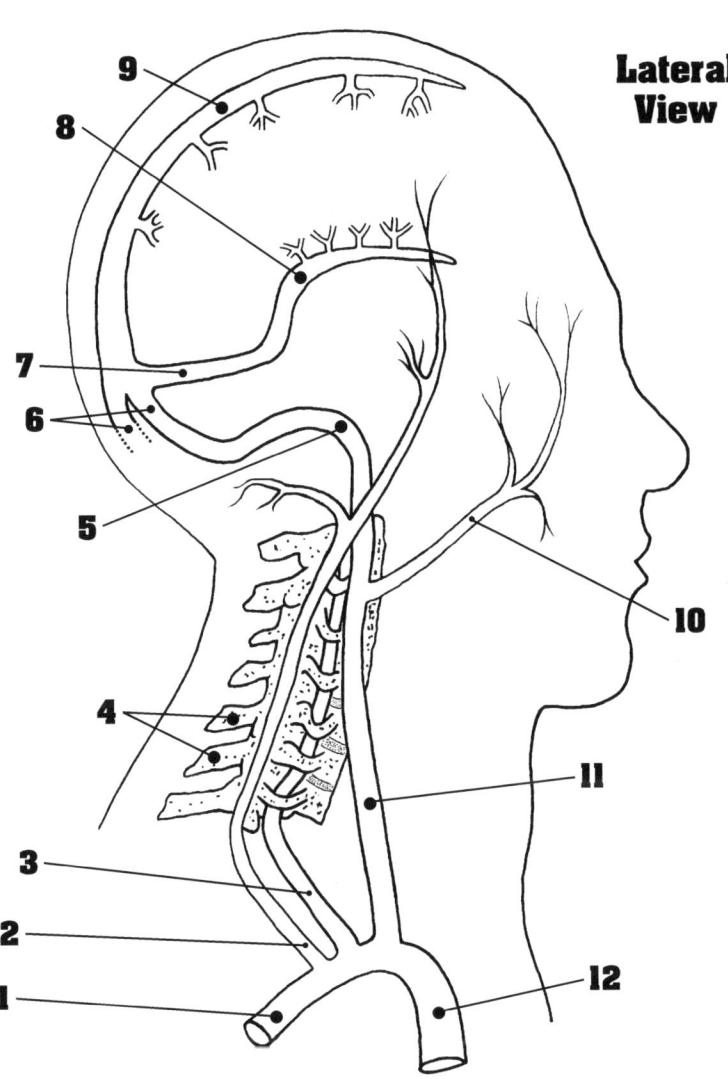

Lateral View

Sagittal Sinus
___ Inferior
___ Superior
___ Brachiocephalic
___ Facial
___ Sigmoid Sinus
___ Cervical Vertebrae
___ Straight Sinus
Jugular
___ Subclavian
___ External
___ Transverse Sinuses
___ Internal
___ Vertebral

© Copyright 2010 Gene Johnson

# Right Arm & Shoulder Arteries

Before we follow the arteries let us note the skeletal components: the **manubrium** (17), **clavicle** (10), **acromion process** (8), **scapula** (20), **humerus** (4), **radius** (1), and **ulna** (25).

Just beyond where the **brachiocephalic** (16) diverges into the **subclavian** (11) and the **common carotid** (15), three arteries ascend and one descends. The three ascending arteries are the **vertebral** (14), **thyrocervical** (13), and **costocervical** (12). The one descending artery is the **internal thoracic** (18).

Just prior to reaching the scapular region the subclavian gives way to the **axillary** (9), and at that region a **thoracoacromial** (19) descends. A little further down the **anterior** (7) and **posterior** (6) **circumflex** arteries surround the neck of the humerus.

Just inferior to the circumflex arteries the axillary gives way to the **brachial** (5), and a **deep brachial** (21) immediately branches off to run a posterior route behind the humerus.

The next two brachial branches shown are the **superior** (22) and **inferior** (23) **ulnar collaterals** which serve the medial aspect of the elbow (olecranal) region.

Below the elbow, the brachial diverges into a lateral **radial** (2) and a medial **ulnar** (24). A **radial recurrent** (3) ascends from the radial to eventually anastomose with the deep brachial.

The radial and ulnar terminate in a **superficial** (27) and **deep** (26) **palmar arch**.

**Metacarpals** (28) and **digitals** (29) extend from the palmar arches into the hands and fingers.

# Right Arm & Shoulder Arteries

235

___ Acromion Process
___ Axillary
___ Brachial
___ Brachiocephalic
Circumflex
___    Anterior
___    Posterior
___ Clavicle
___ Common Carotid
___ Costocervical
___ Deep Brachial
___ Digital
___ Humerus
___ Internal Thoracic
___ Manubrium
___ Metacarpal
Palmar Arch
___    Deep
___    Superficial
___ Radial
___ Radial Recurrent
___ Radius
___ Scapula
___ Subclavian
___ Thoracoacromial
___ Thyrocervical
___ Ulnar
Ulnar Collateral
___    Inferior
___    Superior
___ Ulna
___ Vertebral

Anterior View

© Copyright 2010 Gene Johnson

# Right Arm & Shoulder Veins

While blood from the fingers moves upward through the **digitals** (17) to the **superficial palmar arch** (16), blood from the hands moves upward through the **metacarpals** (1) to the **deep palmar arch** (2).

From the palmar arches the blood flows upward laterally through the **radial** (3) and medially through the **ulnar** (15). The radial and ulnar veins merge in the region of the elbow to form the **brachial** (6). In addition to the radial and ulnar, the **median antebrachial** (4), **cephalic** (5), and **basilic** (13), also provide return routes through the lower arm. The brachial, cephalic, and basilic continue upward through the upper arm.

In the antecubital region (in front of the elbow) a **median cubital** (14) bridges between the basilic and the cephalic. This is the gleeful phlebotomist's favorite target!

The brachial and basilic merge into the **axillary** (12), and the axillary and cephalic merge to form the **subclavian** (7). The subclavian, in turn, gives rise to the **brachiocephalic** (11).

The trunks of the three major neck veins are also shown: the **internal jugular** (10), **vertebral** (9), and **external jugular** (8).

# Right Arm & Shoulder Veins

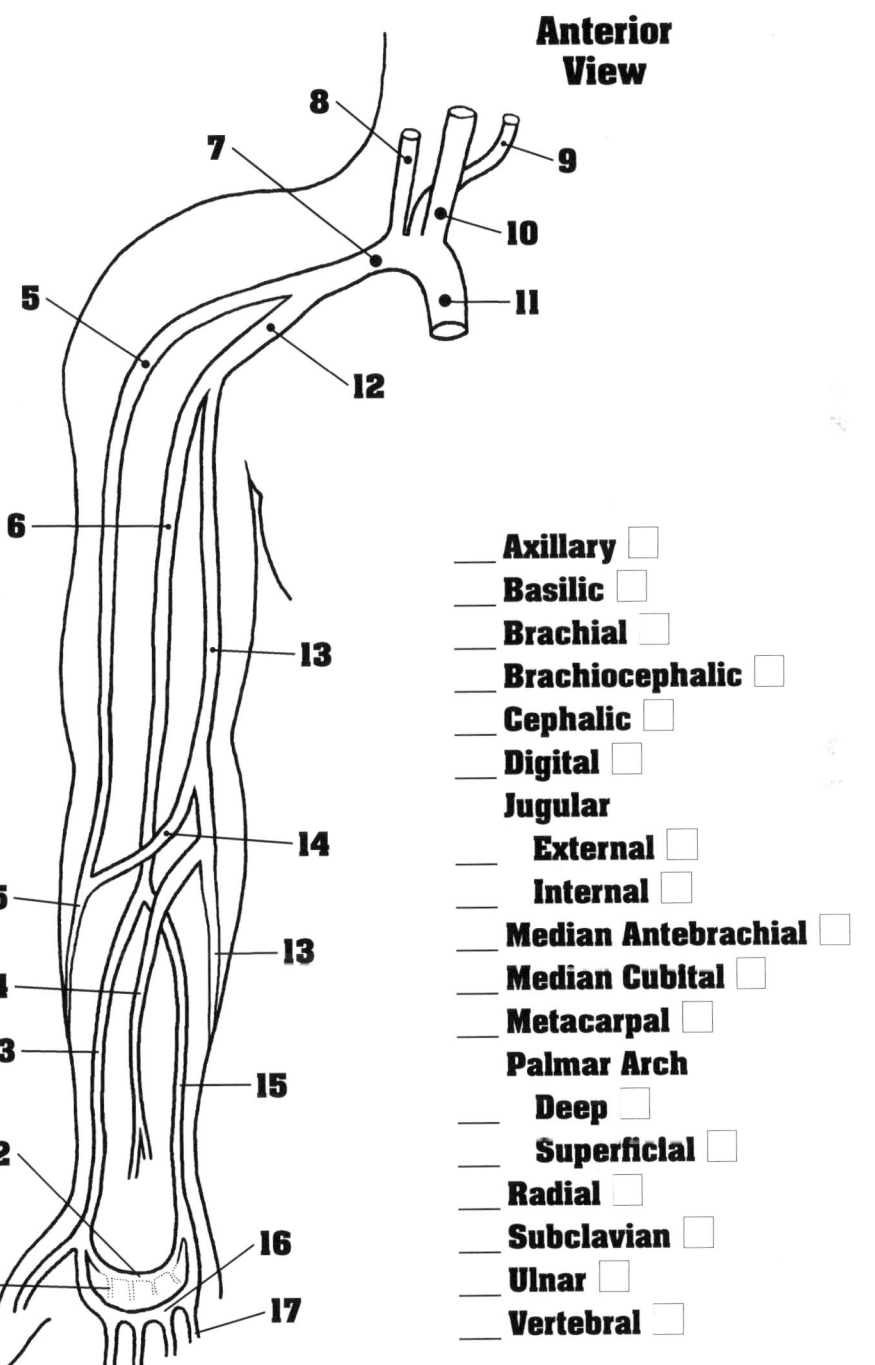

**Anterior View**

___ Axillary ☐
___ Basilic ☐
___ Brachial ☐
___ Brachiocephalic ☐
___ Cephalic ☐
___ Digital ☐
     Jugular
___    External ☐
___    Internal ☐
___ Median Antebrachial ☐
___ Median Cubital ☐
___ Metacarpal ☐
     Palmar Arch
___    Deep ☐
___    Superficial ☐
___ Radial ☐
___ Subclavian ☐
___ Ulnar ☐
___ Vertebral ☐

# Right Leg Arteries

The skeletal components of this drawing are the **coxa** (5), **femur** (3), **fibula** (1), and **tibia** (18).

Following the blood flow, the **abdominal aorta** (8) first divides into the **common iliacs** (7). Each common iliac divides into an **external** (6) and **internal** (10) **iliac**.

The external iliac gives way to the **femoral** (4), and, in the region of the neck of the femur, a **deep femoral** (12) branches laterally from the femoral.

A **lateral** (13) and **medial** (11) **circumflex** artery branch off immediately from the deep femoral, and a **descending branch of the lateral circumflex** (2) runs parallel and lateral to the femur in the upper leg.

In the popliteal region (behind the knee) the femoral becomes the **popliteal** (14). The popliteal, in turn, splits, behind the knee, into the **anterior** (17) and **posterior** (16) **tibials**.

Anastomosing arteries form the **patellar plexus** (15) which surrounds the tibiofemoral joint.

At the common iliac junction a **sacral** (9) descends.

# Right Leg Arteries

___ Abdominal Aorta ☐
___ Coxa ☐
___ Deep Femoral ☐
___ Descending Branch of the Lateral Circumflex ☐
___ Femoral ☐
___ Femur ☐
___ Fibula ☐
   Iliacs
___   Common ☐
___   External ☐
___   Internal ☐
___ Lateral Circumflex ☐
___ Medial Circumflex ☐
___ Patellar Plexus ☐
___ Popliteal ☐
___ Sacral ☐
___ Tibia ☐
   Tibial
___   Anterior ☐
___   Posterior ☐

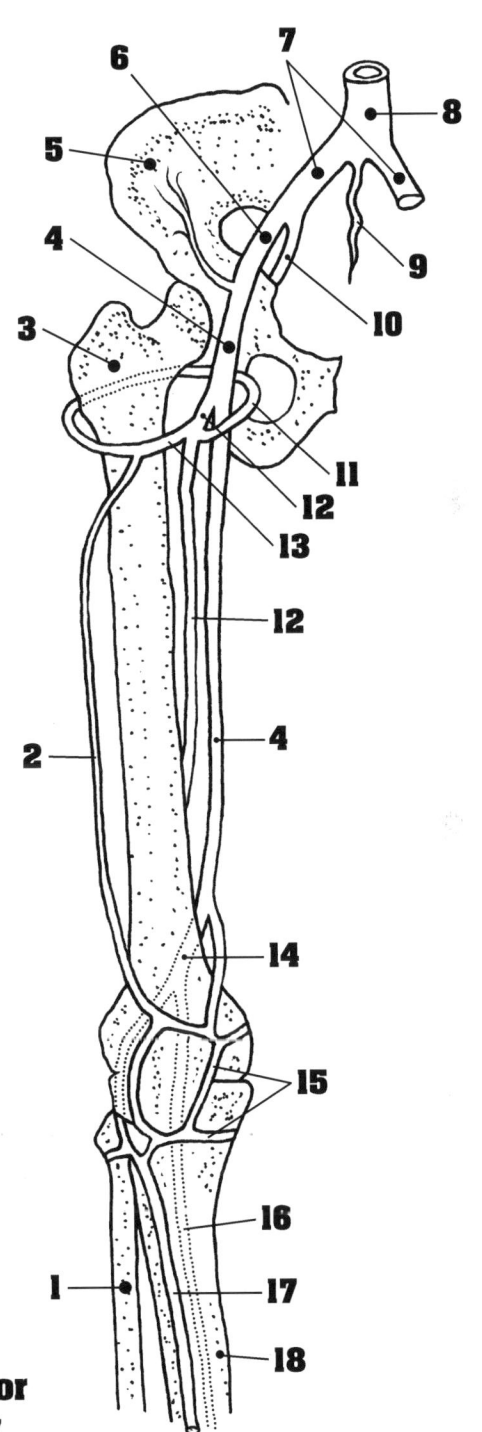

**Anterior View**

# Right Leg Veins

The skeletal components shown are the **coxa** (9), **femur** (5), **fibula** (1), and **tibia** (19).

Five major venous pathways are seen in the lower leg:

- **Fibular** (3)
- **Anterior tibial** (2)
- **Posterior tibial** (18)
- **Greater saphenous** (14)
- **Lesser saphenous** (17)

(Did we not tell you that in a given region there are usually more veins than arteries?)

The fibular and the anterior and posterior tibials merge, behind the knee, into the **popliteal** (4). The lesser saphenous merges, above the knee, with the **femoral** (15).

An **ascending branch of the popliteal** (6) runs upward laterally, and on the medial aspect, the popliteal gives way to the femoral. A **deep femoral** (16) and the greater saphenous, which originated in the lower leg, also run upward in the upper leg.

The femoral, deep femoral, and great saphenous veins merge, near the head of the femur to form the **external iliac** (8). And a **lateral circumflex** (7) also joins the merger.

In the coxal region the **internal iliac** (13) joins the external iliac to form the **common iliac** (10). And the common iliacs merge to form the **inferior vena cava** (11) … and a **sacral** (12) also joins that merger.

# Right Leg Veins

___ Ascending Branch of the Popliteal ▢
___ Coxa ▢
___ Deep Femoral ▢
___ Femoral ▢
___ Femur ▢
___ Fibula ▢
___ Fibular ▢
Iliac
  ___ Common ▢
  ___ External ▢
  ___ Internal ▢
___ Inferior Vena Cava ▢
___ Lateral Circumflex ▢
___ Popliteal ▢
___ Sacral ▢
Saphenous
  ___ Greater ▢
  ___ Lesser ▢
___ Tibia ▢
Tibial
  ___ Anterior ▢
  ___ Posterior ▢

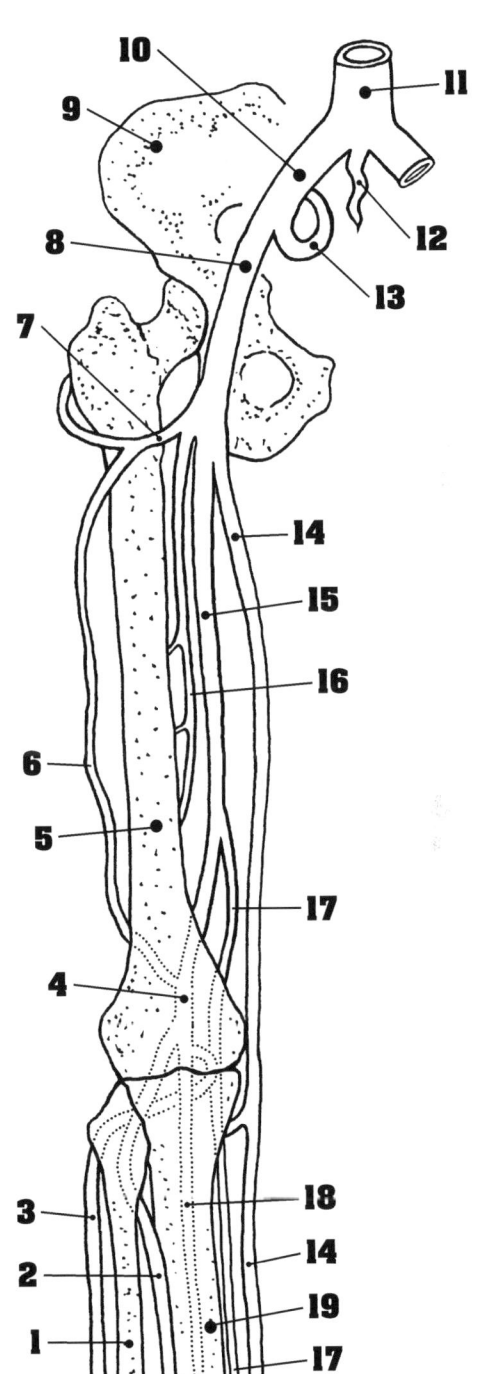

Anterior View

# Hepatic Portal System

Blood from the digestive organs does not return directly to the inferior vena cava and the heart. It is first transported to the liver via the hepatic portal system. And why is it first transported to the liver? Because after eating a large meal — or even a small one — who knows what might have gotten into the blood?

The **liver** (8), with its large repertoire of biochemical orchestrations, has the capacity to monitor, modify, cleanse, and balance the blood because it:

- Facilitates lipid digestion by secreting bile salts.
- Stores excess sugar as glycogen.
- Disarms or modifies toxins, and if it cannot modify them it seals them off in toxic storage vesicles.
- Contains resident populations of macrophages for combating bacteria and other foreign agents.
- Converts lipids to sugar via a process called beta oxidation.
- Converts amino acids of one type, to another type, via transamination.
- Deaminates amino acids and converts them to sugar-like molecules for energy metabolism.
- Disassembles aging and worn-out red blood cells and recycles key elements to the bone marrow for the production of new red blood cells.

And these are but a few of the liver's amazing biochemical orchestrations.

The **hepatic portal vein** (4) receives blood from the:

- **Stomach** (11) — via the **gastric vein** (12).
- **Spleen** (13) and **pancreas** (15) — via the **splenic vein** (14).
- **Small intestine** (5) and the **proximal end of the large intestine** (1) — via a **superior mesenteric vein** (2).
- **Rectum** (18) and the **distal end of the large intestine** (17) — via the **inferior mesenteric vein** (16).

After having passed through the liver's various rigorous inspections and ministrations, the blood is delivered, via the **hepatic veins** (9) to the **inferior vena cava** (10).

The **hepatic ducts** (7) deliver bile from the liver to the **gall bladder** (6). The **common bile duct** (3) delivers bile from the gall bladder to the small intestine.

# Hepatic Portal System

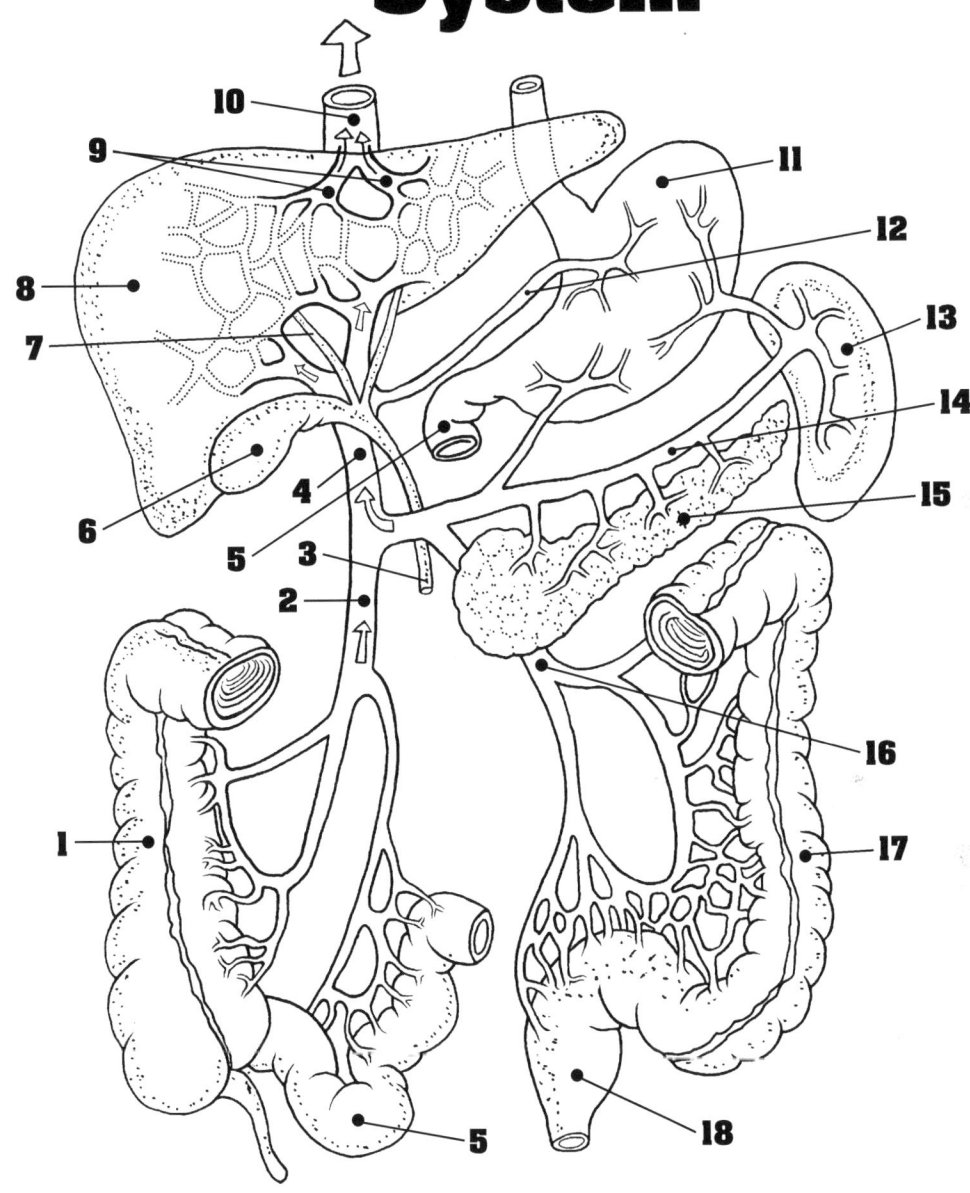

___ Common Bile Duct
___ Gall Bladder
___ Gastric Vein
___ Hepatic Duct
___ Hepatic Portal Vein
___ Hepatic Veins
___ Inferior Vena Cava
___ Large Intestine
  ___ Distal End
  ___ Proximal End
___ Liver
___ Mesenteric
  ___ Inferior
  ___ Superior
___ Pancreas
___ Rectum
___ Small Intestine
___ Spleen
___ Splenic Vein
___ Stomach

243

© Copyright 2010 Gene Johnson

# Skeletal Muscle Blood Pump

Because of the implications it has to your good health we are obliged to include the simple exercise on the opposite page which illustrates how muscular action facilitates venous blood flow.

Because we have a powerful blood pump (heart) in our chest it is easy to understand how blood may be forced outward, upward, and downward, but because we do not have blood pumps in our feet it is more difficult to understand how blood is "forced" upward and back to the heart.

We do know that as a **contracting muscle** (3) impinges on a **vein** (5), it forces **venous blood flow** (1) upward through one-way valves in what is called a "milking action." The **one-way valves**, which are **closed** (4) to prevent blood from "falling" back and downward when there is little or no blood pressure, are forced **open** (2) when muscle contractions squeeze the veins.

*Looking at this diagram should leave you with little doubt about the efficacy of exercise in enhancing circulation. Even when we are not consciously "using" them our skeletal muscles are facilitating blood flow by alternately contracting. Deep breathing, as it raises and lowers the rib cage and diaphragm, also facilitates the return of venous blood to the heart.*

*Because good muscle tone is essential to good circulation you should, as often as possible, get off your ischial tuberosities and put your best metatarsals forward.*

# Skeletal Muscle Blood Pump

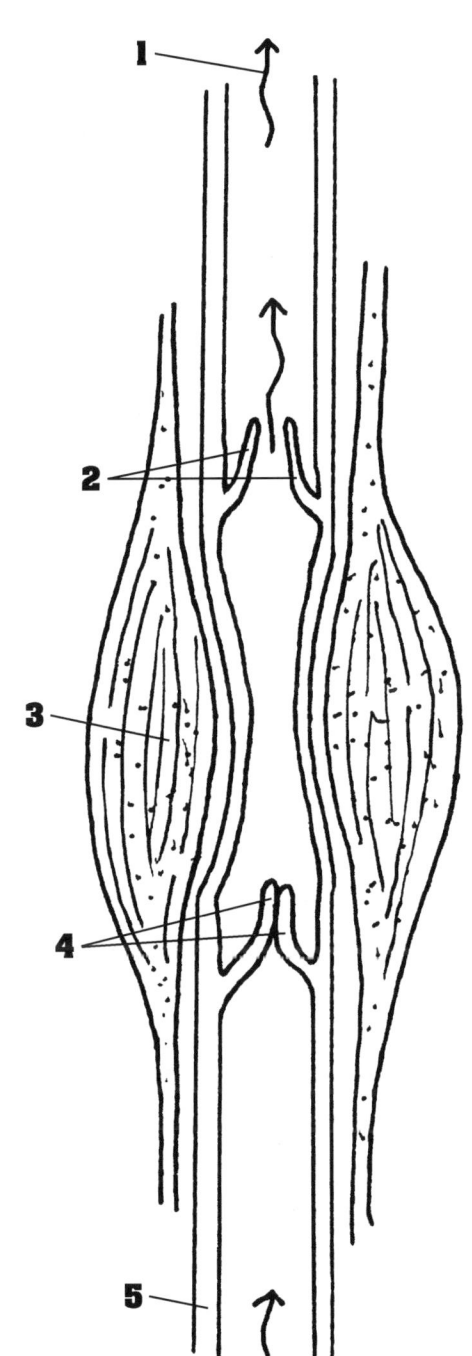

___ **Contracting Muscle** ☐
**One-Way Valves**
___ **Closed** ☐
___ **Open** ☐
___ **Vein** ☐
___ **Venous Blood Flow** ☐

# Immune/Lymphatic System

Although the immune and lymphatic systems are intimately intertwined they are often listed as separate and independent systems. But you know, because we have previously stiffly warned you, that none of the body systems are "separate and independent." As we see, in recent articles, references to a *neuro-immuno-endocrine system* we cannot resist offering a larger systems synthesis: *the integumento-skeleto-musculo-neuro-endocrino-immuno-circulato-respirato-digesto-urino-reproducto system.*

Having satisfied our holistic urge we now turn our attention to the immune system which is primarily devoted to protecting the body from alien microbial pathogens. (If we do hold immunity to be a body system we should note that unlike the other organ body systems, the immune system is a non-organ, or functional, system.

As you can see in the upper chart on the opposite page the immune system is comprised of two defense systems:

## The Innate (Non-Specific) Defense System
Using "two lines" of defense, this system protects the body from all foreign agents:

- The first line is membranous and includes the external skin and the internal mucous membranes.
- The second line uses antimicrobial proteins, phagocytes, and other cells, to inhibit invaders.

## The Adaptive (Specific) Defense System
This system provides a "third line" of defense. Although this defense system takes more time to mount a response than the non-specific innate system, the response is potent and effectual, because it targets specific pathogenic cells.

The innate and adaptive systems work together to orchestrate the body's overall defense against pathogenic invaders. If the first line of defense (skin and mucous membranes) fails to hold out invaders the second line of defense becomes active: phagocytes and natural killer cells (which work with complement proteins) lyse and destroy foreign cells — while, at the same time, fever and inflammation "turn up the heat."

And all the while the adaptive third line of defense is preparing for action by sending out scout cells, called antigen presenting cells (APCs) which, after engulfing and digesting a foreign cell, presents antigen fragments to the specific white blood cells that will actually deal with the invasive cells. Thus two different troops of specialized white blood cells (B cells and T cells) are activated to respond to a specific microbial invader.

# Immune/Lymphatic System

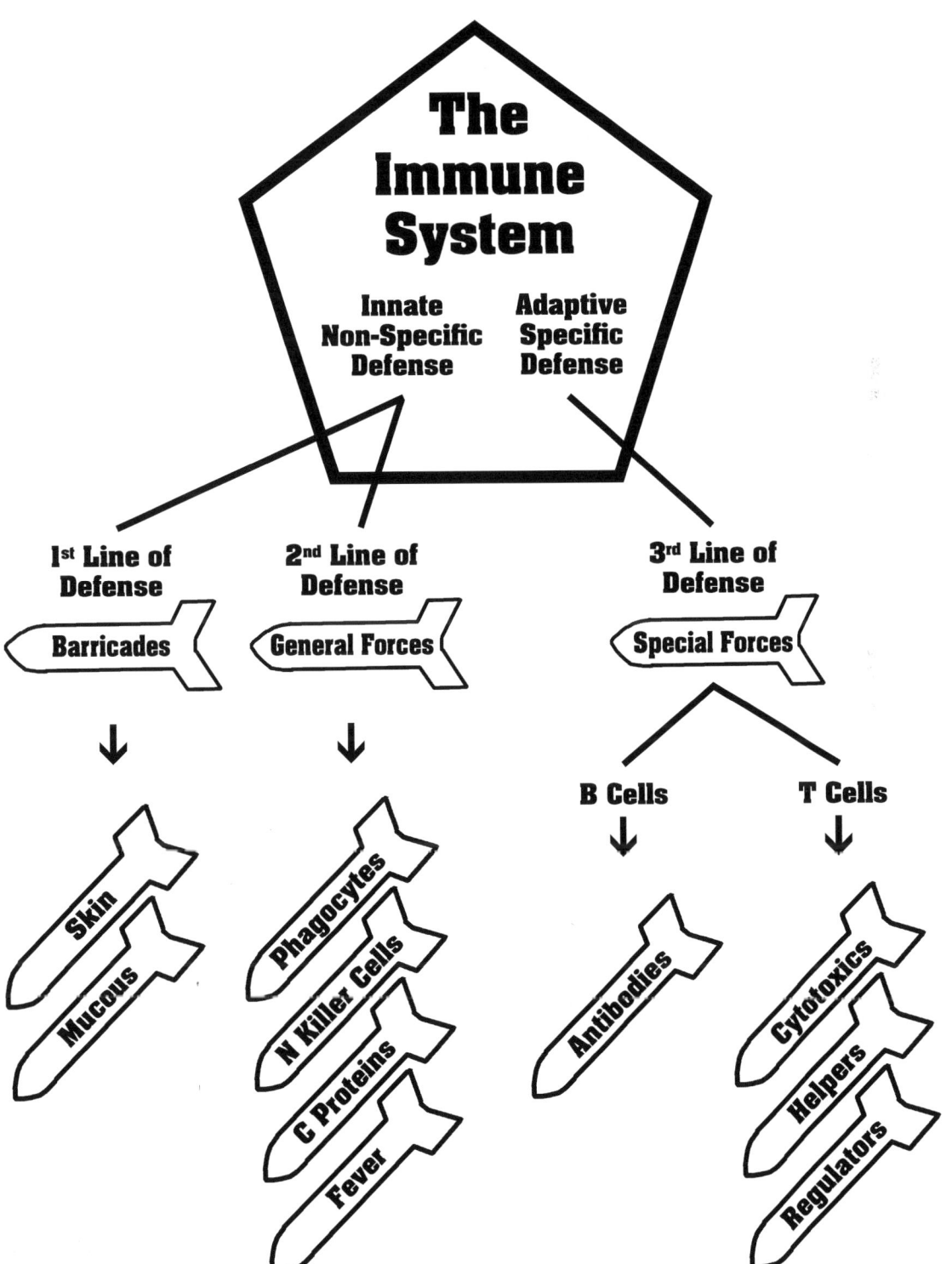

# Lymphatic Vessels

Because more blood plasma is given out at the arterial ends of capillaries than is picked up at the venous ends, there is need for an alternative blood plasma/tissue fluid return system — and that alternative return system is the lymphatic system.

But that is only one of the two main functions of the lymphatic system. The other has to do with body defense and immunity. Lymph nodes, perfused with lymphocytes, are scattered throughout the body, but particularly clustered in the following strategic regions:

- **Inguinal nodes** (1) — groin
- **Axillary nodes** (5) — armpit
- **Cervical nodes** (8) — lateral regions of the neck

Although the lymph nodes in the drawing might seem a little exaggerated, if we assume this guy to be in the midst of a microbial war his lymph nodes would be overwhelmed and swollen up with bacteria or virus particles they are trying to trap and destroy. Such swollen lymph nodes are called buboes. Yes, *buboes*, as in bubonic plague.

A larger lymphoid organ called the **cisterna chyli** (11), located in the abdomen, collects lymph from two large **lumbar ducts** (2).

All the lymph from the lower body regions eventually moves up through **thoracic ducts** (3). Each thoracic duct joins with a **subclavian** (10) and **cervical** (9) **duct** before entering the venous blood flow at the **subclavian jugular junction** (4), where the **subclavian** (6) and **jugular** (7) **veins** merge.

In addition to lymph nodes there are other aggregates of lymphatic tissue called lymphoid organs. These include the spleen, thymus gland, tonsils, and the Peyer's patches of the small intestine. Lymphatic vessels, like veins, have one-way valves, and like veins they also depend upon nearby tissue movements for facilitating good flow. Tissue movements, in turn, are facilitated by exercise and deep breathing. Unlike veins, lymphatic vessels begin as blind pouches.

The importance of the lymphatic return is clearly and quickly noted whenever, and wherever, lymph nodes are destroyed or plugged up. Because the tissue fluids cannot be efficiently returned to the blood, the body regions beyond the incompetent nodes swell up.

A microscopic nematode called the filarial worm infects the lymph nodes of people in some regions of Africa. The filarial worms block the lymph nodes and cause appendages distal to nodes to enlarge enormously, a condition called elephantiasis.

Another function of the lymphatic system has to do with the transport and packaging of certain lipid molecules. Each of the millions of villi in the small intestine has a central lacteal which is a lymphatic blind pouch for picking up certain lipid components (see page 283).

# Lymphatic Vessels

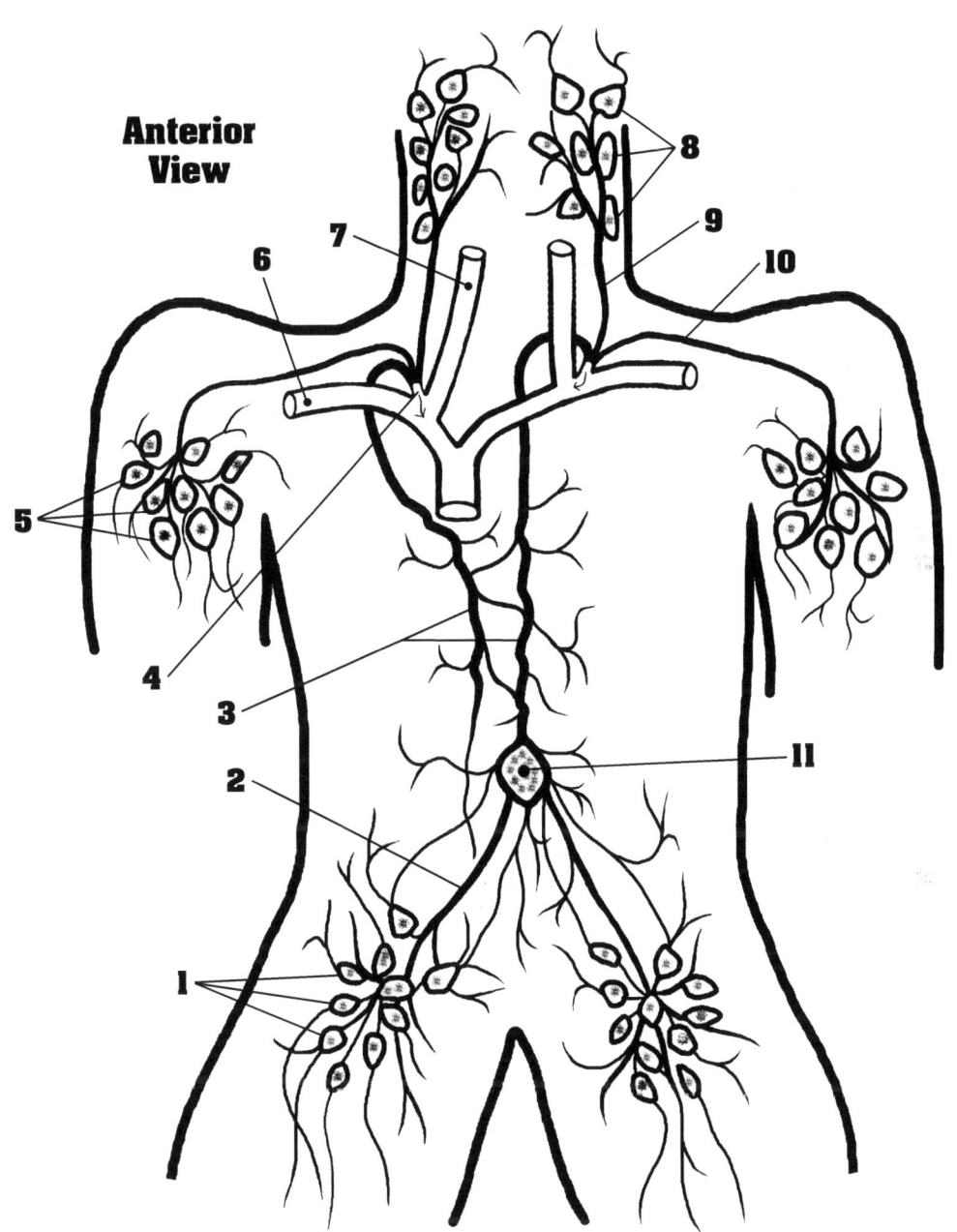

Anterior View

___ Axillary Nodes
___ Cervical Duct
___ Cervical Nodes
___ Cisterna Chyli
___ Inguinal Nodes
___ Jugular Vein
___ Lumbar Duct
___ Subclavian Duct
___ Subclavian Jugular Junction
___ Subclavian Vein
___ Thoracic Ducts

# Lymph Node

Lymph vessels, like veins, have uni-directional **valves** (2). Lymph, arriving at a lymph node via **afferent lymphatic ducts** (1), flows first into the **subcapsular sinus** (15), then inward through **peritrabecular sinuses** (14) to **medullary sinuses** (12).

Connective tissue strands called **trabeculae** (13) extend inward from the fibrous **capsule** (3).

The outer cortex is occupied with **follicles** (4) set in **cortical tissue** (6). Strands of cortical tissue called **medullary cords** (8) extend into the **medullary tissue** (7).

The follicles have **germinal centers** (5) with populations of B cells. The germinal centers can enlarge dramatically (remember those buboes?) when B cells are dividing to produce plasma cells and antibodies.

An invagination called the **hilus** (9) occurs in the region where **efferent lymphatic ducts** (10) exit and **blood vessels** (11) enter and leave.

# Lymph Node

**Longitudinal Section**

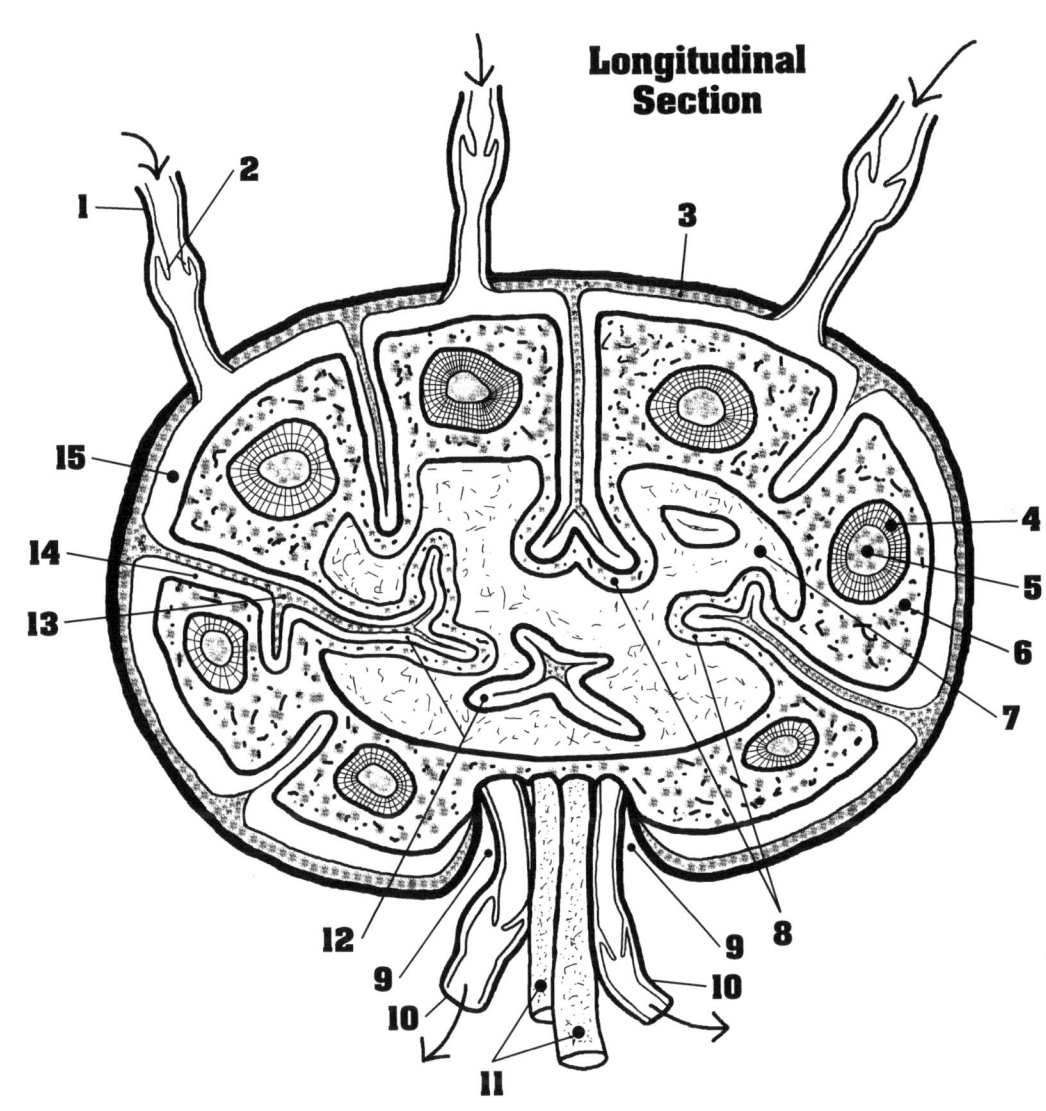

___ Blood Vessels ☐
___ Capsule ☐
___ Cortical Tissue ☐
___ Follicle ☐
___ Germinal Center ☐
___ Hilus ☐
___ Lymphatic Duct
   ___ Afferent ☐
   ___ Efferent ☐

___ Medullary
   ___ Cords ☐
   ___ Sinuses ☐
   ___ Tissue ☐
___ Peritrabecular Sinus ☐
___ Subcapsular Sinus ☐
___ Trabecula ☐
___ Valve ☐

© Copyright 2010 Gene Johnson

# Respiratory System

Each human living cell contains hundreds to tens of thousands of tiny mitochondrial combustion chambers which utilize oxygen to "burn" sugar. The energy released from the combustion of sugar is used to synthesize ATP, which supplies all our energy needs. The energy transforming "fires of metabolism" can only burn so long as they are supplied with oxygen. Hence, twelve times a minute we take a deep breath and fill our lungs with air. Air is approximately 80% nitrogen and 20% oxygen. As air reaches the deepest part of our lungs and fills tiny thin-walled alveoli, oxygen molecules are snatched away by iron atoms on hemoglobin molecules which are embedded in the membranes of red blood cells (iron has a strong affinity for oxygen). The hemoglobin-bound oxygen is then transported to cells and delivered to mitochondria. As oxygen combines with hemoglobin the blood changes from blue, to purple, to red. Conversely, as oxygen dissociates from hemoglobin the blood changes from red, to purple, to blue.

What we call "the respiratory system" covers a broad category of oxygen-related activities, hence, it is often subdivided into the following "four phases of respiration:"

- **Pulmonary respiration** — exchange of gas ($O_2$ & $CO_2$) between air and lungs.

- **External respiration** — exchange of gas between lungs and blood.

- **Internal respiration** — exchange of gas between blood and cells.

- **Cellular respiration** — utilization of $O_2$ in the mitochondria.

*A pulmonologist was once overheard describing a friend by saying, "I'm worried about him, he does not look well-oxygenated." Indeed, only the well-oxygenated enjoy good health. Best breathe deeply and break ranks with the majority who are "shallow breathers." Fortunately there is not yet a cost for air, therefore deep breaths don't cost any more than shallow breaths. Yes, breathe deeply — put a "spring" in your step and a "red" in your cheeks!*

# Respiratory System

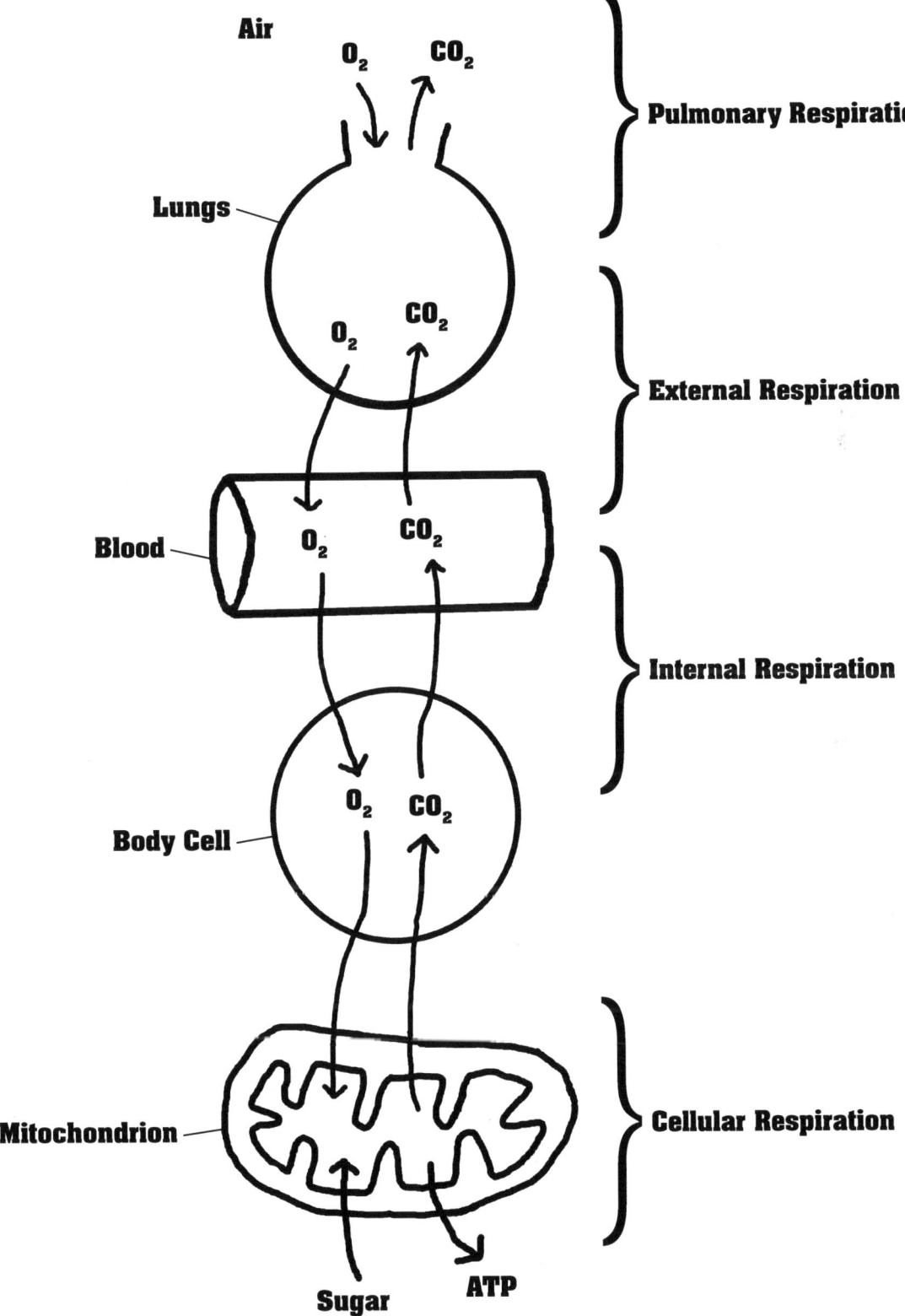

# Thoracic Cavity

The **pleural cavity** (5), is lined with a **parietal pleura** (4) serous membrane. The **lungs** (8), which lie within the pleural cavity, are lined with a **visceral pleura** (6) serous membrane.

The **pericardial cavity** (10), is lined with a **parietal pericardial** (9) serous membrane. The **heart** (14), which lies within the pericardial cavity, is lined with a **visceral pericardial** (11) serous membrane.

The rib cage, which supports the thoracic cavity, is comprised of **thoracic vertebrae** (2) at the posterior, a **sternum** (13) at the anterior, and **ribs** (3) which articulate posteriorly with the vertebrae and connect anteriorly with the sternum via **costal cartilage** (12).

The space behind the heart and between the lungs is called the **mediastinum** (15), which at this plane contains only the **esophagus** (7). The **spinal cord** (1) runs through the vertebrae.

And here is a serious, serous membrane, summary:

- Parietal pleura lines the pleural (lung) cavity.
- Visceral pleura covers the lungs.

- Parietal pericardium lines the pericardial (heart) cavity.
- Visceral pericardium covers the heart.

- Parietal peritoneum lines the peritoneal (abdominal) cavity.
- Visceral peritoneum covers the organs within the peritoneal cavity.

*Keep this serous membrane vocabulary clear so your teacher cannot trip you up with any tricky questions ... but surely you do not have a "tricky teacher" do you?*

# Thoracic Cavity

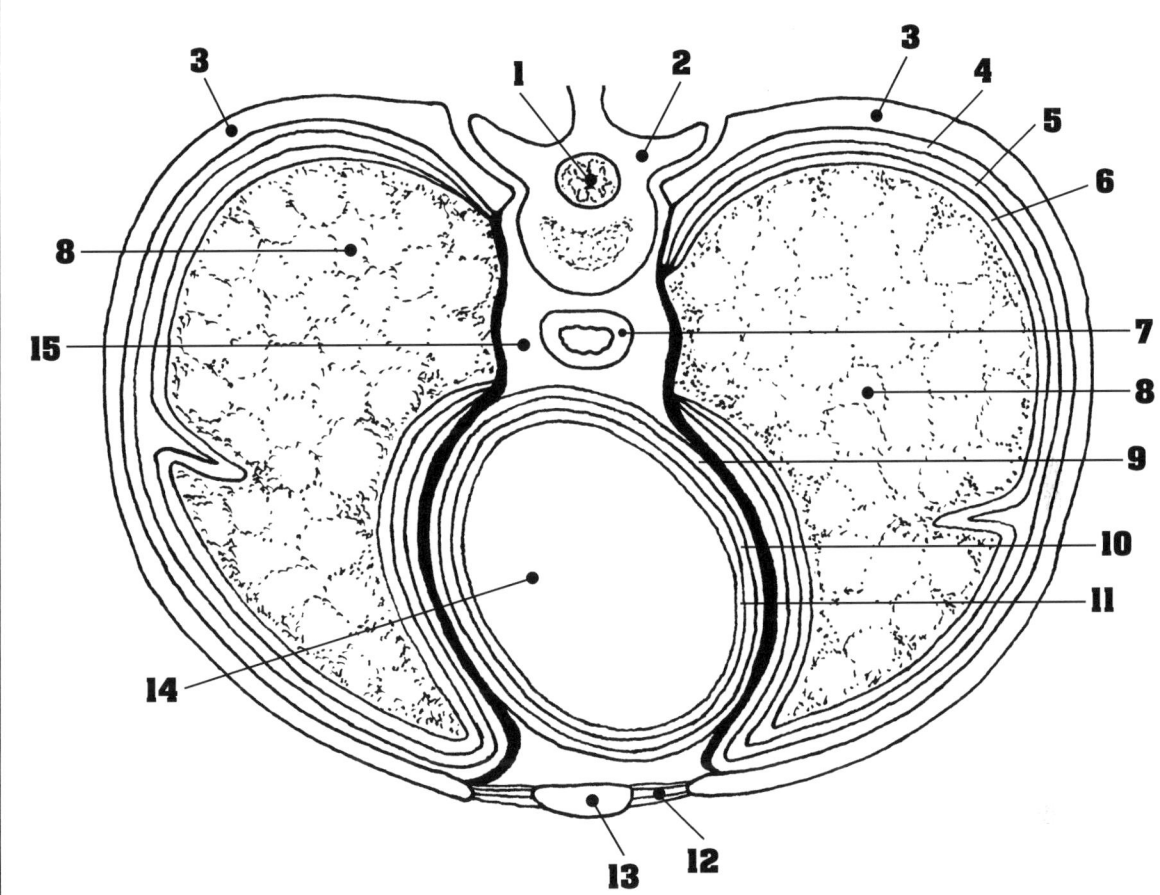

**Transverse Section**

___ Costal Cartilage
___ Esophagus
___ Heart
___ Lungs
___ Mediastinum
___ Pericardial Cavity
___ Pericardium
   ___ Parietal
   ___ Visceral

___ Pleura
   ___ Parietal
   ___ Visceral
___ Pleural Cavity
___ Ribs
___ Spinal Cord
___ Sternum
___ Thoracic Vertebra

# Heart & Lungs
## With Lungs Reflected Laterally

The intimate intertwining of heart and lungs, often referred to as the pulmo-circulatory system, is shown here with the **lungs** (3) reflected laterally.

Unoxygenated blood returns to the heart from the head via **jugular veins** (12), and from the arms via **subclavian veins** (10). Jugular and subclavian veins converge to form **brachiocephalic veins** (9). The right and left brachiocephalic veins converge to form the **superior vena cava** (7).

Blood from the lower body regions returns to the heart via the **inferior vena cava** (1). The **right ventricle** (2) pumps blood to the lungs via the **pulmonary trunk** (21), which divides into the **right** (8) and **left** (19) **pulmonary arteries**. Freshly oxygenated blood returns from the lungs to the heart via the **right** (6) and **left** (20) **pulmonary veins**.

The **left ventricle** (23) pumps blood up through the **aortic trunk** (5) to the **aortic arch** (18). At the top of the aortic arch the **brachiocephalic artery** (14) divides into the **right common carotid artery** (13) — which takes blood up through the neck to the head, and the **right subclavian artery** (11) — which takes blood toward the right shoulder region.

Because the **left common carotid artery** (16) and the **left subclavian artery** (17) arise directly from the aortic arch, there is no brachiocephalic artery on the left side.

The **right** (4) and **left** (22) **coronary arteries** derive from the extreme base of the aortic arch. The **thoracic aorta** (24) emerges just beneath the heart.

Note the posteriorly positioned **trachea** (15).

# Heart & Lungs
## With Lungs Reflected Laterally

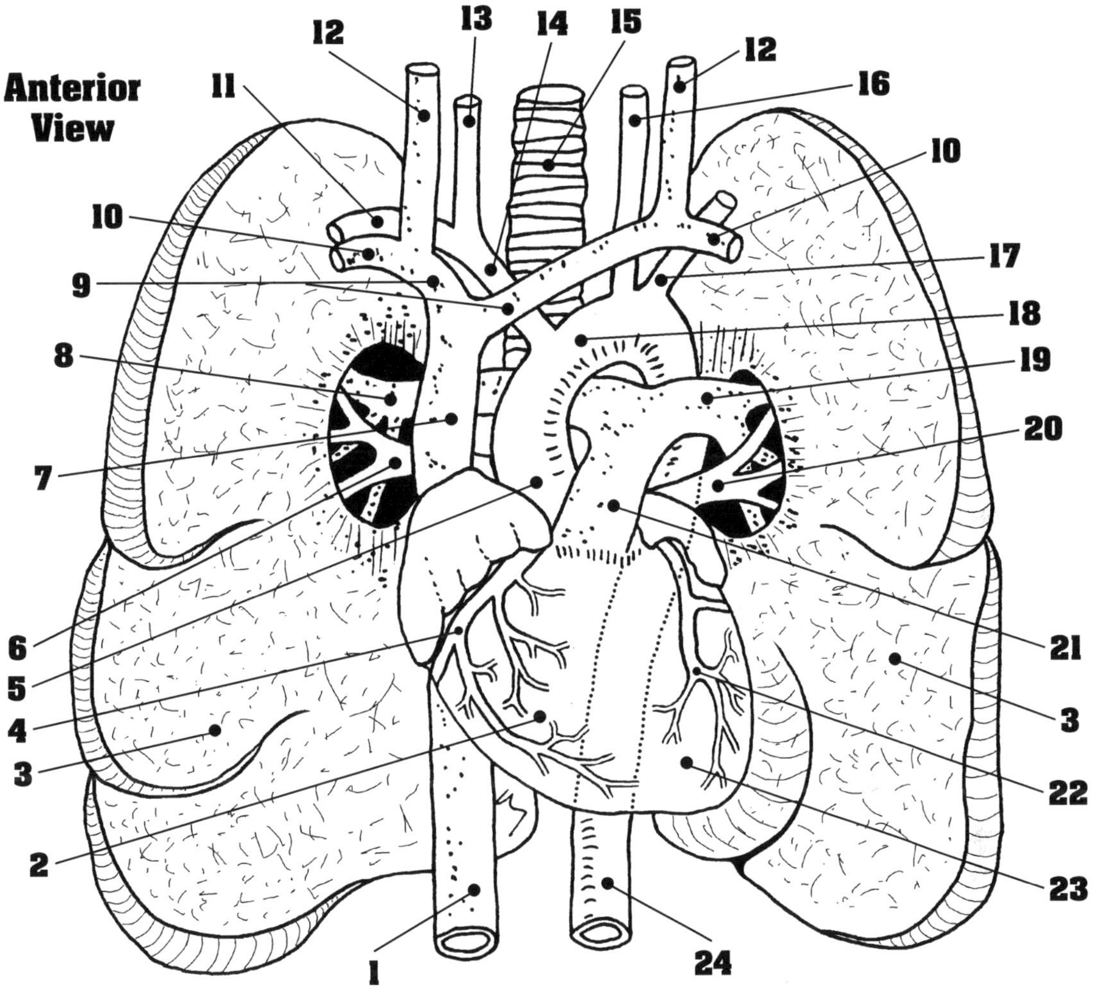

**Anterior View**

___ Aortic Arch ☐
___ Aortic Trunk ☐
___ Brachiocephalic
   Artery ☐
___ Veins ☐
___ Common Carotid Artery
   ___ Left ☐
   ___ Right ☐
___ Coronary Artery
   ___ Left ☐
   ___ Right ☐
___ Jugular Veins ☐
___ Lungs ☐
___ Pulmonary Artery
   ___ Left ☐
   ___ Right ☐
___ Pulmonary Trunk ☐
___ Pulmonary Vein
   ___ Left ☐
   ___ Right ☐
___ Subclavian Artery
   ___ Left ☐
   ___ Right ☐
___ Subclavian Veins ☐
___ Thoracic Aorta ☐
___ Trachea ☐
___ Vena Cava
   ___ Inferior ☐
   ___ Superior ☐
___ Ventricle
   ___ Left ☐
   ___ Right ☐

# Respiratory System

Air enters and leaves the body through either the **oral cavity** (8) or the **nasal cavity** (7), or both. Air enters the nasal cavity through the **external nares** (6) and moves from the nasal cavity into the pharynx via the **internal nares** (9).

The **pharynx** (13) is subdivided into three anatomical regions:

- **Nasopharynx** (10)
- **Oropharynx** (11)
- **Laryngopharynx** (12)

The larynx is supported by the **hyoid bone** (5), and two cartilaginous components: **thyroid cartilage** (15) and **cricoid cartilage** (16).

From the larynx, air flows downward through the **trachea** (17), into the **bronchi** (18). Each bronchus divides into many smaller **bronchioles** (19). The bronchioles terminate in **alveolar clusters** (2).

The right lung has a **superior** (4), **median** (3), and **inferior lobe** (1). The left lung has only a superior and inferior lobe.

The left lung has a **cardiac notch** (20) to accommodate the heart, which is directed toward the left side of the body. The **esophagus** (14) is posterior to the trachea.

*Ciliated epithelial cells lining the trachea act synchronously to create what is called the "tracheal escalator." The escalator assists in the removal of particulate matter that comes in with the air flow. One of the many, many ... many negative side effects of smoking is that nicotine poisons and disables the tracheal escalator.*

*How much particulate matter enters your lungs each day depends not only on how you live, but upon where you live.*

# Respiratory System

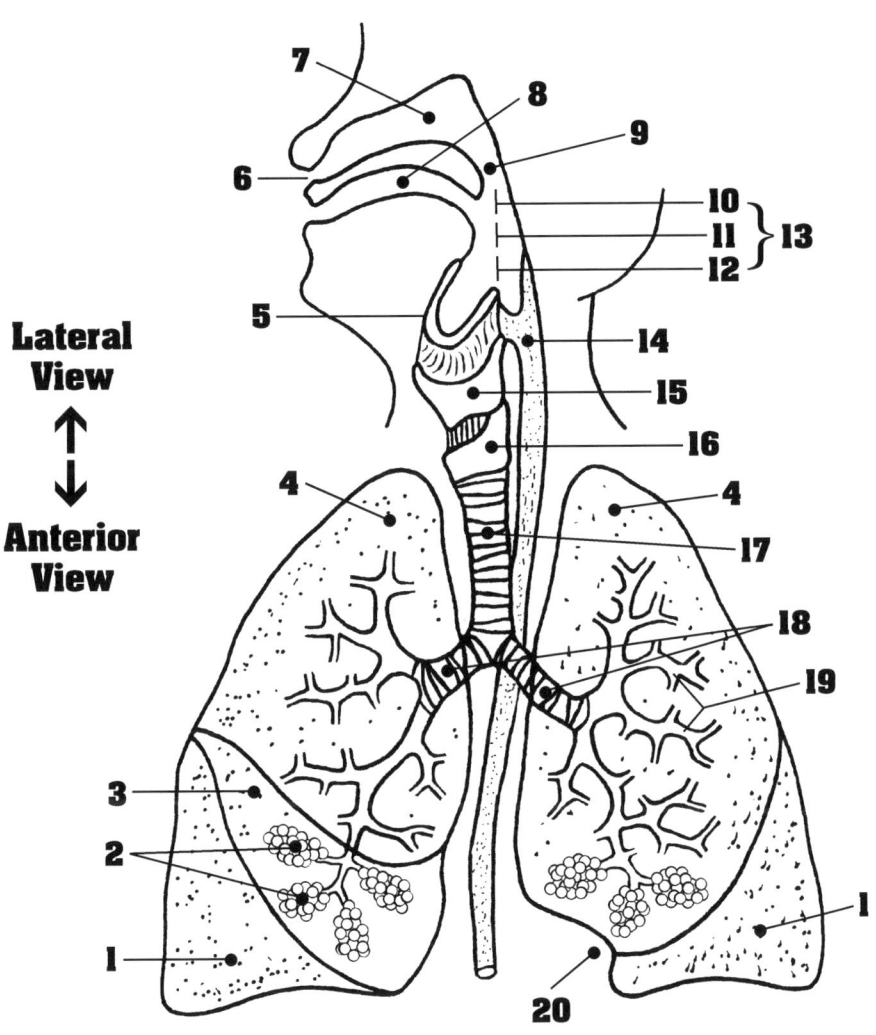

**Lateral View ↕ Anterior View**

| | | |
|---|---|---|
| ___ Alveolar Clusters ☐ | **Lobes** | ___ Nasal Cavity ☐ |
| ___ Bronchi ☐ | ___ Inferior ☐ | ___ Nasopharynx ☐ |
| ___ Bronchioles ☐ | ___ Median ☐ | ___ Oral Cavity ☐ |
| ___ Cardiac Notch ☐ | ___ Superior ☐ | ___ Oropharynx ☐ |
| ___ Cricoid Cartilage ☐ | **Nares** | ___ Pharynx ☐ |
| ___ Esophagus ☐ | ___ External ☐ | ___ Thyroid Cartilage ☐ |
| ___ Hyoid Bone ☐ | ___ Internal ☐ | ___ Trachea ☐ |
| ___ Laryngopharynx ☐ | | |

© Copyright 2010 Gene Johnson

# Alveolar Cluster

The air flow through a terminal **bronchiole** (2) is partially regulated by **smooth muscle** (1) sphincters.

A vascular net of blood **capillaries** (6) surrounds each **alveolus** (5).

Pulmonary **arterioles** (4) carry oxygen-depleted blood to the alveoli, while pulmonary **venules** (3) carry oxygen-rich blood away from the alveoli.

*15 million Americans suffer from bronchial asthma. Asthma is characterized by episodes of coughing, labored breathing, wheezing, and chest tightening.*

*Hypersensitivity to foreign materials (allergens) often causes the bronchial airways to become inflamed. The inflammation amplifies the effect of bronchial spasms and dramatically reduces the airflow.*

*Chronic obstructive pulmonary disease (COPD) afflicts an additional 10 million people in the United States. Most COPD patients develop obstructive emphysema, which is characterized by permanent enlargements of the alveoli and accompanied by a deterioration of the alveolar walls. Chronic inflammation leads to lung fibrosis and a loss of lung elasticity.*

*Psst! Psst! COPD patients almost invariably have a history of smoking or living around smoke.*

*Psst! Psst! Lung cancer is responsible for a third of cancer deaths in the United States. More than 90% of lung cancer patients were smokers.*

# Alveolar Cluster

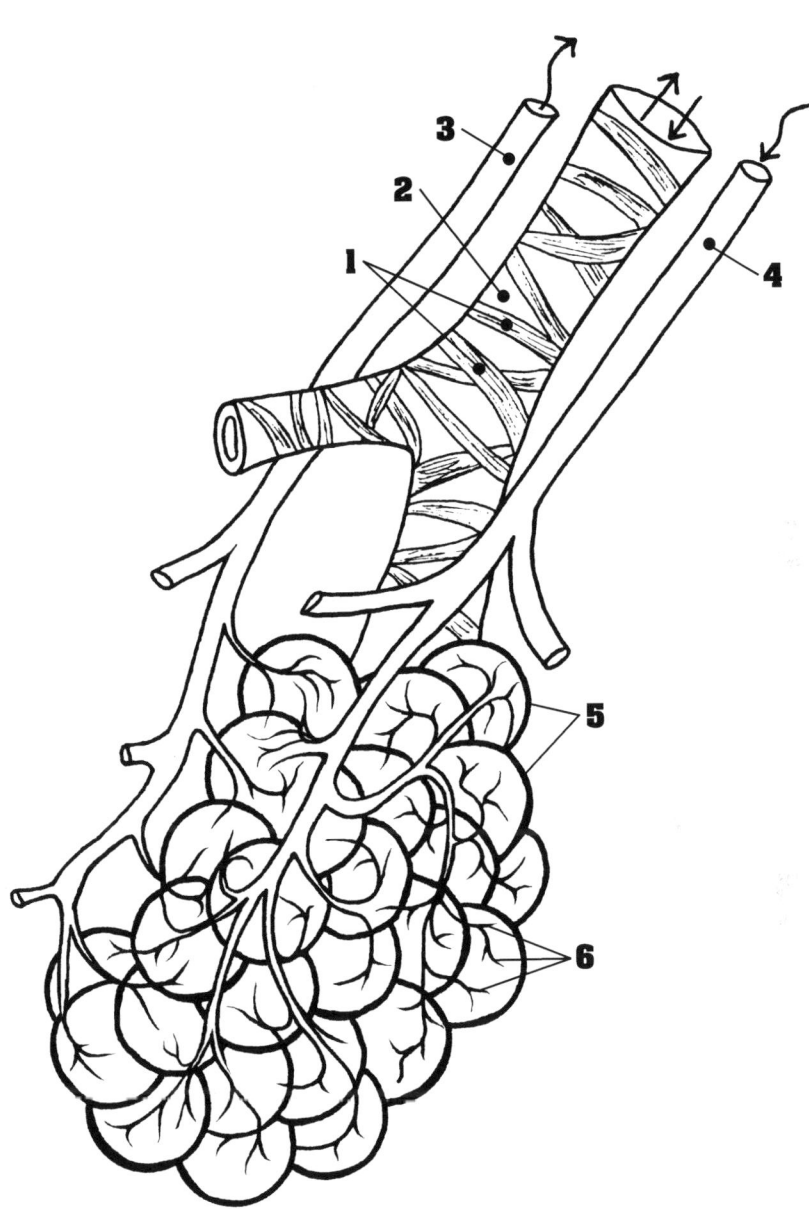

___ **Alveoli**
___ **Arteriole**
___ **Bronchiole**
___ **Capillaries**
___ **Venule**
___ **Smooth Muscle**

# Larynx

The infrastructure of the voice box (larynx) is largely comprised of three cartilaginous components — **thyroid cartilage (Adam's apple)** (4), **cricoid cartilage** (6), and **epiglottis** (1) — along with one osseous component — the **hyoid bone** (2).

The thyroid cartilage is knit, inferiorly, to the cricoid cartilage by the **cricothyroid membrane** (5). The thyroid cartilage is knit, superiorly, to the hyoid bone by the **thyrohyoid membrane** (3).

The cartilaginous components of the **trachea** (7) are located inferior to the larynx.

During the swallowing process the tongue-like, cartilaginous epiglottis depresses downward to close the entrance to the larynx and trachea so that all food particles will be shunted down the esophagus (food pipe) rather than the trachea (air pipe).

*Under certain circumstances, the epiglottis does not close fast enough to prevent small food particles — particularly dry, "dusty" food particles — from going down the wrong pipe.*

# Larynx

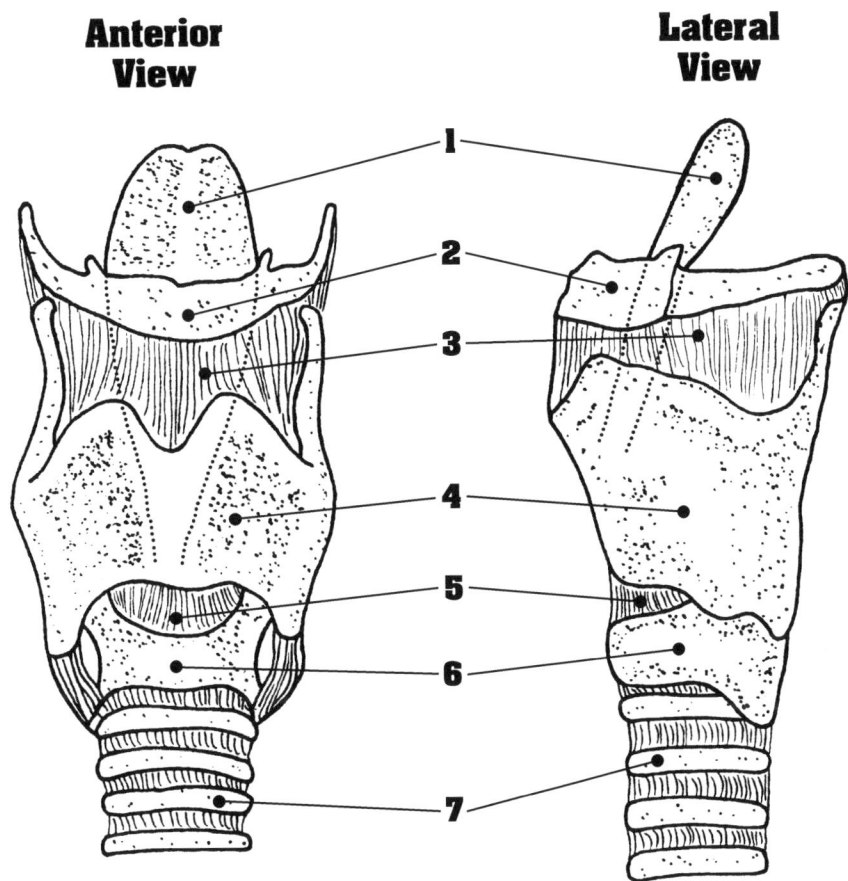

**Anterior View**     **Lateral View**

___ Cricoid Cartilage
___ Cricothyroid Membrane
___ Epiglottis
___ Hyoid Bone
___ Thyrohyoid Membrane
___ Thyroid Cartilage
___ Trachea

# Respiratory Center

The **medullary rhythmicity center** (14), located in the medulla of the brain, contains an **inspiratory** (16) and **expiratory** (15) **center.**

Because expiration usually passively follows inspiration there is little activity at the expiratory center. In the case of forced expirations impulses are sent via the **internal intercostal nerve** (20) to the **internal intercostal muscles** (2). The internal intercostal muscles pull the ribs above them downward, thereby lowering the rib cage and forcing air out of the lungs. However, in the normal breathing process, the rib cage needs no stimulus and falls back passively.

By contrast, many factors and circumstances impinge on the inspiratory center:

A **pneumotaxic area** (12), located in the **pons respiratory center** (11), acts to inhibit the inspiratory center in the medulla, while an **apneustic area** (13), also located in the pons respiratory center, stimulates the medullary inspiration center.

An **emotional center** (10) in the hypothalamus can also affect the breathing rate.

**Central chemoreceptors** (17) — located in the brain stem, and **peripheral chemoreceptors** (6) — located in the **aorta** (5) and **carotid arteries** (7), all respond to increases in carbon dioxide and hydrogen ions by stimulating the inspiratory center. Impulses from the carotid receptors travel via the **glossopharyngeal nerve** (8), whereas impulses from the aortic chemoreceptor travel via the **vagus nerve** (9).

**Stretch receptors** (4) in the lungs send impulses to the inspiratory center via the vagus nerve.

In response to the stimulations or inhibitions that impinge on it, the inspiratory center sends a lesser or greater number of impulses to the muscles that effect inspiration. The **diaphragm** (3) is innervated by the **phrenic nerve** (18), and the **external intercostal muscles** (1) are innervated by the **external intercostal nerve** (19).

Because origins and insertions of the external intercostal muscles are reversed, as compared with the internal intercostal muscles, they pull the ribs upward, thereby lifting the rib cage and facilitating inspiration.

# Respiratory Center

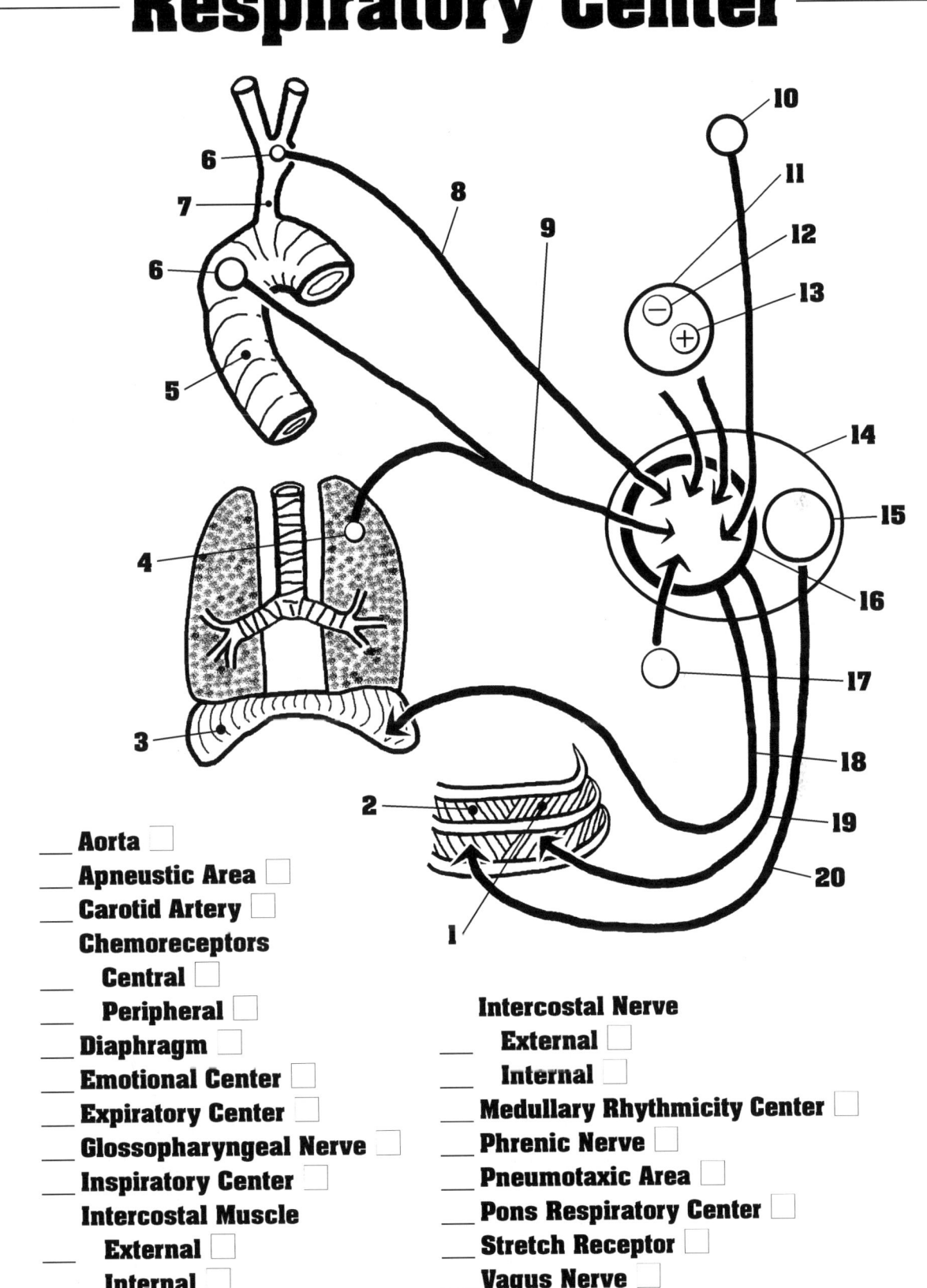

___ Aorta
___ Apneustic Area
___ Carotid Artery
___ Chemoreceptors
   ___ Central
   ___ Peripheral
___ Diaphragm
___ Emotional Center
___ Expiratory Center
___ Glossopharyngeal Nerve
___ Inspiratory Center
___ Intercostal Muscle
   ___ External
   ___ Internal

Intercostal Nerve
   ___ External
   ___ Internal
___ Medullary Rhythmicity Center
___ Phrenic Nerve
___ Pneumotaxic Area
___ Pons Respiratory Center
___ Stretch Receptor
___ Vagus Nerve

# Digestive System

Whereas the respiratory system delivers oxygen to body cells, the digestive system is devoted to delivering nutrients to body cells. But whereas oxygen needs only to be transported, most nutrients, because they come in rather large, complex chunks, need not only to be transported but to also be digested. Fortunately, the "complex chunks" are of only three major types, all of which are polymers which can be quickly depolymerized into monomers. More fortunate still (at least for beginning anatomy & physiology students) the chemical mechanism, hydrolysis (hydration decomposition), by which they are depolymerized (digested), is the same for each of the three food types:

- **Carbohydrate polymers (polysaccharides)** — depolymerized into monosaccharide monomers (glucose, fructose, galactose).
- **Lipid (fat) polymers** — depolymerized into glycerol and fatty acid monomers.
- **Protein polymers (polypeptides)** — depolymerized into amino acid monomers.

Although some polymerization (digestion) occurs in the oral cavity and stomach, it mostly occurs inside the approximately 20-foot-long small intestine. Enzymes, which assist the depolymerizations (decompositions) at every step along the way, are primarily contributed by the salivary glands, pancreas, and small intestine. As will be shown in later exercises, the stomach and liver also play critical roles in the processing of nutrients.

When depolymerization is completed and the monomers are delivered to cells, they can then be used as:

- **Fuel** — for the mitochondria (ATP).
- **Building blocks** — for constructing, or replacing, structural components.
- **Building blocks** — for constructing regulatory molecules, such as enzymes or hormones.

Thus we can ingest carbohydrate, lipid, and protein polymers obtained from other organisms, depolymerize them into monomers, and rearrange (polymerize) the monomers into polymers suited to our particular needs. In such a fashion one animal or plant can be handed on to another animal or plant. Thus grass becomes antelope, and antelope becomes lion. The living world is, in a sense, a constantly shifting symphony of dehydration syntheses and hydration decompositions — a symphony in three polymers! Well, we do need vitamins and minerals too.

# Digestive System

**Vitamins**

A — retinol
D — calciferal
E — tocopheral
K₁ — phylloquinone
K₂ — menatetranone
C — ascorbic acid
B₁ — thiamine
B₂ — riboflavin
B₃ — niacin
B₅ — pantothenic acid
B₆ — pyridoxine
B₁₂ — cyanocobalemin
Biotin
Folic Acid

**Minerals**

Ca — calcium
Cl — chlorine
Co — cobalt
Cr — chromium
Cu — copper
F — flourine
Fe — iron
I — iodine
K — potassium
Mg — magnesium
Mn — manganese
Na — sodium
P — phosphorus
S — sulphur
Se — selenium
Zn — zinc

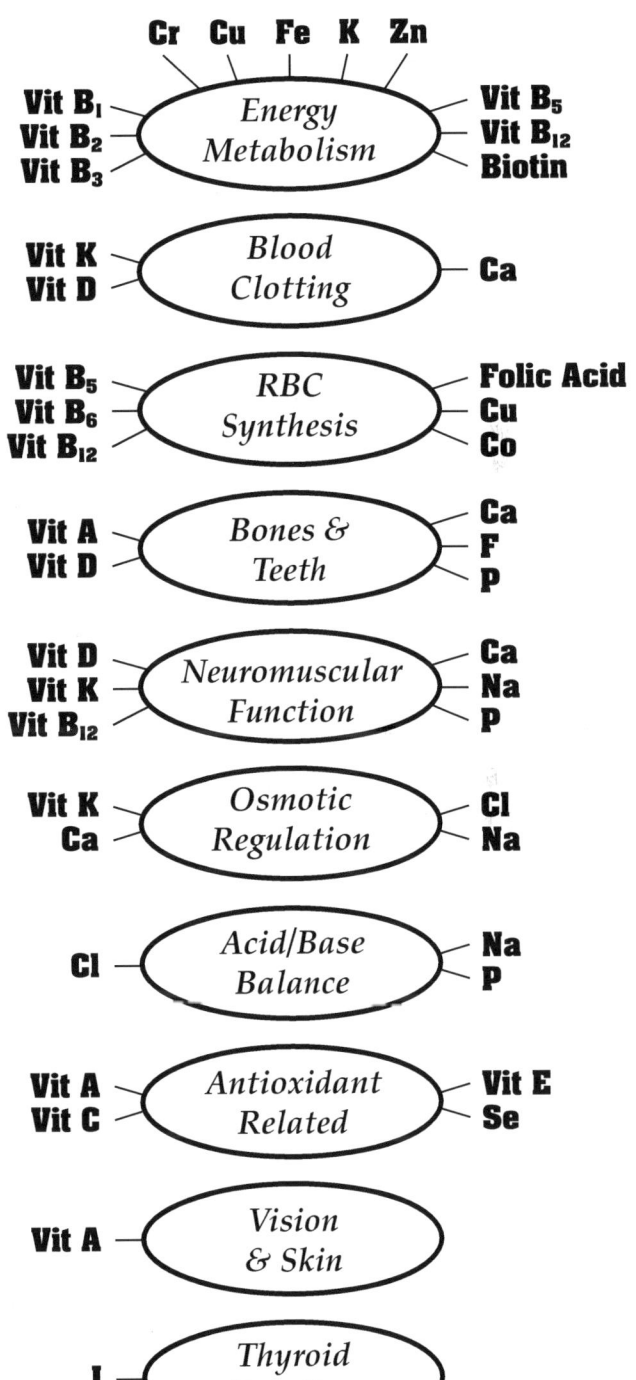

# Salivary Glands

The **oral cavity** (3) receives saliva from:

- **Parotid glands** (4) — via **parotid ducts** (5).
- **Submaxillary glands** (1) — via **submaxillary ducts** (8).
- **Sublingual glands** (2) — via **sublingual ducts** (7).

The sublingual gland is embedded inferiorly in the **tongue** (6).

1,000 to 1,500 ml of saliva are produced each day. Each of the three pairs of salivary glands contains a different mix of saliva. The following are among the many known active ingredients in saliva:

- **Mucin** — lubricates the oral cavity and hydrates foodstuffs.

- **Amylase** — digestive enzyme which breaks down (depolymerizes) carbohydrates.

- **Lysozyme** — bacteriostatic enzyme which inhibits bacterial growth.

- **IgA antibodies** — protect against pathogenic microorganisms.

- **Defensins** — chemical messengers which assist the immune response.

*In addition to the above salvic contributions of saliva, we now know that symbiotic (friendly) bacteria, living on the back of the tongue, also act to hold alien (pathogenic) bacteria in check. No wonder animals are always licking their wounds or babies!*

*Great hosts of friendly, symbiotic bacteria live almost everywhere on and in our bodies. We could live but a few hours without these "significant others." Didn't we suggest earlier that "there is no such thing as an individual?"*

# Salivary Glands

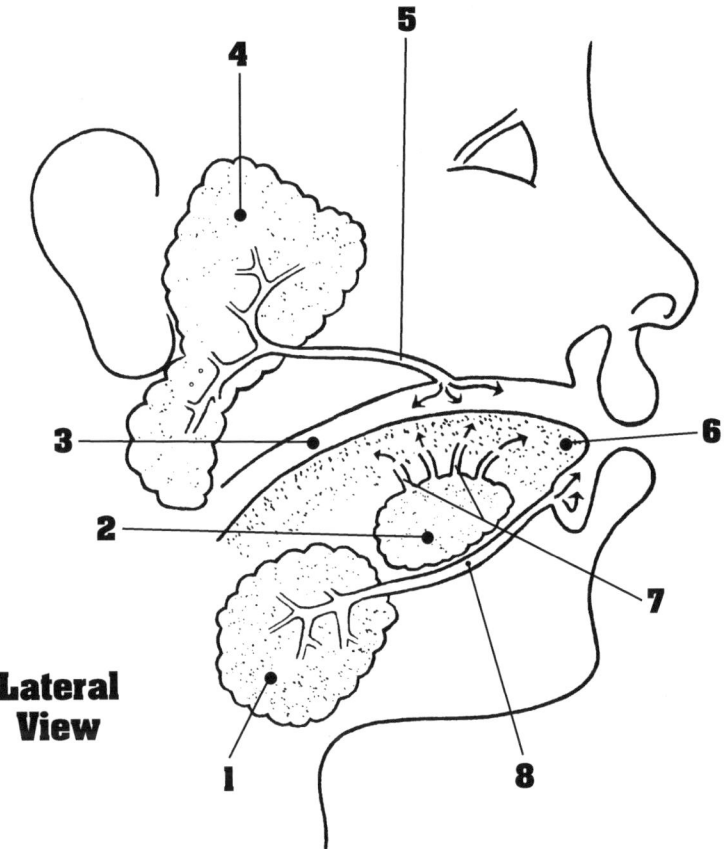

**Lateral View**

___ Oral Cavity
___ Parotid Duct
___ Parotid Gland
___ Sublingual Ducts
___ Sublingual Gland
___ Submaxillary Duct
___ Submaxillary Gland
___ Tongue

# Tooth
## Typical Molar

The **crown** (11) of the tooth is covered by **enamel** (9). The **pulp cavity** (10), located in the region of the **neck** (12), is surrounded by **dentin** (8), and contains **capillaries** (5) and **nerve endings** (6).

The **root canal** (1), with its associated blood and nerve supply, runs through the **root** (13).

The tooth is anchored to the **bone** (4) by a **periodontal ligament (membrane)** (3) and a layer of **cementum** (2). The **gingiva (gum)** (7) is located at the base of the crown where the tooth emerges from the alveolar socket in the bone.

The dental formula for the normal adult human is:

2I, 1C, 2PM, 3M (two incisors, one canine, two premolars, three molars).

The above formula is per quadrant, so multiply by four to get the total number of teeth (8 x 4 = 32).

*Many people have, during their lifetime, three different sets of teeth:*

*First, baby (deciduous) teeth.*
*Later, permanent teeth.*
*And still later, false teeth.*

# Tooth
## Typical Molar

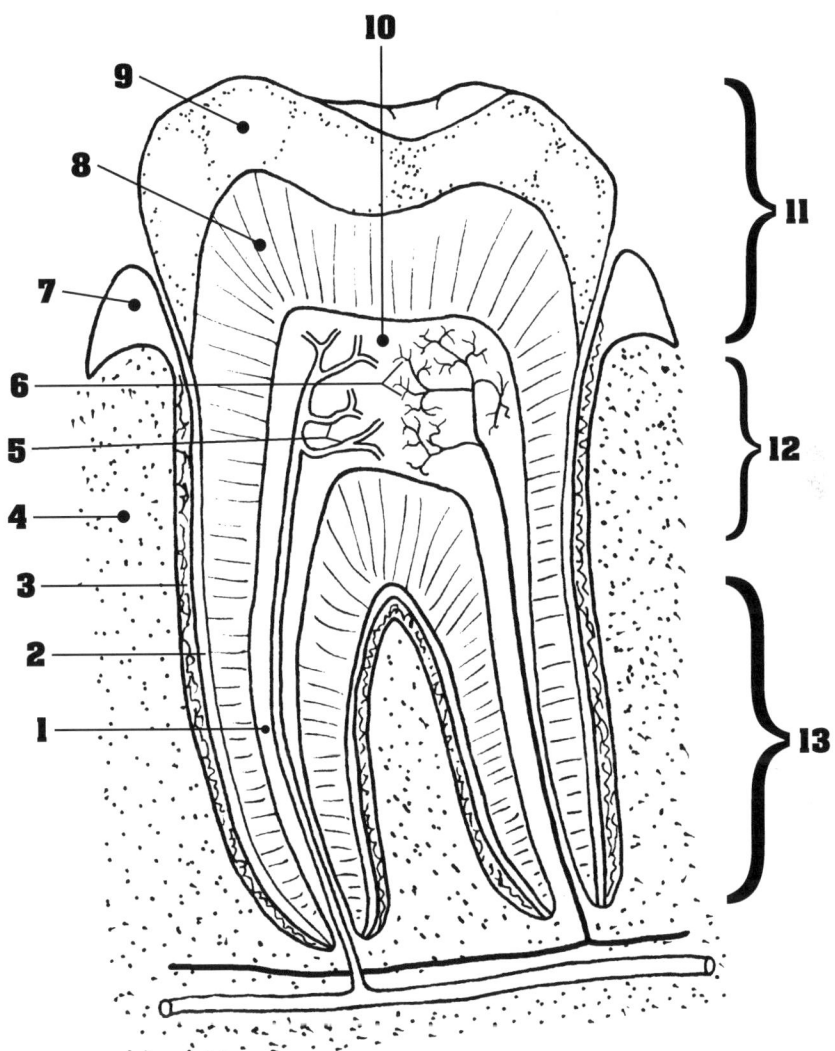

___ Bone
___ Capillaries
___ Cementum
___ Crown
___ Dentin
___ Enamel
___ Gingiva
___ Neck
___ Nerve Endings
___ Periodontal Ligament
___ Pulp Cavity
___ Root
___ Root Canal

# Alimentary Canal
## (Gastrointestinal Tract)

From superior to inferior, the alimentary canal consists of the:

- **Esophagus** (1)
- **Stomach** (2)
- Small intestine
- Large intestine
- **Rectum** (8)

The small intestine is anatomically subdivided into the:

- **Duodenum** (12)
- **Jejunum** (5)
- **Ileum** (7)

The large intestine is anatomically subdivided into the:

- **Ascending colon** (10)
- **Transverse colon** (11)
- **Descending colon** (6)

The **appendix** (9) arises at the beginning of the ascending colon.

The duodenum receives secretions from the **pancreas** (3) via the **pancreatic duct** (4), and from the **liver** (16) via the **hepatic ducts** (17) and **common bile duct** (13).

Bile, secreted by the liver, is stored in the **gall bladder** (14) and periodically released via the **cystic duct** (15) to the common bile duct.

# Alimentary Canal
## (Gastrointestinal Tract)

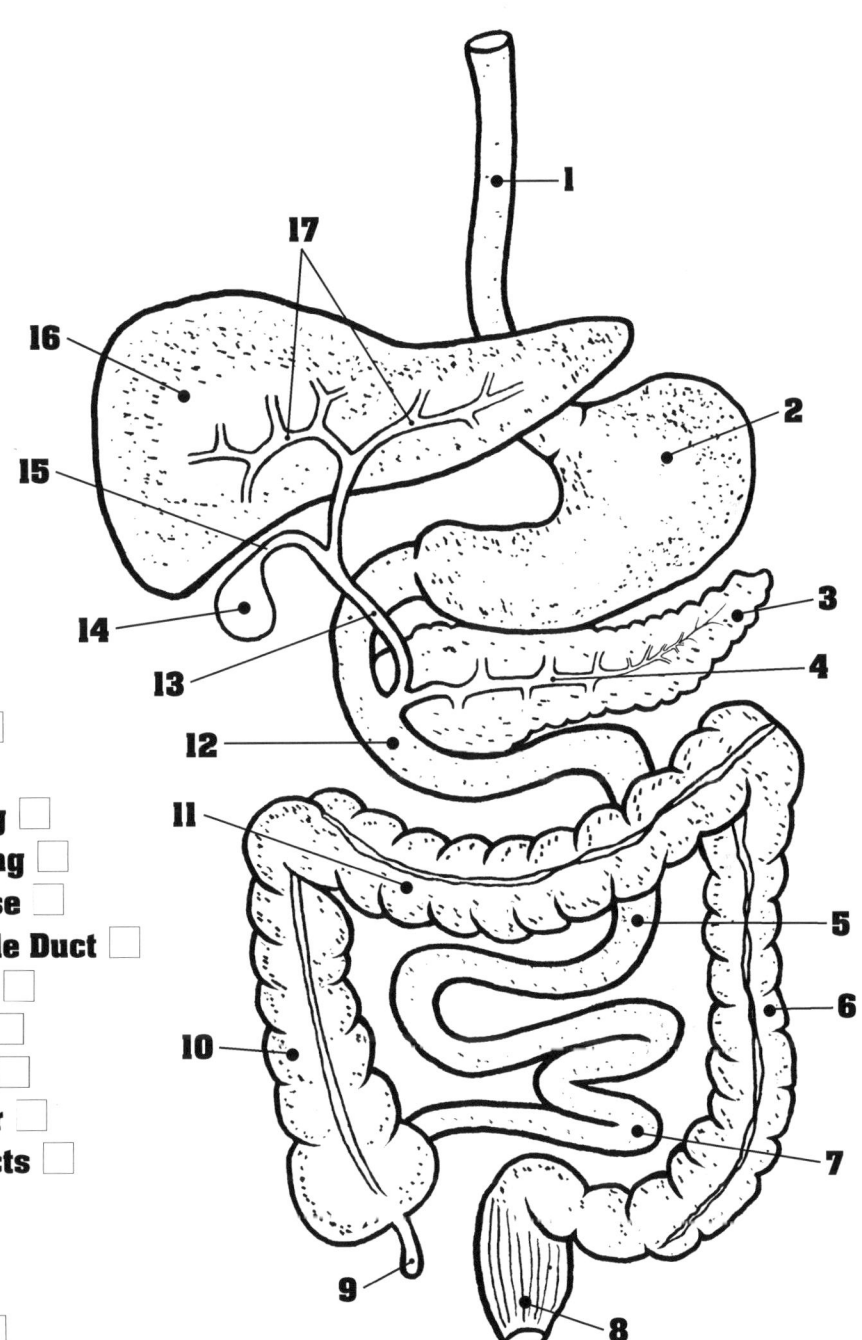

___ Appendix ☐
Colon
___ Ascending ☐
___ Descending ☐
___ Transverse ☐
___ Common Bile Duct ☐
___ Cystic Duct ☐
___ Duodenum ☐
___ Esophagus ☐
___ Gall Bladder ☐
___ Hepatic Ducts ☐
___ Ileum ☐
___ Jejunum ☐
___ Liver ☐
___ Pancreas ☐
___ Pancreatic Duct ☐
___ Rectum ☐
___ Stomach ☐

# Stomach

The **esophagus** (7) delivers food to the stomach through the **cardiac (esophageal) sphincter** (6). Food, in the form of chyme, leaves the stomach via the **pyloric sphincter** (2) and enters the **duodenum** (1).

The stomach is subdivided into four anatomical regions:

- **Cardia** (8) — the region closest to the heart.
- **Pylorus** (12) — the region nearest the pyloric valve.
- **Fundus** (9) — the expanded upper region.
- **Body** (10) — the central region.

The stomach has three muscular layers which run in three different directions, providing for the diverse mixing actions of the stomach:

- **Longitudinal muscle** (5)
- **Circular muscle** (4)
- **Oblique muscle** (3)

The mixing actions are further facilitated by rough ridges called **rugae** (11), which line the internal surface of the stomach.

The human stomach mechanically breaks food down into a homogenous chyme. The chyme is delivered, bolus by bolus, to the small intestine.

The stomach cleanses chyme by subjecting it to an "acid bath." Hydrochloric acid, produced by parietal cells in the stomach lining, maintains the gastric juice at a pH of 1.5 to 2, which is, indeed, a very strong acid.

Because most enzymes (including digestive enzymes) operate at a pH range close to neutral, the contents from the stomach must be immediately neutralized in the small intestine by buffer solutions from the pancreas.

*As is the case with most systems, the buffering system does not work well when you are upset, angry or anxious. This is one of many good reasons not to eat with anyone who might irritate or upset you. Hide, run ... do anything and everything to avoid such people — especially at meal time!*

# Stomach

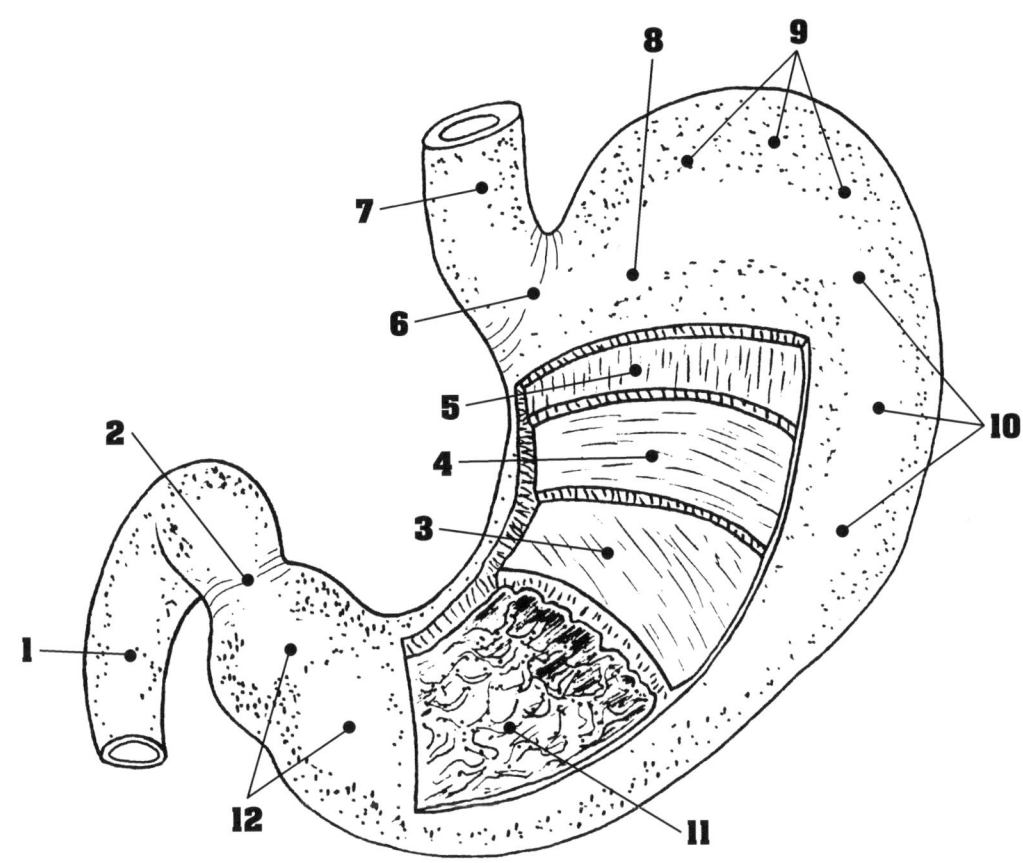

___ Body ▢
___ Cardia ▢
___ Cardiac (Esophageal) Sphincter ▢
___ Duodenum ▢
___ Esophagus ▢
___ Fundus ▢
    Muscle
___   Circular ▢
___   Longitudinal ▢
___   Oblique ▢
___ Pyloric Sphincter ▢
___ Pylorus ▢
___ Rugae ▢

# Upper Abdominal Organs
## With Stomach Removed

With the stomach removed we see the **pancreas** (11) "cradled in the arms of the **duodenum**" (1).

Prior to entering the duodenum the **pancreatic duct** (15) and **common bile duct** (2) merge. An alternative route is available for the pancreatic juices via an **accessory duct** (14).

The liver, which is positioned up against the **diaphragm** (8), is divided by a **falciform ligament** (7) into a **right** (5) and **left** (9) **lobe**.

The **hepatic ducts** (6) bring bile to the **gall bladder** (3), which delivers the bile, via the **cystic duct** (4), to the common bile duct.

The **kidneys** (12), with **ureters** (13), are also shown.

… And goodness gracious, let's throw in the **spleen** (10)!

# Upper Abdominal Organs
## With Stomach Removed

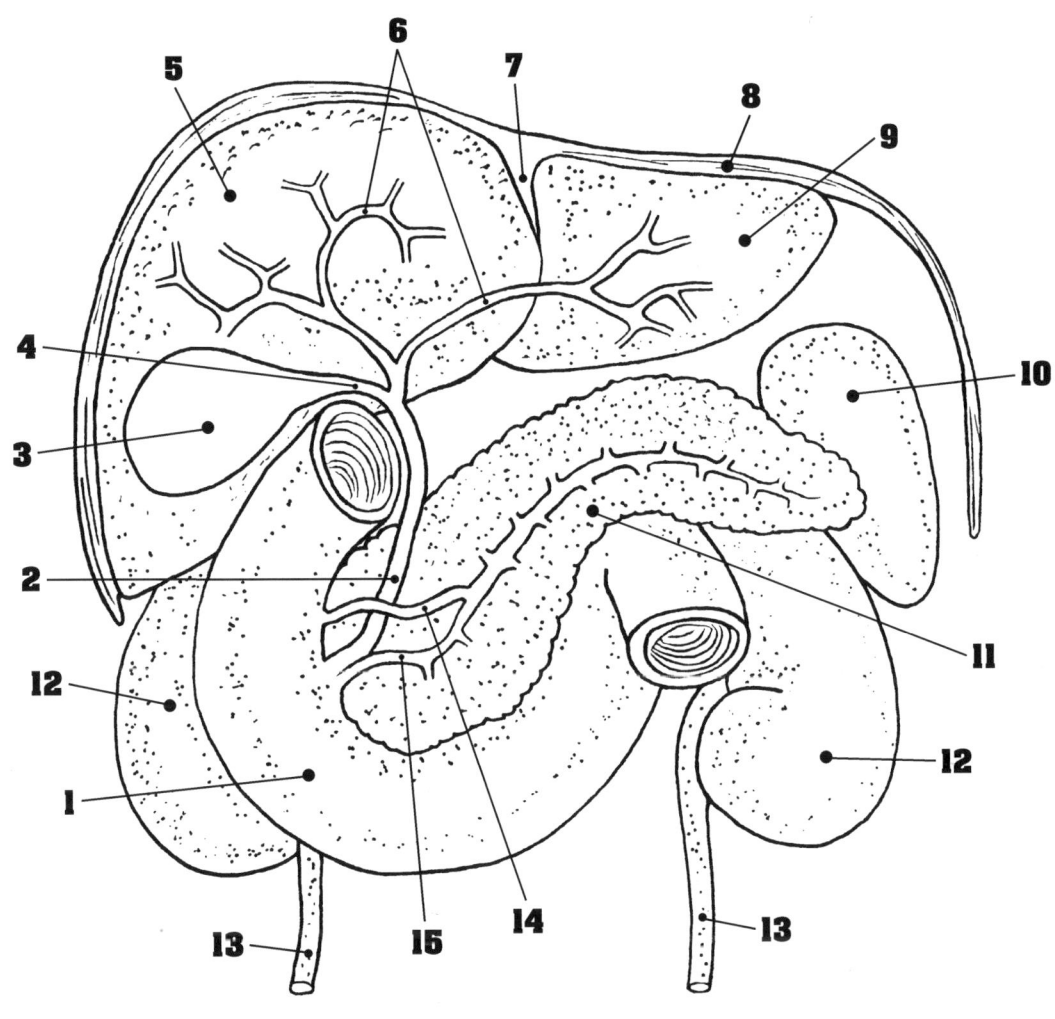

___ Accessory Duct
___ Common Bile Duct
___ Cystic Duct
___ Diaphragm
___ Duodenum
___ Falciform Ligament
___ Gall Bladder
___ Hepatic Ducts
___ Kidneys
___ Liver
   ___ Left Lobe
   ___ Right Lobe
___ Pancreas
___ Pancreatic Duct
___ Spleen
___ Ureters

# Liver Lobule

Stacked rows of **hepatocytes** (3) form the thousands of tiny hexagonal **liver lobules** (1) that comprise the liver.

At each of the six corners of a liver lobule there is a **portal triad** (7) comprised of:

- **Bile duct** (4)
- **Portal venule** (5)
- **Portal arteriole** (6)

Venous and arterial blood converge to flow through **liver sinusoids** (2) to a **central vein** (9). Blood from the central veins will eventually join the inferior vena cava via the hepatic veins.

The bile flows outward, through **bile canaliculi** (10), in the opposite direction of the blood flowing inward in the sinusoids. The bile proceeds on through hepatic bile ducts to the gall bladder and small intestine.

Fixed (resident) **hepatic macrophages (Kupfer cells)** (8) patrol the sinusoids and help fight bacteria, remove debris, recycle worn-out red blood cells, and perhaps perform many other numinous and yet unknown functions.

We acknowledge the supreme importance of the liver by repeating its functions:

- Facilitates lipid digestion by secreting bile salts.
- Stores excess sugar as glycogen.
- Disarms or modifies toxins, and if it cannot modify them it seals them off in toxic storage vesicles.
- Contains resident populations of macrophages for combating bacteria and other foreign agents.
- Converts lipids to sugar via a process called beta oxidation.
- Converts amino acids of one type, to another type, via transamination.
- Deaminates amino acids and converts them to sugar-like molecules for energy metabolism.
- Disassembles aging and worn-out red blood cells and recycles key elements to the bone marrow for the production of new red blood cells.

# Liver Lobule

___ Bile Canaliculus
___ Bile Duct
___ Central Vein
___ Hepatic Macrophages
___ Hepatocytes
___ Liver Lobule
___ Liver Sinusoids
___ Portal Arteriole
___ Portal Triad
___ Portal Venule

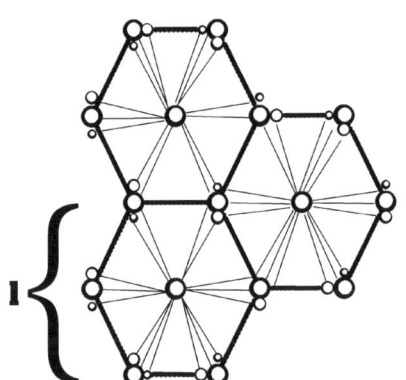

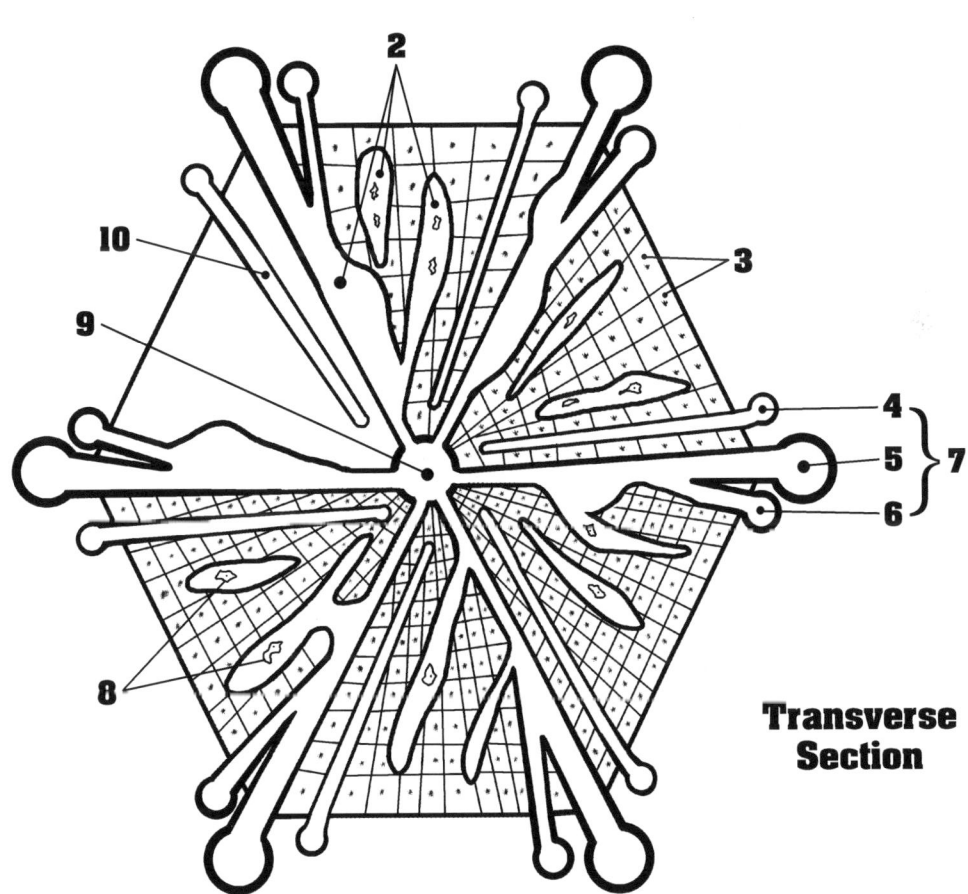

**Transverse Section**

# Small Intestine
## Histology

The small intestine is surrounded by **serosa** (2) (visceral peritoneal) membrane, and held in place by **mesentery** (3).

The intestine is comprised of three major tissue layers:

- **Mucosa** (8)
- **Submucosa** (7)
- **Muscularis** (6)

The muscularis layer is subdivided into **longitudinal muscle** (4) and **circular muscle** (5).

The **lumen** (9) of the intestine has an **epithelial lining** (10). External **accessory glands** (1) — like the pancreas and liver — and **submucosal glands** (11), bring secretions to the lumen of the intestine.

The submucosa is perfused with blood vessels. Larger, more collapsed **veins** (13), contrast with smaller, more rounded **arteries** (12).

The small intestine, which is approximately 20 feet in length, is the major organ for digestion and absorption. The remarkable micro-anatomical details of the lining of the small intestine are considered in the next two exercises.

# Small Intestine
## Histology

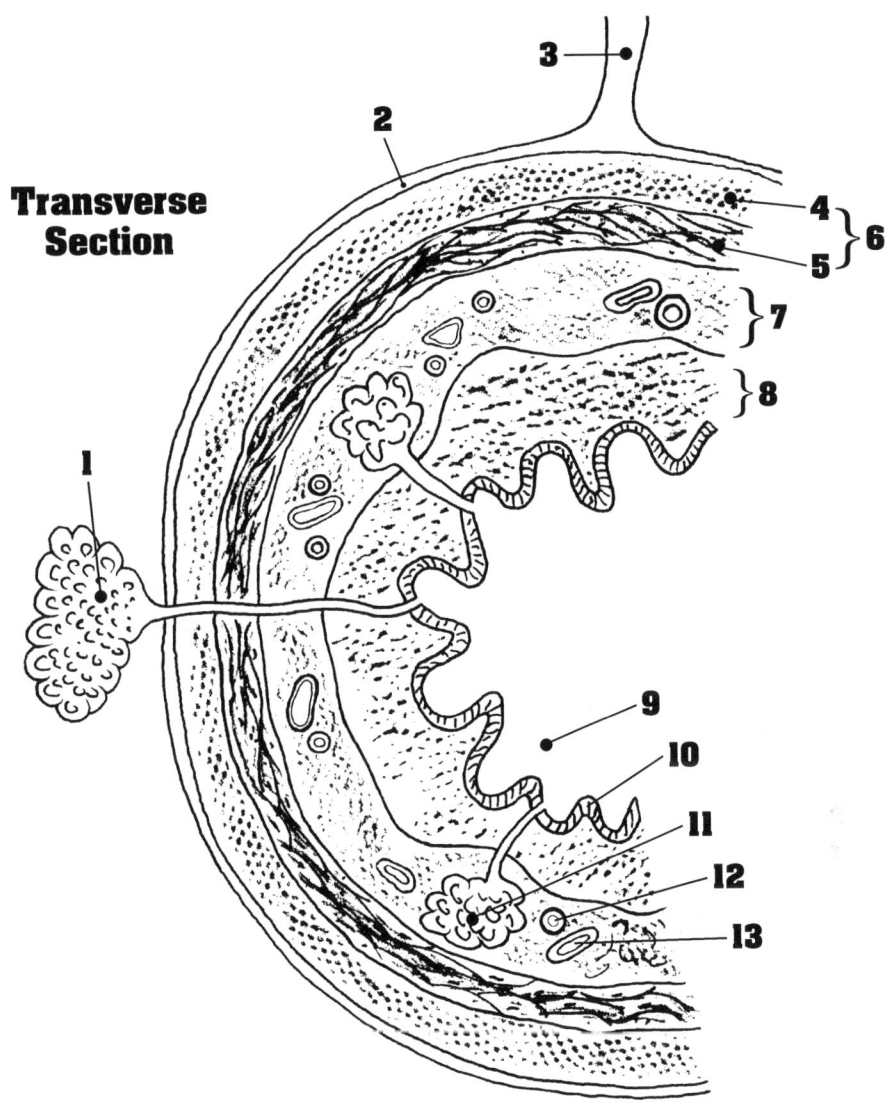

Transverse Section

___ Accessory Gland
___ Artery
___ Circular Muscle
___ Epithelial Lining
___ Longitudinal Muscle
___ Lumen
___ Mesentery
___ Mucosa
___ Muscularis
___ Serosa
___ Submucosa
___ Submucosal Gland
___ Vein

# Villus

A single villus is not much longer than 1 mm, yet, the many millions that collectively line the lumen of the small intestine do Herculean work in digestion and absorption. In this drawing we emphasize their role in absorption.

Digestion, as noted earlier, is primarily a process of depolymerization. When we eat we take on large protein, carbohydrate and lipid polymers. The digestive enzymes sequentially break these large polymers down into the individual building-block monomers from which they were originally polymerized.

- **Protein polymers** (7) are depolymerized into **amino acid monomers** (8).

- **Polysaccharide (carbohydrate) polymers** (9) are depolymerized into **monosaccharide (sugar) monomers** (10).

- **Lipid polymers** (12) are depolymerized into **short-chain** (11) and **long-chain** (13) **fatty acid monomers**.

Amino acids, monosaccharides and short-chain fatty acids are absorbed directly into the **capillaries** (3), whereas the long-chain fatty acids are absorbed into the **lacteal** (4). From the lacteal long-chain fatty acids travel through the **lymphatic vessels** (1) where they are modified, repackaged and eventually returned to the blood at the subclavian-jugular junction (see page 249).

**Columnar epithelial cells** (5), with a microvilli **brush border** (6), constitute the absorptive surface of the villus.

Thin strands of **smooth muscle cells** (2) endow the villi with contractile properties, thus, the villi can alternatively shorten and lengthen in a pulsating fashion. This not only facilitates contact with foodstuffs in the lumen of the intestine, but also acts to "milk" lymph through the lacteals. Rather like the rhythmic movements of breathing help "milk" venous blood upward through veins.

# Villus

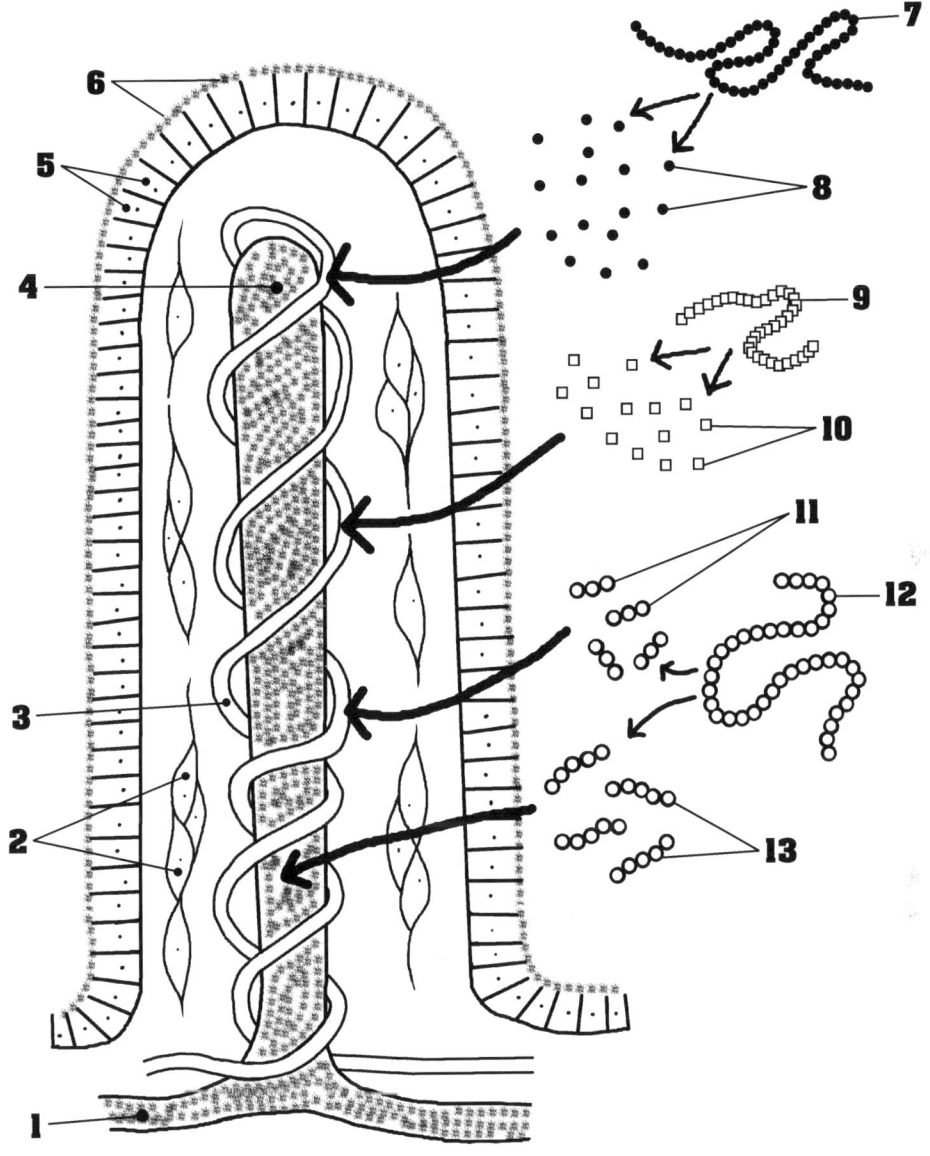

___ Amino Acid Monomers
___ Brush Border
___ Capillary
___ Columnar Epithelial Cells
___ Fatty Acids
    ___ Long-Chain Monomers
    ___ Short-Chain Monomers
___ Lacteal
___ Lipid Polymer
___ Lymphatic Vessel
___ Monosaccharide
    (Sugar) Monomers
___ Polysaccharide
    (Carbohydrate) Polymer
___ Protein Polymer
___ Smooth Muscle Cells

# Brush Border

Five anatomical specializations for increasing surface area in the small intestine are shown:

- The **small intestine** (1) is a very long tube (approximately 20 feet).

- Ridges called **plicae circularae** (2) occur all along the inside of the tube.

- Microscopic, finger-like **villi** (3) line the lumen of the intestine.

- Protoplasmic "fingers" called **microvilli** (6) extend from the outer edges of the **columnar epithelial cells** (4) forming a **brush border** (5) around each villus.

- **Macromolecular "fingers"** (7) extend out from between the microvilli "fingers."

*Because most digestion and absorption occur in the brush border, keeping the lining of the small intestine clean is prerequisite to good health. The typical American diet of soft, mushy, greasy, refined foods would seem almost purposefully calculated to coat and plug-up the brush border. American foods tend to be over-cooked and over-greased!*

*One doctor, when asked, "what's wrong with all these people anyway?" replied, "they all have dirty bowels!" "How then shall we eat?" someone once asked a food guru. The guru replied, "The sooner you get it out of the dirt and into your mouth the better it will be for you. Eat close to the earth! Eat close to the earth!"*

*The brush border, like your teeth, needs a good daily brushing. Stringy beans and celery! Peelings! Roughage! No decent potato comes to the table without its coat on! And never, ever, peel a kiwi fruit, for the prickly peel of the kiwi has the capacity to tickle your GI tract from top to bottom! ... and if all else fails, tie the kiwi on a string, swallow it, and plumb it up and down in your small intestine!*

# Brush Border

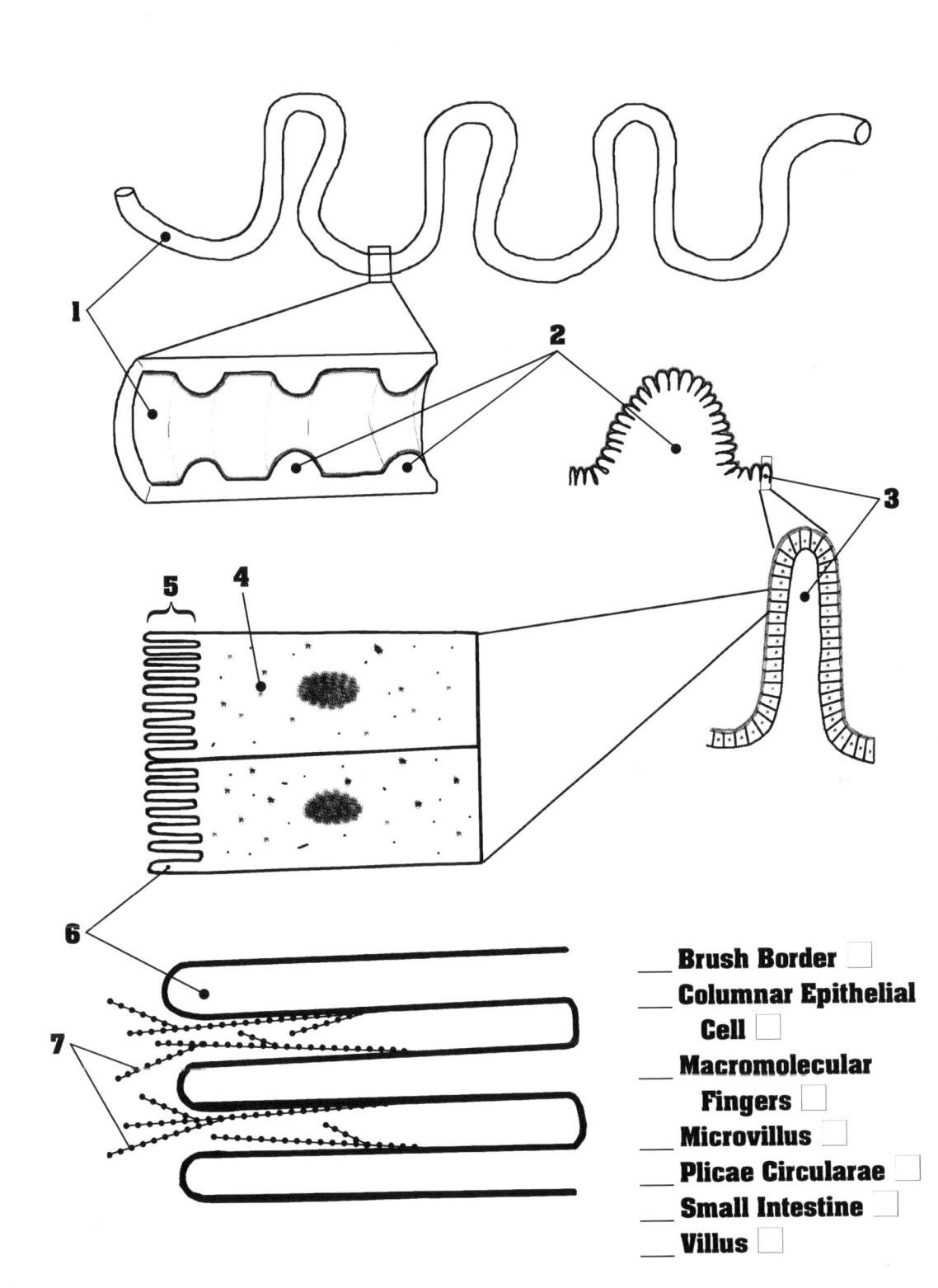

___ Brush Border
___ Columnar Epithelial Cell
___ Macromolecular Fingers
___ Microvillus
___ Plicae Circularae
___ Small Intestine
___ Villus

# Large Intestine

The **ileum** (4) of the small intestine joins the large intestine at the **cecum** (3). The **vermiform appendix** (1) extends out from the cecum. The large intestine extends upward on the right side of the body via the **ascending colon** (6), then across the top of the abdomen via the **transverse colon** (8), and finally downward along the left side of the body via the **descending colon** (11). The descending colon gives way to the "S-shaped" **sigmoid colon** (13), which in turn merges into the **rectum** (14). Fecal matter takes leave through the **anus** (15).

Each segmentation of the colon is called a **haustrum** (12). Longitudinal muscular strips called **taenia coli** (5) assist muscular movements of the large intestine.

The colon is stabilized by the **mesocolon** (7). The vermiform appendix is supported by the **mesoappendix** (2).

The proximal region of large intestine is supplied with blood by the **superior mesenteric** (9) artery while the distal region receives blood from the **inferior mesenteric** (10) artery.

*The large intestine is home to a diverse community of wee beasties (microbes). This bacterial flora synthesizes B-complex vitamins and most of the vitamin K needed by the liver to synthesize clotting proteins.*

*Bacterial flora ferment indigestible foodstuffs, releasing a mixture of gases — some, like dimethyl sulfide, are quite odorous. About 500 ml of floral flatus (gas) is produced each day — a small price to pay for good vitamins? Yes? No?*

# Large Intestine

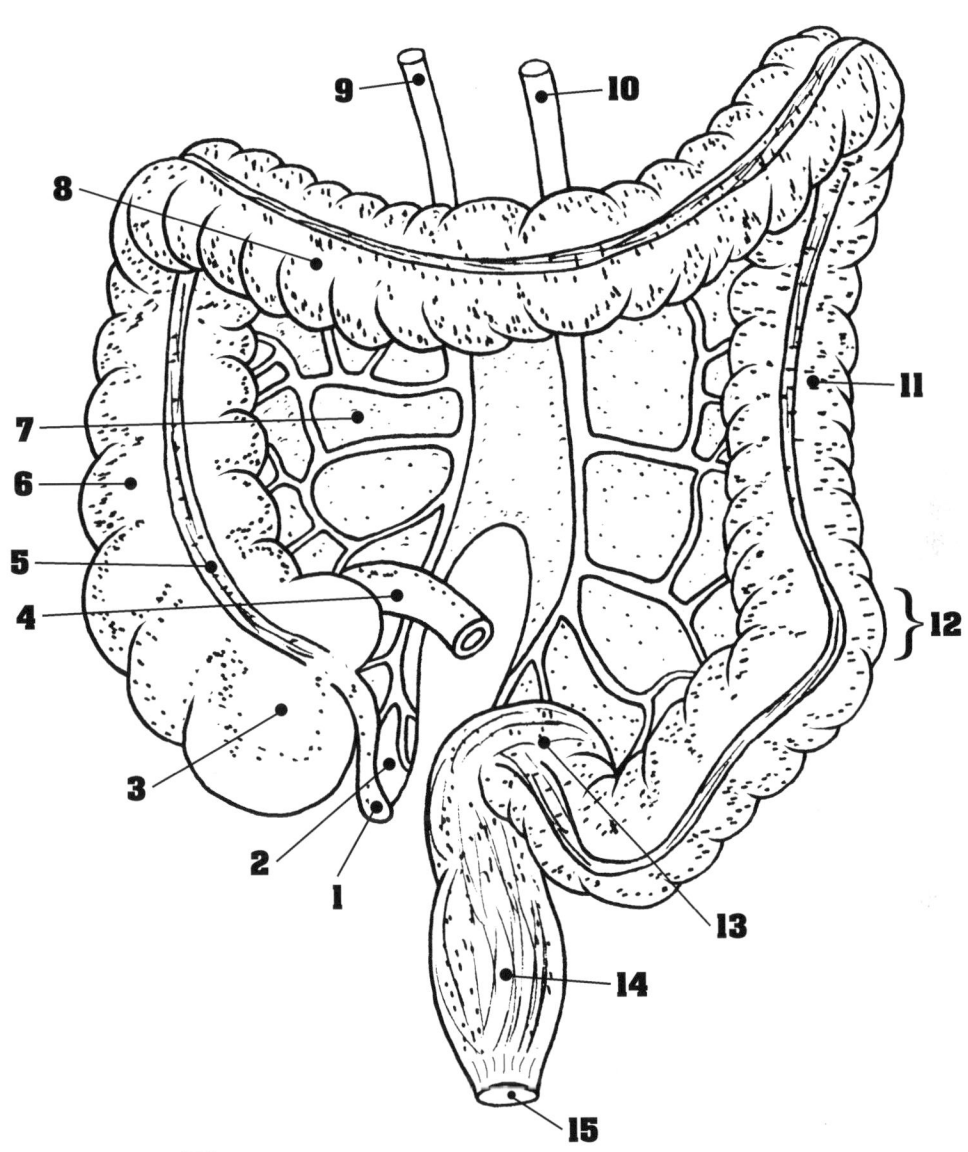

___ Anus
___ Cecum
Colon
   ___ Ascending
   ___ Descending
   ___ Sigmoid
   ___ Transverse
___ Haustrum
___ Ileum
Mesenteric
   ___ Inferior
   ___ Superior
___ Mesoappendix
___ Mesocolon
___ Rectum
___ Taenia Coli
___ Vermiform Appendix

# Defecation Reflex

Thinking you might be up to at least one anal exercise we delineate matters relating to the elimination of the feces, a process referred to as defecation.

Stretching of the rectal walls, due to the presence of feces, initiates the defecation reflex wherein signals are sent via a **sensory nerve** (3) to the **spinal cord** (2). The spinal cord in turn sends signals back via the **involuntary motor nerve** (4). These signals cause the **sigmoid colon** (5) and the **rectum** (6) to contract, the **internal anal sphincter** (8) to relax, and the **external anal sphincter** (7) to contract.

Later, as feces are forced into the anal canal, messages are sent to the **brain** (1) thus allowing us to consciously decide whether the external anal sphincter should be opened or remain constricted. (to allow us time to go to ... wherever it is we might be headed). Messages are sent from the brain to the sphincter muscles via a **voluntary motor nerve** (9).

*Defecation is aided by what is called Valsalva's maneuver which involves closing the glottis, contracting the diaphragm and abdominal muscles, and grunting — thereby increasing intra-abdominal pressure and facilitating defecation.*

# Defecation Reflex

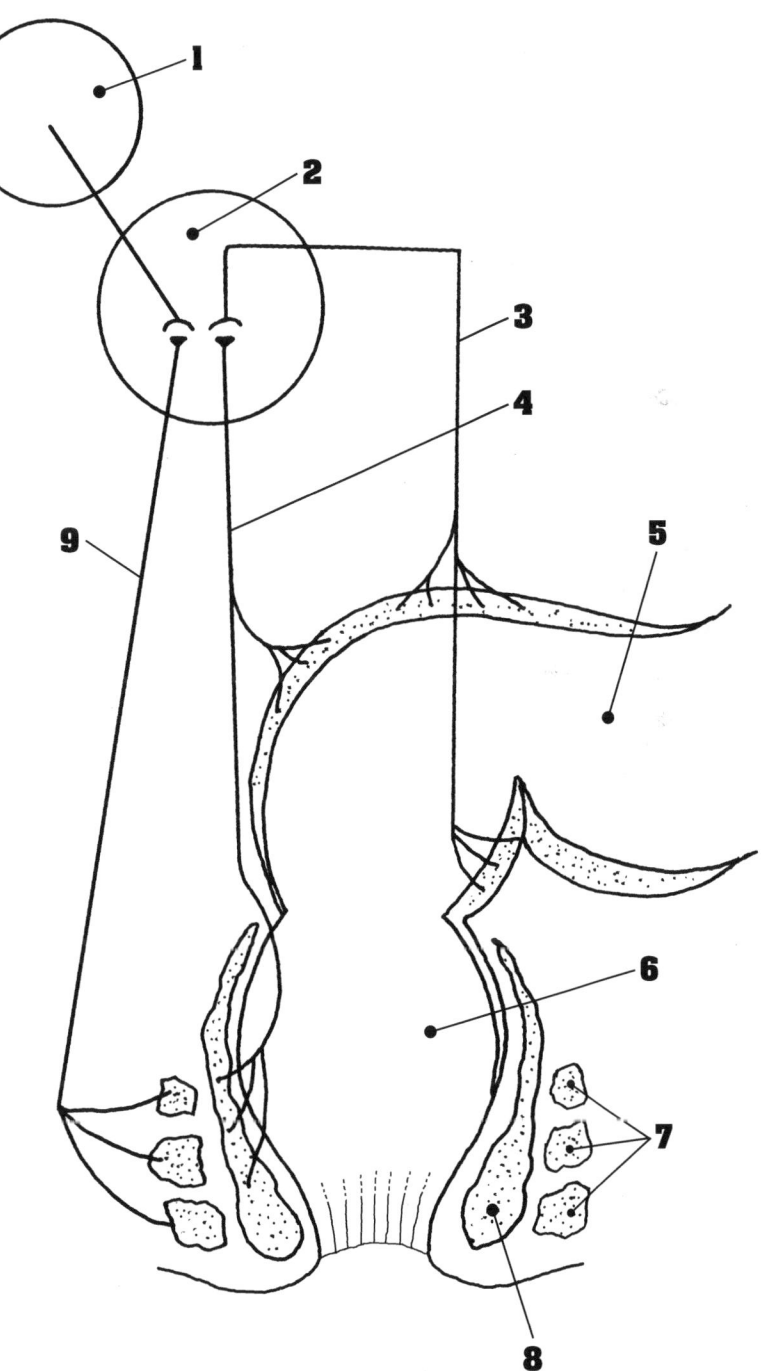

**Anal Sphincter**
___ External ☐
___ Internal ☐
___ Brain ☐
**Motor Nerve**
___ Involuntary ☐
___ Voluntary ☐
___ Rectum ☐
___ Sensory Nerve ☐
___ Sigmoid Colon ☐
___ Spinal Cord ☐

# Urinary System

The total volume of blood plasma — approximately five liters — is processed through the kidneys about forty times every 24 hours (once every 22 minutes!). The cleansing, balancing, and volume-regulating functions of the kidneys are so vital that if our kidneys shut down we would live for only a few hours. Perhaps nothing you do is more vital to your good health than urinating. Although it is called other things — in other places — in the urologist's office it is called "the golden, cleansing stream of purification."

The multifarious, complex processes involved in urine formation make renal physiology one of the most challenging of all "systems studies." You can be sure that your instructor will be, or at least should be, coming to you with fear and trembling, when he, or she, comes to the podium to lecture on renal physiology. Be assured that even now we are trembling as we write this brief introductory summary.

The functional unit of the kidney is called a nephron (see page 301). There are approximately one million nephrons in the kidneys.

Mechanisms of urine formation are usually set forth in three steps:

- **Glomerular filtration** — this ultra-filtration process, which produces about 200 liters of glomerular filtrate per day, occurs in the glomerulus. The glomeruli are located at the proximal ends of the nephrons. Details relating to glomerular structure and function are shown on page 303.

- **Tubular reabsorption** — approximately 99% of the glomerular filtrate is selectively reabsorbed back into the blood via the renal tubules and collecting ducts. Most textbooks provide a table showing what substances are reabsorbed in the various tubular regions.

- **Tubular secretion** — while some substances are reabsorbed back into the blood, other substances are secreted from the blood into the urine.

One of the kidney's most striking features is its ability to create and maintain a remarkable osmotic gradient, which increases from 300 milliosmols at the cortex to 1,200 milliosmols at the base of the medullary region. As glomerular filtrate is carried down and up through this gradient (via u-shaped renal tubules), osmotic pressure, in conjunction with a variety of ingenious structural devices in the walls of the tubules, is used to selectively reabsorb 99% of the filtrate back into the blood.

# Urinary System

Because dehydration jeopardizes so many physiologic processes the kidneys are especially equipped to assist in maintaining the proper level of body hydration. The osmotic sensors (osmoreceptors), however, are located, not in the kidneys but in the hypothalamic region of the brain. When body fluids become too concentrated osmoreceptors send impulses to hypothalamic neurons, which synthesize and release antidiuretic hormone (ADH) from the posterior lobe of the pituitary gland. ADH, via a second messenger, causes the insertion of aquaporins (water pores) into the walls of the collecting ducts in the kidneys, thereby directing water back into the tissues rather than allowing it to pass out with the urine.

In a dehydrated state we pass only about a half liter of concentrated urine per day as compared with one and a half liters of not-so-concentrated urine in a hydrated state. There are several significant disadvantages related to passing concentrated urine:

- **It puts a heavy burden on the kidneys.**
- **It can contribute to the formation of kidney stones (renal calculi).** (And that's a big hurt you don't want to have to deal with!)
- **It portends a plethora of other metabolic struggles related to the consequences of dehydration.**
- **And it smells bad.**

So be kind to both yourself and your kidneys and drink a lot of pure water — not juice, not coffee, not soda, not beer, but water!

The kidneys also monitor proper blood pressure. At a point where blood enters the glomerulus (via the afferent arteriole — see page 301), a sensor, called the juxtaglomerular apparatus (JGA), monitors the blood pressure. If blood pressure is too low a sequential series of events called the **renin-angiostenson pathway** occur:

- **The JGA releases renin.**
- **Renin converts angiotensinogen to angiotensin.**
- **Angiotensin causes the adrenal cortex to release aldosterone.**
- **Aldosterone acts on the renal tubules, causing more sodium (salt) to be reabsorbed into the blood.**
- **"Water follows salt" and the additional water increases blood volume.**
- **An increase in blood volume causes an increase in blood pressure.**
- **And presto! Our pressure problem is solved.**

# Abdominal & Retro-peritoneal Cavities

The **kidneys** (2) and **spinal cord** (5) are located in the **retroperitoneal cavity** (10). Other abdominal organs are located in the **peritoneal (abdominal) cavity** (9), which is vacated (eviscerated) in this drawing. A **peritoneal membrane** (11) lines the abdominal cavity.

**Renal arteries** (6) transport blood from the **abdominal aorta** (8) to the kidneys, while **renal veins** (14) carry blood from the kidneys to the **inferior vena cava** (12).

While **lumbar arteries** (7) carry blood to **lumbar muscles** (3), **lumbar veins** (13) carry blood away from lumbar muscles.

The spinal cord is shown within the vertebral foramen of a **lumbar vertebra** (4).

The kidneys are surrounded and protected by **adipose tissue** (1).

# Abdominal & Retroperitoneal Cavities

**Transverse Section**

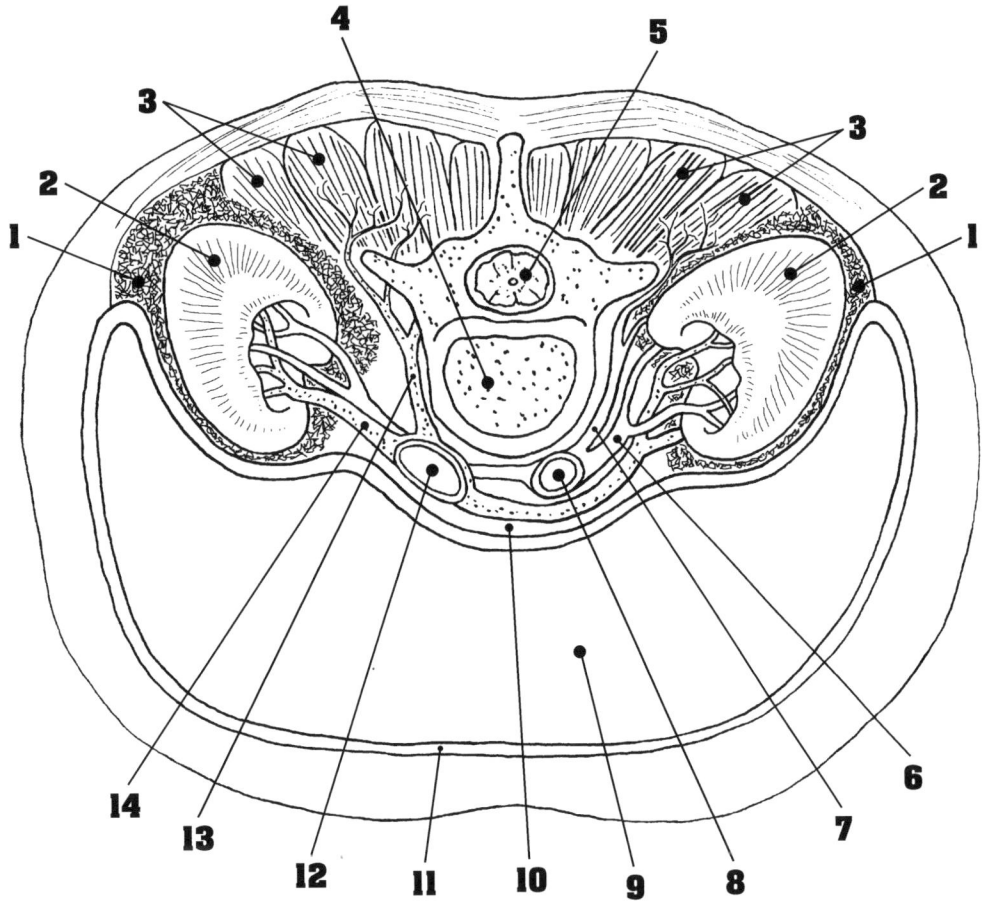

___ Abdominal Aorta ▢
___ Adipose Tissue ▢
___ Inferior Vena Cava ▢
___ Kidneys ▢
___ Lumbar Artery ▢
___ Lumbar Muscles ▢
___ Lumbar Vein ▢
___ Lumbar Vertebra ▢
___ Peritoneal Cavity ▢
___ Peritoneal Membrane ▢
___ Renal Artery ▢
___ Renal Vein ▢
___ Retroperitoneal Cavity ▢
___ Spinal Cord ▢

# Urinary System

Blood flows from the **abdominal aorta** (12) into the **kidneys** (7) via **renal arteries** (8), and flows out of the kidneys into the **inferior vena cava** (4) via **renal veins** (6).

Urine, formed within the kidneys, flows out through the **ureters** (1) and down to the **urinary bladder** (19). The urine exits the body via the **urethra** (20).

**Adrenal glands** (9) receive blood via **adrenal arteries** (10).

The **celiac artery** (13) trifurcates into the:

- **Gastric artery** (14)
- **Splenic artery** (15)
- **Hepatic artery** (11)

We're trying to help you remember those beautiful arterial maps you learned earlier. You did learn them didn't you?

**Gonadal arteries** (17) are found midway between the **superior** (16) and **inferior** (18) **mesenteric arteries**. The abdominal aorta bifurcates into the **iliac arteries** (3).

The inferior vena cava bifurcates into the **iliac veins** (2).

Ahh! ... **gonadal veins** (5) are also featured.

# Urinary System

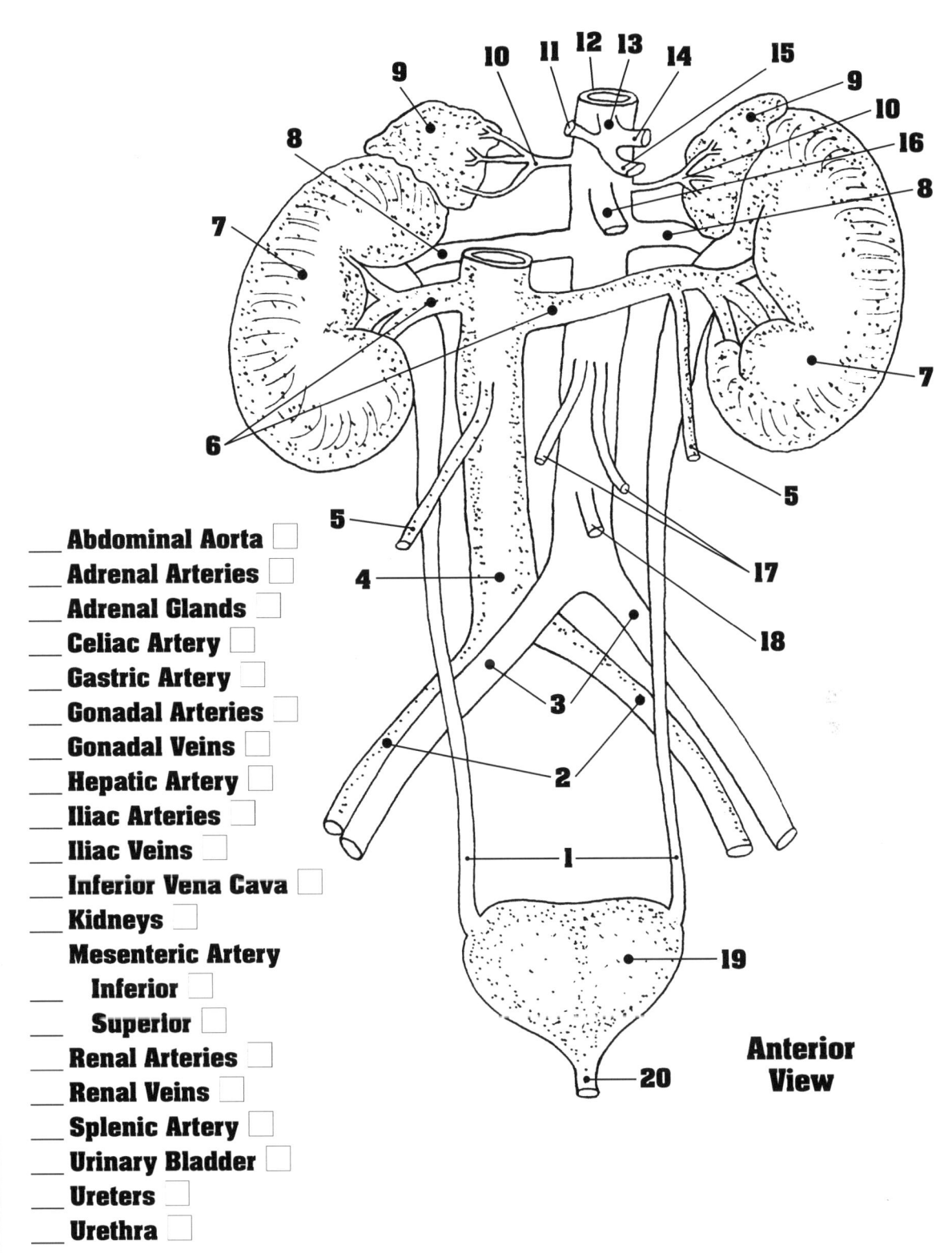

- Abdominal Aorta
- Adrenal Arteries
- Adrenal Glands
- Celiac Artery
- Gastric Artery
- Gonadal Arteries
- Gonadal Veins
- Hepatic Artery
- Iliac Arteries
- Iliac Veins
- Inferior Vena Cava
- Kidneys
- Mesenteric Artery
   - Inferior
   - Superior
- Renal Arteries
- Renal Veins
- Splenic Artery
- Urinary Bladder
- Ureters
- Urethra

Anterior View

# Renal Arterial Circulation

Within the kidneys:

- A **renal artery** (3) divides into **segmental arteries** (2).

- Segmental arteries, in turn, branch into **interlobar arteries** (1).

- Interlobar arteries give rise to **arcuate arteries** (6).

- Arcuate arteries give rise to **interlobular arteries** (7).

- Interlobular arteries give way at last to **afferent arterioles** (8).

- Afferent arterioles deliver blood to the **renal corpuscles** (9), where the blood will be filtered.

The venous pattern, which closely parallels the arterial pattern, is not shown except for the **renal vein** (4).

In this drawing we also see the **renal calyses** (10), **renal pelvis** (11), and **collecting ducts** (12). Urine collected in the renal pelvis will later exit through the **ureter** (5).

Calyses and collecting ducts will be seen in more detail in the next drawing.

*Note the subtle difference in nomenclature between "interlobar" and "interlobular" in the above sequence — and don't be tricked by a tricky test.*

# Renal Arterial Circulation

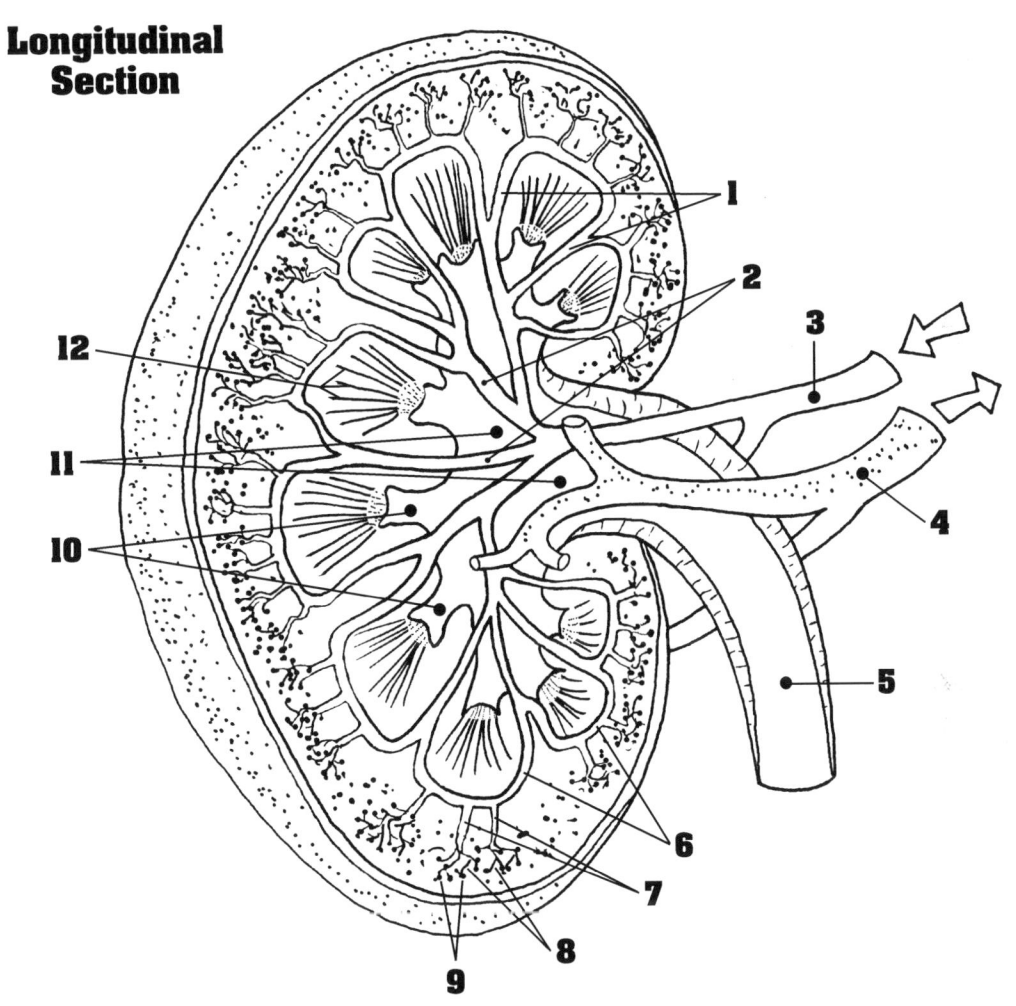

**Longitudinal Section**

___ Afferent Arterioles
___ Arcuate Arteries
___ Collecting Ducts
___ Interlobar Arteries
___ Interlobular Arteries
___ Renal Artery
___ Renal Calyses
___ Renal Corpuscles
___ Renal Pelvis
___ Renal Vein
___ Segmental Arteries
___ Ureter

# Kidney

The kidney, which is surrounded by a tough, fibrous **renal capsule** (6), is regionally subdivided into:

- **Renal cortex** (4)
- **Renal medulla** (5)

The medullary region is comprised of **renal pyramids** (2) alternating with **renal columns** (1). The cortical region is besprinkled with **renal corpuscles** (7). Renal corpuscles are proximal ends of nephrons (see the next exercise).

Renal pyramids are comprised of urine-collecting ducts that deliver urine through **renal papillae** (3) to **minor calyces** (8). Urine then flows into **major calyces** (9), on to the **renal pelvis** (10), and out of the kidney into the **ureter** (11).

Following the pathway of urine from its collection to its elimination, we have the following anatomical sequence:

- Collecting ducts
- Renal papillae
- Minor calyces
- Major calyses
- Renal pelvis
- Ureter
- Urinary bladder
- Urethra

*Do your excretory systems, and your close friends, a kindly favor. Drink, drink, drink! Water, water, water! If your body is well-watered, your metabolic wastes will be more diluted, and both your armpits and your urine will be less smelly.*

*You might also, thereby, avoid kidney stones, which have a greater chance of developing in the context of more concentrated urine ... not to speak of a dozen other major health benefits enjoyed by the "well-watered."*

# Kidney

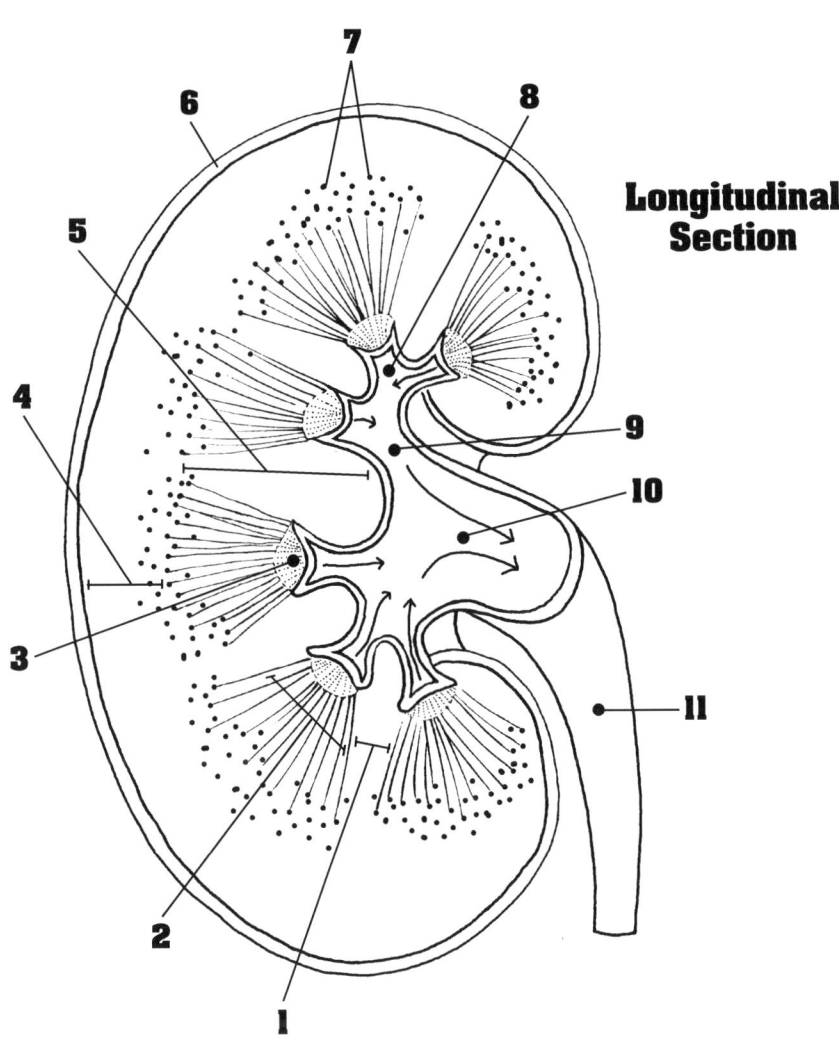

**Longitudinal Section**

___ Major Calyx
___ Minor Calyx
___ Renal Capsule
___ Renal Column
___ Renal Corpuscles
___ Renal Cortex
___ Renal Medulla
___ Renal Papilla
___ Renal Pelvis
___ Renal Pyramid
___ Ureter

# Nephron

Nephrons are the functional units of the urinary system. There are approximately one million nephrons in the kidneys. The nephron in this drawing is shown in connection with the blood vessels that service it and the **collecting duct** (2) that transports urine to the ureter.

Blood comes to the nephron via an **interlobar artery** (17), **arcuate artery** (15), **interlobular artery** (13), and, finally, an **afferent arteriole** (10).

Blood is carried away from the nephron via an **interlobular vein** (14), **arcuate vein** (16), and **interlobar vein** (18).

As blood enters a nephron the afferent arteriole breaks up into a yarn-like ball of capillaries called the **glomerulus** (7). The glomerulus is contained within a **glomerular capsule** (6). The glomerulus, together with the glomerular capsule, comprise the **renal corpuscle** (8) — all shown in more detail in the next exercise.

Blood plasma filtered out of the renal corpuscle into the renal tubule is called glomerular filtrate. The renal tubule is subdivided into three anatomical regions:

- **Proximal convoluted tubule** (5)
- **Loop of Henle (thin segment)** (3)
- **Distal convoluted tubule** (12)

The convoluted tubules are intimately intertwined with **peritubular capillaries** (4). The loop of Henle is enmeshed in a capillary net called the **vasa recta** (1).

As blood enters the nephron, a **juxtaglomerular apparatus** (11) monitors blood pressure.

An **efferent arteriole** (9) carries blood away from the renal corpuscle.

# Nephron

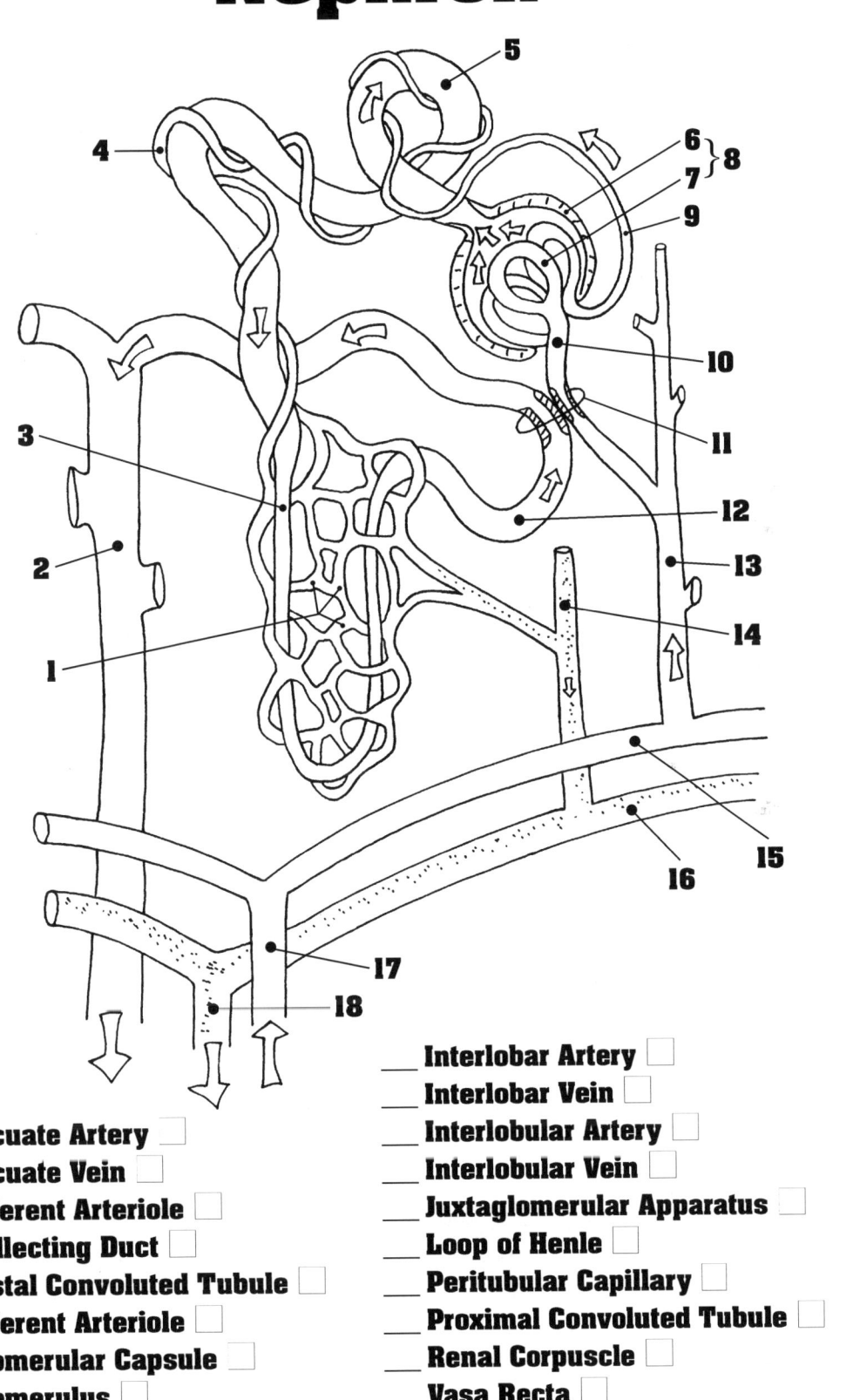

___ Arcuate Artery
___ Arcuate Vein
___ Afferent Arteriole
___ Collecting Duct
___ Distal Convoluted Tubule
___ Efferent Arteriole
___ Glomerular Capsule
___ Glomerulus
___ Interlobar Artery
___ Interlobar Vein
___ Interlobular Artery
___ Interlobular Vein
___ Juxtaglomerular Apparatus
___ Loop of Henle
___ Peritubular Capillary
___ Proximal Convoluted Tubule
___ Renal Corpuscle
___ Vasa Recta

# Renal Corpuscle

The **glomerulus** (9), a densely packed ball of capillaries, is surrounded by a **glomerular capsule** (8). The glomerulus, with its surrounding glomerular capsule, comprise the **renal corpuscle** (10) — recall, there are approximately one million of these within the kidneys. Unsurprisingly, the space between the glomerulus and the glomerular capsule is called the **glomerular capsule space** (5).

Collectively, renal corpuscles filter approximately 150 liters of blood plasma per day. The filtered plasma, when it enters the **renal tubule** (6) is called **glomerular filtrate (GF)** (7). Blood enters the renal corpuscle via an **afferent arteriole** (1) and exits through an **efferent arteriole** (11).

Four anatomical specializations greatly assist the ultrafiltration process:

- Because the incoming afferent arteriole is larger in diameter than of the outgoing efferent arteriole exceptionally high capillary pressures are created within the glomerular capillaries. This high pressure greatly facilitates the ultrafiltration process.

- **Podocytes** (4), which surround and intertwine with the glomerular capillaries, create **filtration slits** (3) which enhance filtration.

- **Fenestrated glomerular capillaries** (2), because they are more porous than regular, non-fenestrated capillaries, further assist filtration.

- Finally, ultrafiltration is facilitated by the sheer density of the glomerular capillaries, which intertwine so closely that the glomerulus appears as a small ball of "vascular yarn." No where else in the body are capillaries so densely packed.

As the 150 liters of glomerular filtrate passes up and down through the renal tubules a variety of complex osmotic "cleansing" processes facilitate the tubular reabsorption of 99% of the filtrate.

*Because the wastes and toxins removed in the urine would quickly kill us if the kidneys should fail to operate, we should always "urinate with gratitude!" — or, as one urologist put it, "pee with appreciation!"*

# Renal Corpuscle

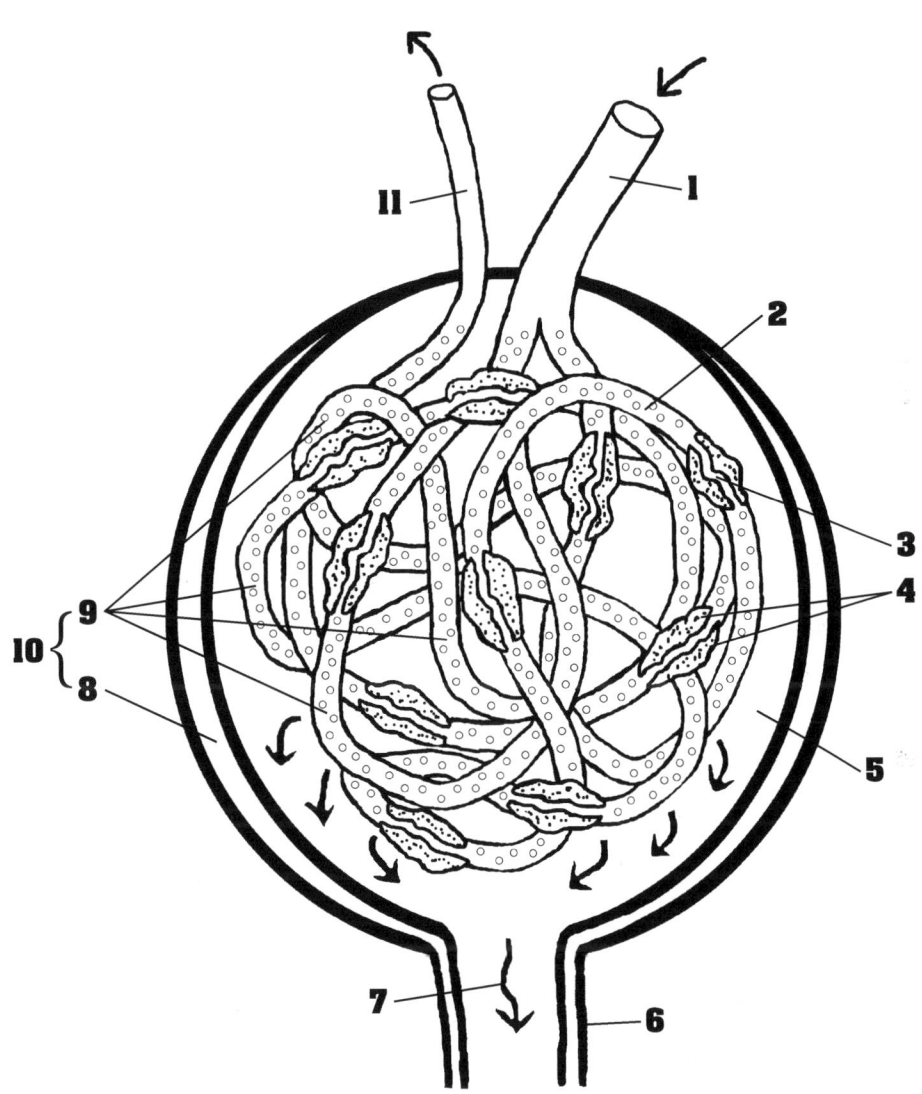

___ Afferent Arteriole
___ Efferent Arteriole
___ Fenestrated Glomerular Capillary
___ Filtration Slit
___ Glomerulus
___ Glomerular Capsule
___ Glomerular Capsule Space
___ Glomerular Filtrate
___ Podocyte
___ Renal Corpuscle
___ Renal Tubule

# Reproductive System

With regard to function, the common purpose of the male and female reproductive organs is to reproduce (produce offspring). The future resides in **gonads** (testes in the male, and ovaries in the female), where genetic blueprints, in the form of DNA molecules (chromosomes) are sorted and packaged into **gametes** (sperm and eggs).

Thus, the reproductive process begins with **gametogenesis** — **spermatogenesis** in the case of the sperm (see page 311) and oogenesis in the case of the egg (ovum). A central feature of gametogenesis is a type of cell division called meiosis wherein the chromosome number is reduced by one half — from a 46-chromosome cell type (diploid) (2n) to a 23-chromosome (haploid) (n) cell type (see page 35). When a haploid sperm fertilizes a haploid egg the diploid (46-chromosome) state will be restored.

Several striking features of the male reproductive anatomy are explained by the fact that developing sperm are temperature-sensitive and cannot properly develop at normal body temperature:

- To accommodate the need for an approximately 3°C lower temperature, the scrotum, which houses the testes, hangs awkwardly (and oftentimes painfully) outside the body.
- Depending on the temperature the scrotal sac, via **cremaster muscles** acting on spermatic cords, can be retracted more closely to the body or released so as to allow it to hang more distant from the body.
- Warm blood coming to the testes via the testicular arteries is cooled before arrival by passing through a network of veins called the **pampiniform plexus** (one of several blood-mediated heat exchange systems in the body).

Psst! New methods of birth control may now occur to the ladies:
- Seduce him into a hot tub before bedtime.
- Encourage him to wear tight jeans on a daily basis.
- And in the near future you may be able to buy him thermally regulated jockey underwear which periodically raise the temperature just enough to prevent mature sperm development (our patent is pending).

# Reproductive System

The following developmental events (processes) are essential prerequisites for "continuing the line:" We offer you the informative task of describing each of the processes and also specifying the anatomical location where each occurs — also an opportunity to use the glossary in your very expensive textbook.

**Spermatogenesis**

**Oogenesis**

**Ovulation**

**Ejaculation**

**Fertilization**

**Cleavage**

**Blastulation**

**Implantation**

**Placentation**

**Gestation**

**Partuition**

**Lactation**

# Male Reproductive System
## Lateral View

The **testis** (20), enclosed in the **scrotum** (19), is subdivided into numerous **lobules** (21). Each lobule contains a coiled **seminiferous tubule** (18), and each seminiferous tubule has the capacity to produce millions of **sperm** (1).

The sperm, after they are produced and have undergone maturation in the **epididymis** (17), are carried up the **vas deferens** (9). At the distal end of the vas deferens the sperm congregate in the **ampulla of the vas deferens** (13) where they will await a forceful ejection through the **ejaculatory duct** (8).

At the point of ejaculation, fluids from three different accessory glands are added to the sperm:

- **Seminal vesicles** (14) secrete **seminal fluid** (3).

- The **prostate gland** (15) secretes **prostate fluid** (2).

- **Bulbourethral glands** (16) secrete **bulbourethral fluid** (4).

Sperm, together with fluids from the accessory glands, comprise semen. Semen exits the **penis** (6) and the **glans penis** (5) via the **urethra** (7).

**Urine** (10), which enters the **urinary bladder** (12) via the **ureter** (11), also exits the body via the urethra.

*When a man is sexually aroused the arterioles which supply blood to the spongy regions of the penis are dilated, thus allowing blood to enter and enlarge the erectile bodies — thus, also making it more difficult for the sexually aroused man to urinate.*

# Male Reproductive System

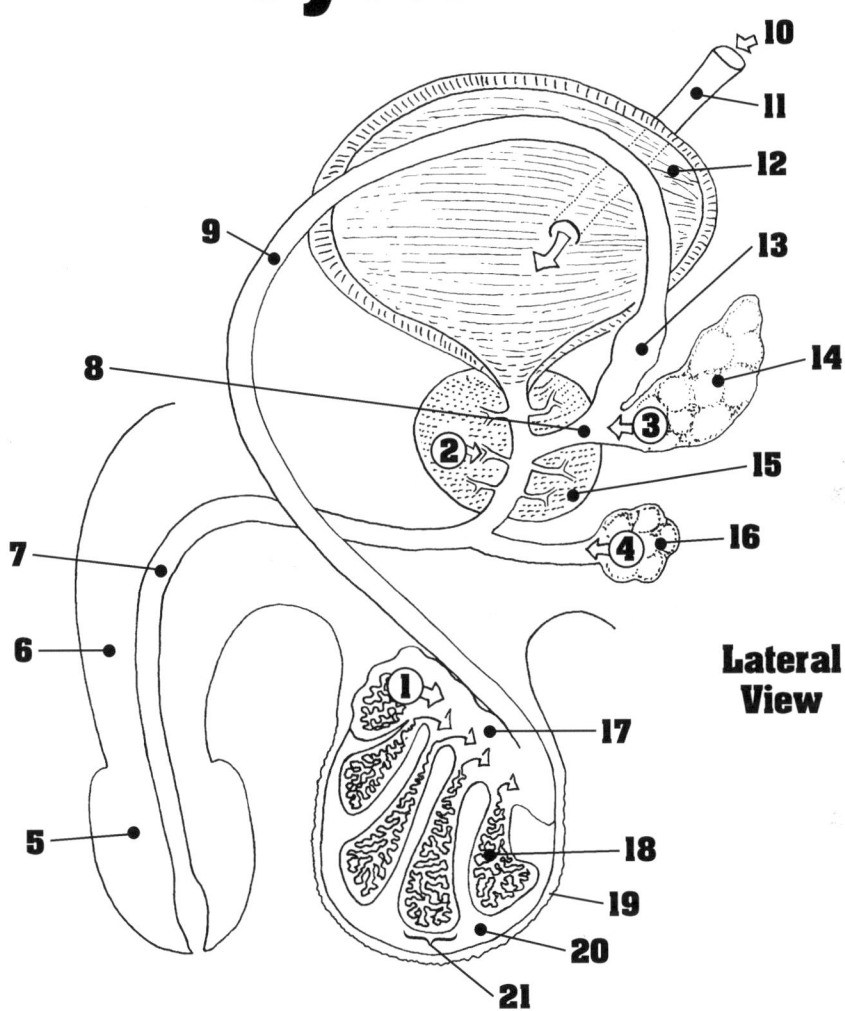

Lateral View

___ Ampulla of the Vas Deferens
___ Bulbourethral Fluid
___ Bulbourethral Gland
___ Ejaculatory Duct
___ Epididymis
___ Glans Penis
___ Lobule
___ Penis
___ Prostate Fluid
___ Prostate Gland
___ Scrotum
___ Seminal Fluid
___ Seminal Vesicle
___ Seminiferous Tubule
___ Sperm
___ Testis
___ Ureter
___ Urethra
___ Urinary Bladder
___ Urine
___ Vas Deferens

# Male Reproductive System
## Anterior View

In this drawing we review the anatomy from the anterior view and look more closely at the internal anatomy of the penis.

Again, following the pathway of sperm from point of production to point of emission, we have the **testis** (1), **epididymis** (3) — with an associated **appendix testis** (2), **vas deferens** (15), **ampulla of the vas deferens** (12), **ejaculatory duct** (10), **prostatic urethra** (16), and finally the **penile urethra** (18).

The accessory glands, from superior to inferior, are the **seminal vesicles** (11), **prostate** (9), and **bulbourethral glands** (8). The **urinary bladder** (14) with its two connecting **ureters** (13), is immediately superior to the prostate gland.

At the proximal end of the **body of the penis** (6) there is a **bulb of the penis** (17). At the distal end there is the **glans of the penis** (19).

Three spongy, expandable regions comprise the body of the penis: a central **corpus spongiosum** (4), and on each side a **corpus cavernosum** (5). The proximal ends of the two corpora cavernosa form the **crus of the penis** (7).

In the transverse section of the penis the major elements of the blood supply are also shown. Dorsally we see the **dorsal veins** (20) and **dorsal arteries** (21). **Deep arteries** (22) are seen within the two corpora cavernosa.

# Male Reproductive System

**Anterior View**

___ Ampulla of the Vas Deferens
___ Appendix Testis
___ Bulbourethral Gland
___ Corpus Cavernosum
___ Corpus Spongiosum
___ Deep Arteries
___ Dorsal Veins
___ Dorsal Arteries
___ Ejaculatory Duct
___ Epididymis
___ Penile Urethra
    Penis
___    Body
___    Bulb
___    Crus
___    Glans
___ Prostate
___ Prostatic Urethra
___ Seminal Vesicle
___ Testis
___ Ureter
___ Urinary Bladder
___ Vas Deferens

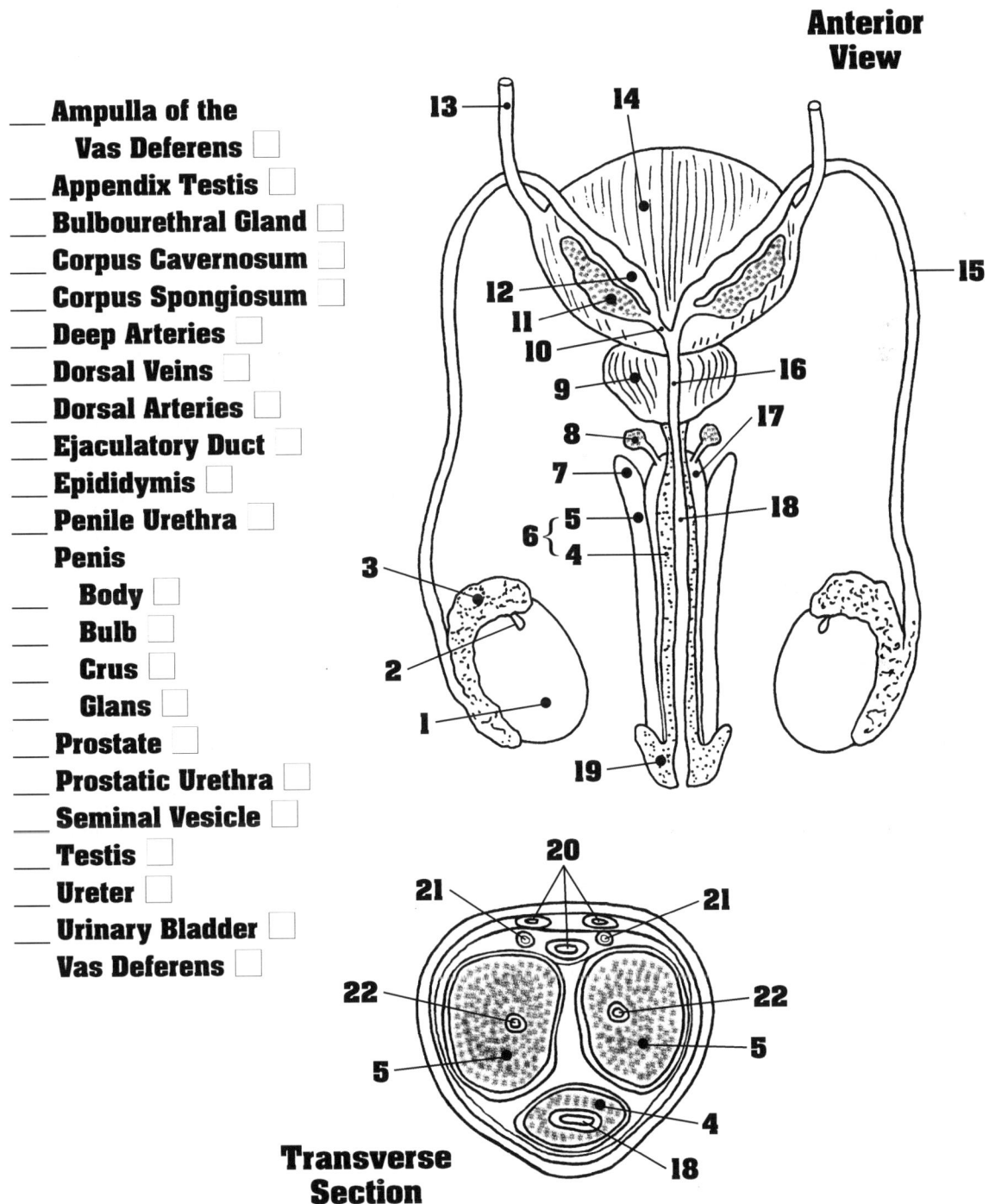

**Transverse Section**

# Spermatogenesis

At male puberty a **spermatogonium** (2) moves downward from the **basal lamina** (1) of a seminiferous tubule and becomes a **primary spermatocyte** (13). The primary spermatocyte then undergoes a process called **meiosis** (6) which involves two meiotic divisions, referred to as **meiosis I** (4) and **meiosis II** (5).

In meiosis I the chromosome number is reduced from 46 (diploid) to 23 (haploid), resulting in two haploid **secondary spermatocytes** (12).

Meiosis II is a mitosis-like, cloning division which produces identical new cells. Thus, from the one original, diploid spermatogonium, we now have four, haploid **spermatids** (11).

Next, in a process called **spermiogenesis** (7), the non-motile spermatids transform into **immature sperm** (10). The immature sperm are then released, as **spermatozoa** (9) into the **lumen of the seminiferous tubule** (8), and soon thereafter are carried upward toward the epididymus (see page 307) where their maturation will continue.

The developing sperm are embedded in super-sized nursing (sustentacular) cells. Two **sustentacular cell nuclei** (3) are shown. Sustentacular cells:

- Provide essential signals and nutrients for sperm development.
- Assist in moving sperm toward the lumen.
- Secrete testicular fluid which aids locomotion.
- Phagocytize cytoplasm sloughed off as sperm transform.
- Produce chemical mediates essential for spermatogenesis.

*Starting at about 14 years of age a normal male produces about 400 million sperm/day, and continues to do so until about 70 years of age. Well, let's see ... 70-14 = 56 years of sperm production. 56 years x 365 days/year x 400 million/day = over 8 trillion!*

# Spermatogenesis

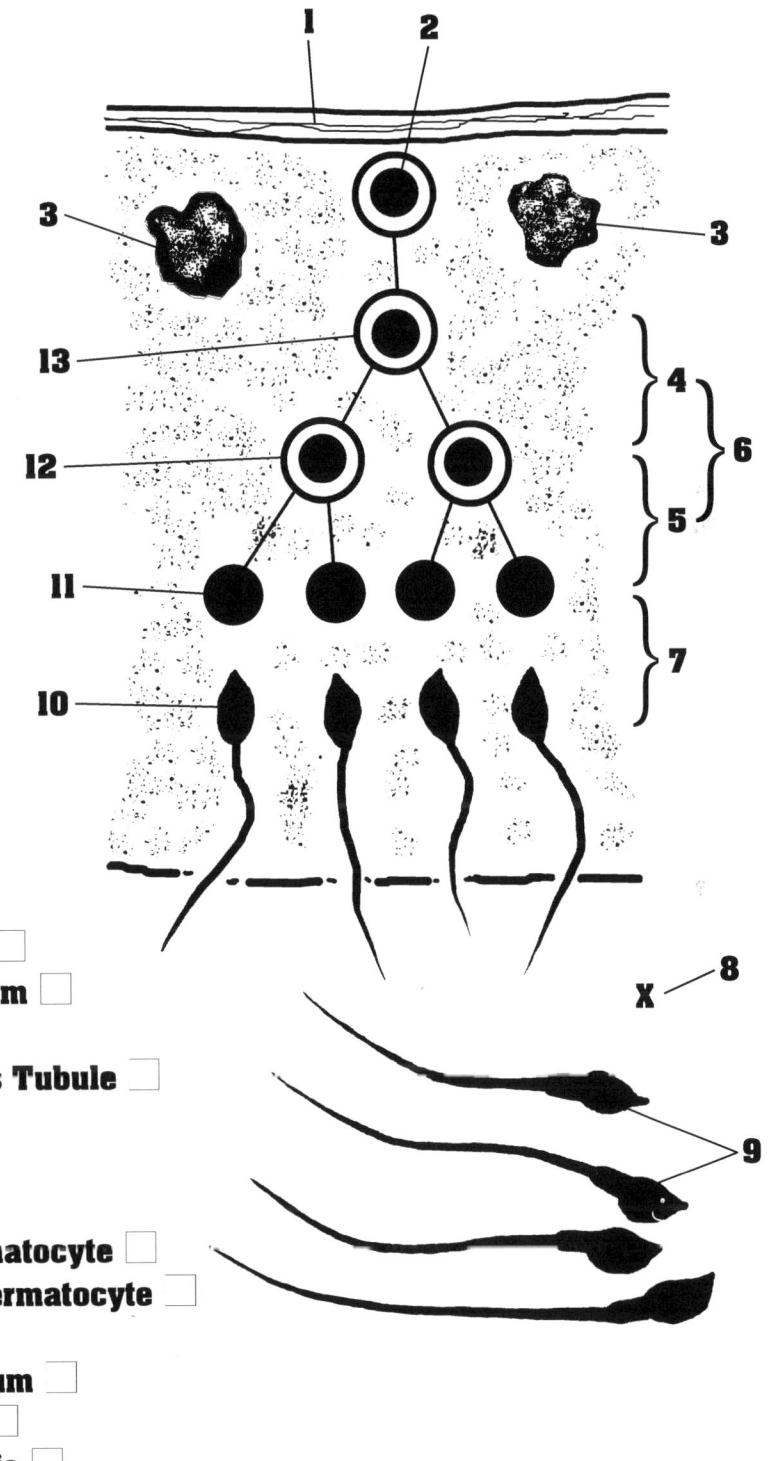

___ Basal Lamina
___ Immature Sperm
___ Lumen of the
      Seminiferous Tubule
___ Meiosis
___ Meiosis I
___ Meiosis II
___ Primary Spermatocyte
___ Secondary Spermatocyte
___ Spermatids
___ Spermatogonium
___ Spermatozoa
___ Spermiogenesis
___ Sustentacular Cell Nuclei

© Copyright 2010 Gene Johnson

# Sperm

The sperm, a superb swimming machine, is magnificently designed not only to carry an all-important genetic payload, but also to deliver that payload to a precise and difficult location. A sperm is anatomically and functionally subdivided into three distinct regions:

- **Head** (1) — carries **DNA** (5) in a **nucleus** (6) and has a nose called an **acrosome** (8) which contains **hydrolytic enzymes** (7) needed to digest the outer cell membrane of the ovum so as to create a fertilization channel for the entrance of a sperm nucleus.

- **Midpiece** (2) — contains numerous **mitochondria** (4) for supplying ATP energy for locomotive power.

- **Tail (flagellum)** (3) — comprised of microtubules (see page 25).

*In the case of sperm arriving at the scene of ovulation the early bird does not get the worm, for only after hundreds or thousands of sperm have already arrived and released their acrosomal enzymes, will a fertilization channel be ready for receiving the sperm nucleus of a later-arriving sperm.*

*So why does it take millions of sperm to fertilize one egg? Because the journey is long, hard, hazardous, and wrought with many obstacles and uncertainties, and because, as previously noted, many survivors are needed at the scene, that one may enter the egg.*

*… And because, being males, they won't ask for directions.*

# Sperm

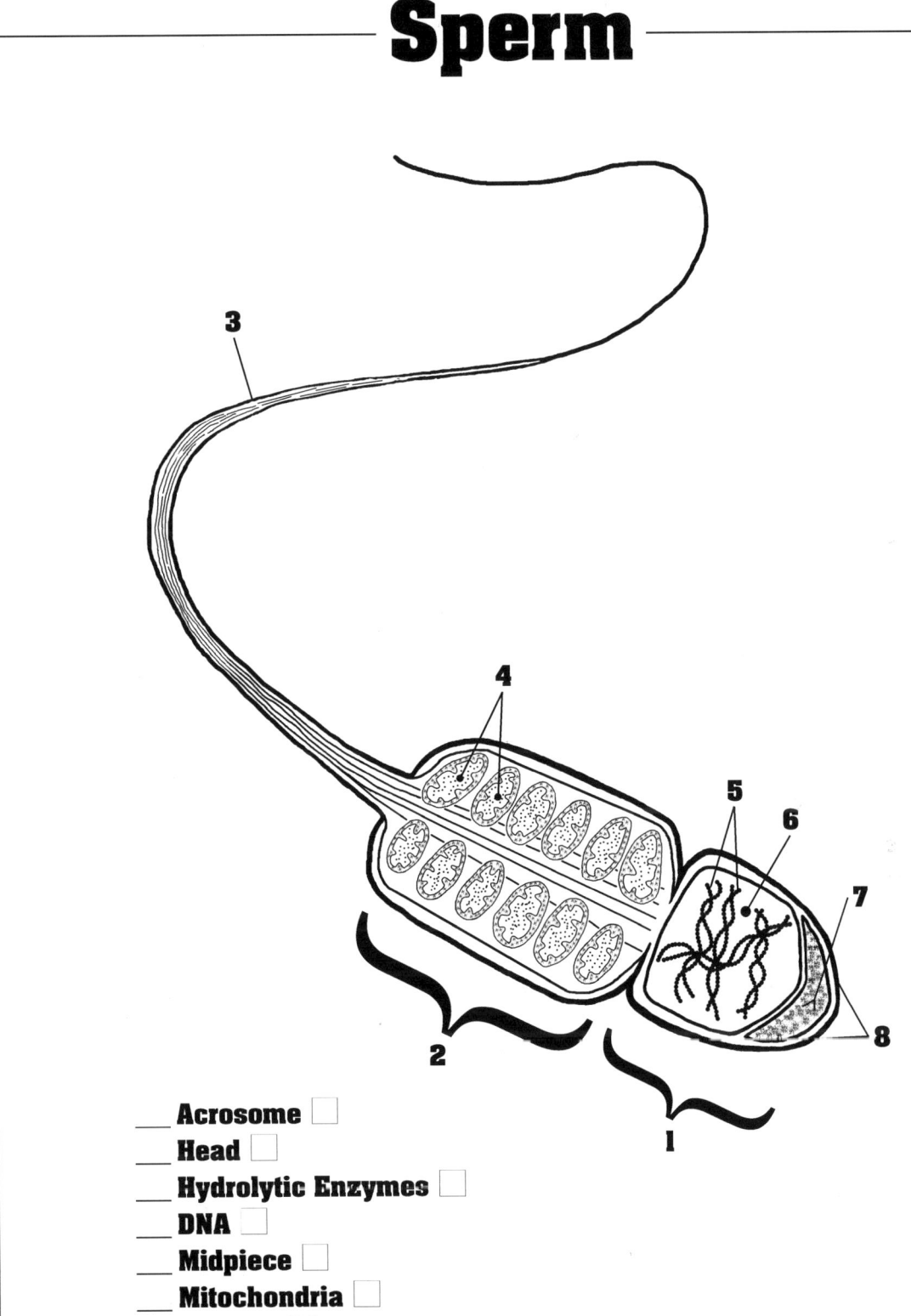

___ Acrosome
___ Head
___ Hydrolytic Enzymes
___ DNA
___ Midpiece
___ Mitochondria
___ Nucleus
___ Tail (Flagellum)

# Female Reproductive System
## Anterior View

The **ovaries** (4) are supported and stabilized:

- Medially by the **ovarian ligament** (9)
- Laterally by the **suspensory ligaments** (1)
- Superiorly by the **mesovarium** (3)
- Inferiorly by the **broad ligament** (5)

The **oviduct** (2) is subdivided into three anatomical regions:

- **Infundibulum** (11)
- **Ampulla** (10)
- **Isthmus** (8)

Finger-like structures called **fimbriae** (27) project out over the ovary from the infundibulum. At the time of ovulation the fimbriae actively search the surface of the ovary for the recently ovulated egg.

The uterus is also subdivided into three anatomical regions:

- **Fundus** (6)
- **Body of the uterus** (12)
- **Isthmus of the uterus** (13)

The **cervix** (14) is immediately inferior to the uterus, and the **vagina** (16) is immediately inferior to the cervix. The cervix protrudes into the vaginal cavity, and the vaginal cavity which surrounds the protuberance is called the **fornix** (22).

During copulation the penis enters the vagina via the **vaginal orifice** (19), and penetrates the **vestibule** (18), **hymen** (20), and **vaginal cavity** (17). In order to fertilize an egg, sperm must migrate through the **external os** (15), **cervical canal** (25), **internal os** (26), **uterine cavity** (7), and on up through the oviduct. Fertilization usually occurs in the ampullar region of the oviduct.

The broad ligaments and **uterosacral ligaments** (23) stabilize the uterus. **Ureters** (24) connect with the urinary bladder (not shown) behind the uterus.

At the entrance to the vaginal cavity **vestibular bulbs** (21) secrete accessory fluids.

# Female Reproductive System

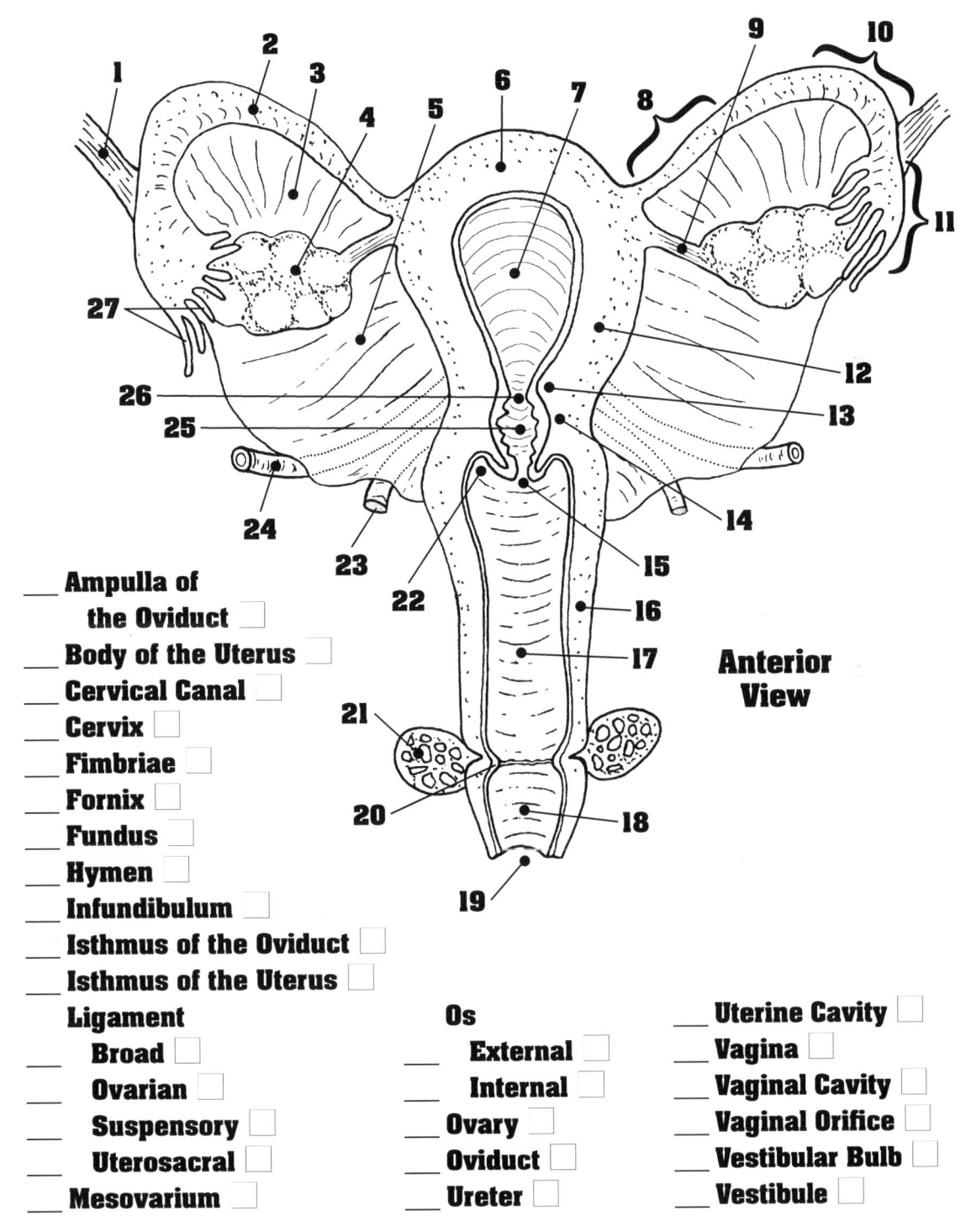

Anterior View

___ Ampulla of the Oviduct ☐
___ Body of the Uterus ☐
___ Cervical Canal ☐
___ Cervix ☐
___ Fimbriae ☐
___ Fornix ☐
___ Fundus ☐
___ Hymen ☐
___ Infundibulum ☐
___ Isthmus of the Oviduct ☐
___ Isthmus of the Uterus ☐
___ Ligament
  ___ Broad ☐
  ___ Ovarian ☐
  ___ Suspensory ☐
  ___ Uterosacral ☐
___ Mesovarium ☐
___ Os
  ___ External ☐
  ___ Internal ☐
___ Ovary ☐
___ Oviduct ☐
___ Ureter ☐
___ Uterine Cavity ☐
___ Vagina ☐
___ Vaginal Cavity ☐
___ Vaginal Orifice ☐
___ Vestibular Bulb ☐
___ Vestibule ☐

© Copyright 2010 Gene Johnson

# Female Reproductive System
## Lateral View

We review, from a lateral view, major features of the female reproductive system.

The **oviduct** (10) and **ovary** (9) are supported and stabilized by the:

- **Suspensory ligament** (11)
- **Ovarian ligament** (13)
- **Mesovarium** (8)

The **uterus** (7) with its **uterine cavity** (6), and the **cervix** (16) with its **cervical canal** (15), are stabilized by the **uterosacral ligament** (14).

The **urinary bladder** (3) with its **bladder cavity** (5) is supported by an **umbilical ligament** (4). Urine enters the bladder cavity via the **ureter** (12) and exits the bladder via the **urethra** (2).

Ventrally, from anterior to posterior, we see the:

- **Clitoris** (1)
- Urethra
- **Vagina** (17)
- **Rectum** (18)

*"Inter faeces et urinam nascimur."* — St. Augustine

# Female Reproductive System

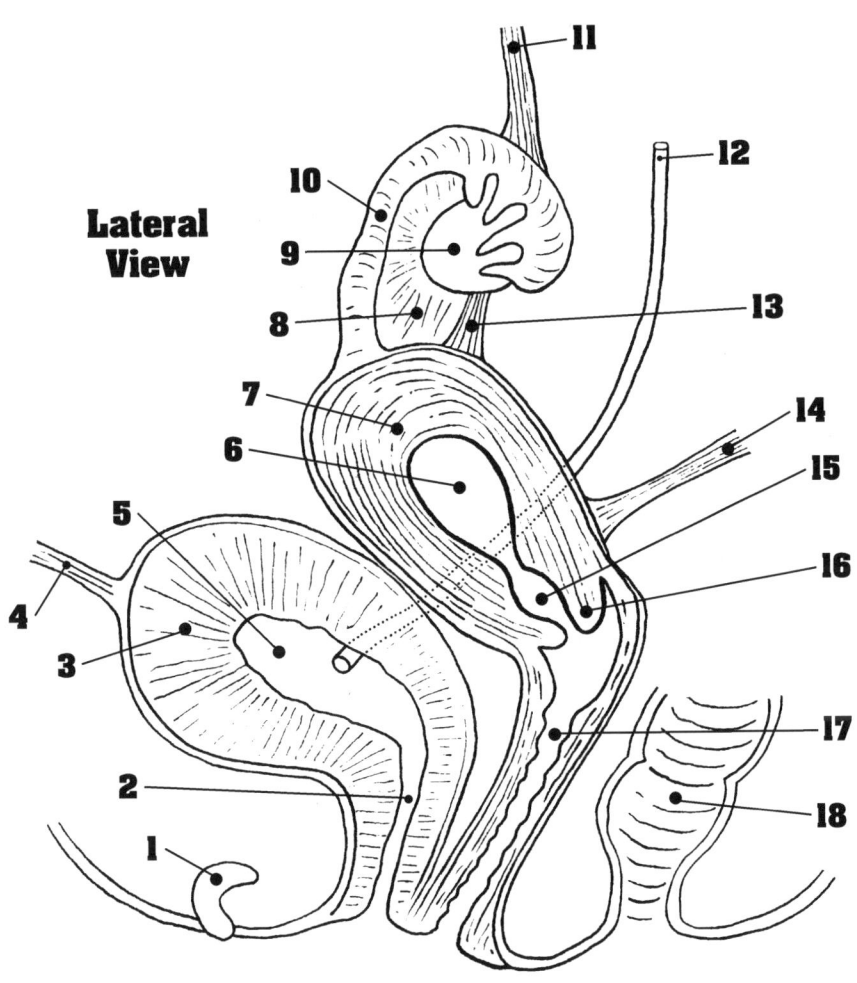

Lateral View

___ Bladder Cavity
___ Clitoris
___ Cervical Canal
___ Cervix
___ Ligament
   ___ Ovarian
   ___ Suspensory
   ___ Umbilical
   ___ Uterosacral
   ___ Mesovarium
___ Ovary
___ Oviduct
___ Rectum
___ Ureter
___ Urethra
___ Urinary Bladder
___ Uterine Cavity
___ Uterus
___ Vagina

© Copyright 2010 Gene Johnson

# Mammary Gland

Toward the end of pregnancy, prolactin from the pituitary stimulates the mammary glands to prepare for milk production. After birth, mechanoreceptors in the **nipple** (4) trigger the release of oxytocin from the pituitary. The oxytocin, in turn, triggers the let-down reflex in the mammary glands with the subsequent ejection of milk.

Milk is produced by glandular alveoli in **lactiferous lobules** (7). The lobules are contained within **lactiferous lobes** (6). Fifteen to 25 lobes embedded in **adipose tissue** (1) radiate behind the nipple.

Milk, secreted into **lactiferous ducts** (2), collects in **lactiferous sinuses** (3), and is then ejected through the nipple. A ring of densely pigmented skin surrounding the nipple is called the **areola** (5).

Posterior to the mammary gland we see the **pectoralis major** (9) and **intercostal** (10) **muscles**, together with the **ribs** (8).

Certain cells and molecules in breast milk are designed to develop, strengthen, and protect the infant:

- IgA antibodies provide broad spectrum immunity.
- Other antibodies (immunoglobulins) act more specifically.
- Lysozyme, interferon, complement, and lactoperoxidase, provide additional protection against foreign agents.
- Interleukins and prostaglandins help prevent hyper-inflammatory responses.
- A glycoprotein acts to mitigate the growth of the bacterium H. pylori.
- A natural laxative helps rid the infant bowels of meconium, a tarry green-black waste paste.
- Breast milk also encourages the normal and needed bacterial flora to colonize the large intestine.

*The above is only a partial listing of the known benefits of breast milk and does not include a plenitude of putative benefits and a multitude of numinous benefits yet to be discovered.*

*Perhaps only those who would be so bold and foolish as to dam(n) a native river or plow a native prairie, would be so bold and foolish as to attempt to concoct a synthetic infant milk formula and try to fob it off onto the world as being as good, if not better than, native (natural) mother's milk.*

# Mammary Gland

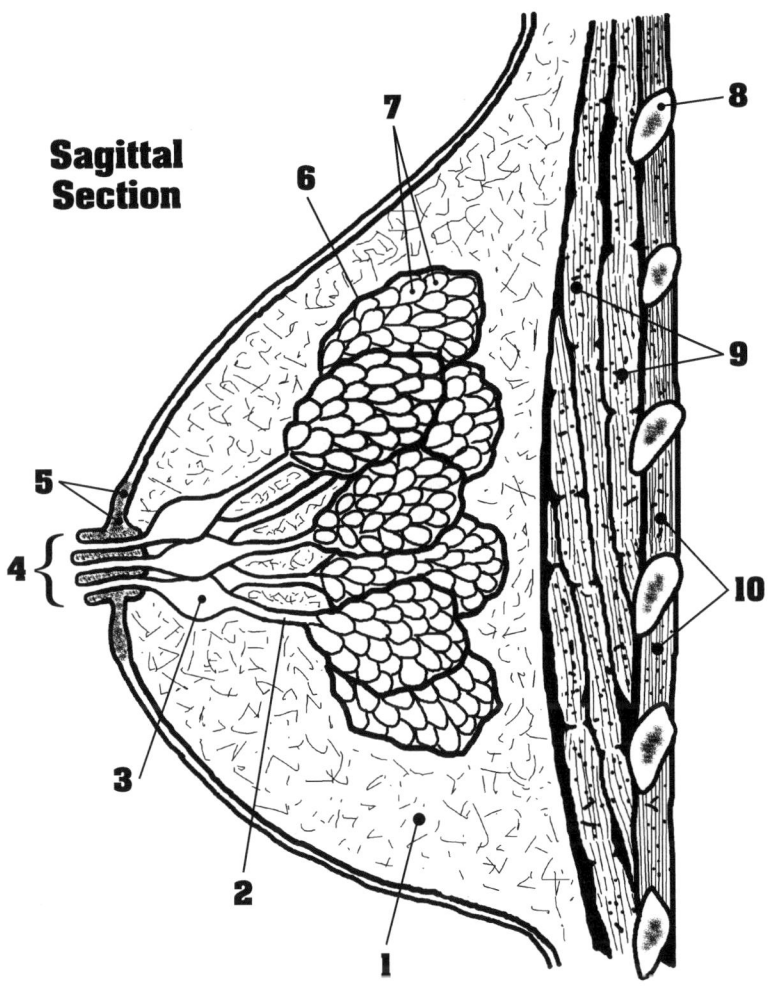

Sagittal Section

___ Adipose Tissue
___ Areola
___ Intercostal Muscles
___ Lactiferous Duct
___ Lactiferous Lobe
___ Lactiferous Lobules
___ Lactiferous Sinus
___ Nipple
___ Pectoralis Major
___ Rib

# Ovarian Cycle

### Follicular Phase

Follicle stimulating hormone (FSH) and leutenizing hormone (LH) from the **pituitary** (1) induce **primordial follicles** (3) to develop into **primary follicles** (4). Primary follicles subsequently develop into **secondary follicles** (5). Once formed, the secondary follicles immediately begin secreting estrogen, which has two major effects:

- It stimulates the pituitary to secrete increasingly greater amounts of FSH and LH.
- It stimulates the development of the endometrial layer of the uterus, thereby preparing the uterus for the possibility of an implantation.

The secondary follicles later develop into fluid-filled **Graafian (vesicular) follicles** (6), which in turn secrete increasingly greater amounts of estrogen. Then, in response to the elevated estrogen levels, the pituitary secretes greater amounts of FSH and LH. At this point, the high level of LH is called the LH surge, and it is the LH surge which causes the **oocyte** (7) to be released from the Graafian follicle — a process called **ovulation** (8). The LH surge, and ovulation, occur on about the 14th day of the approximately 28-day ovarian cycle.

### Luteal Phase

The LH surge also transforms the ruptured Graafian follicle into a **developing corpus luteum** (11) which begins immediately to act as a potent endocrine gland by secreting high levels of estrogen and progesterone. The high levels of estrogen and progesterone produced by the **mature corpus luteum** (12) have two major effects:

- It stimulates the later stages of endometrial development in the uterus.
- It inhibits the secretion of FSH and LH from the pituitary.

And so long as that "luteal inhibition" is present there will be no LH surge, and so long as there is no LH surge there will be no ovulation. Hence "the pill" which, because it contains high levels of estrogen, inhibits the pituitary secretion of LH, and thus prevents ovulation. If pregnancy occurs the corpus luteum is maintained, via hormonal influences from the placenta. At the end of the cycle, or at birth, the corpus luteum shuts down, atrophies, and becomes a star-like **corpus albicans** (13).

The ovary is supported by an **ovarian ligament** (2) and richly supplied with **blood vessels** (10) which enter and leave through the **mesovarium** (9).

*If we had drawn all the follicular and luteal phases in the one ovarian site in which they sequentially developed it would have made a very messy drawing, hence we spread the phases around the ovary even though they do not physically migrate or move.*

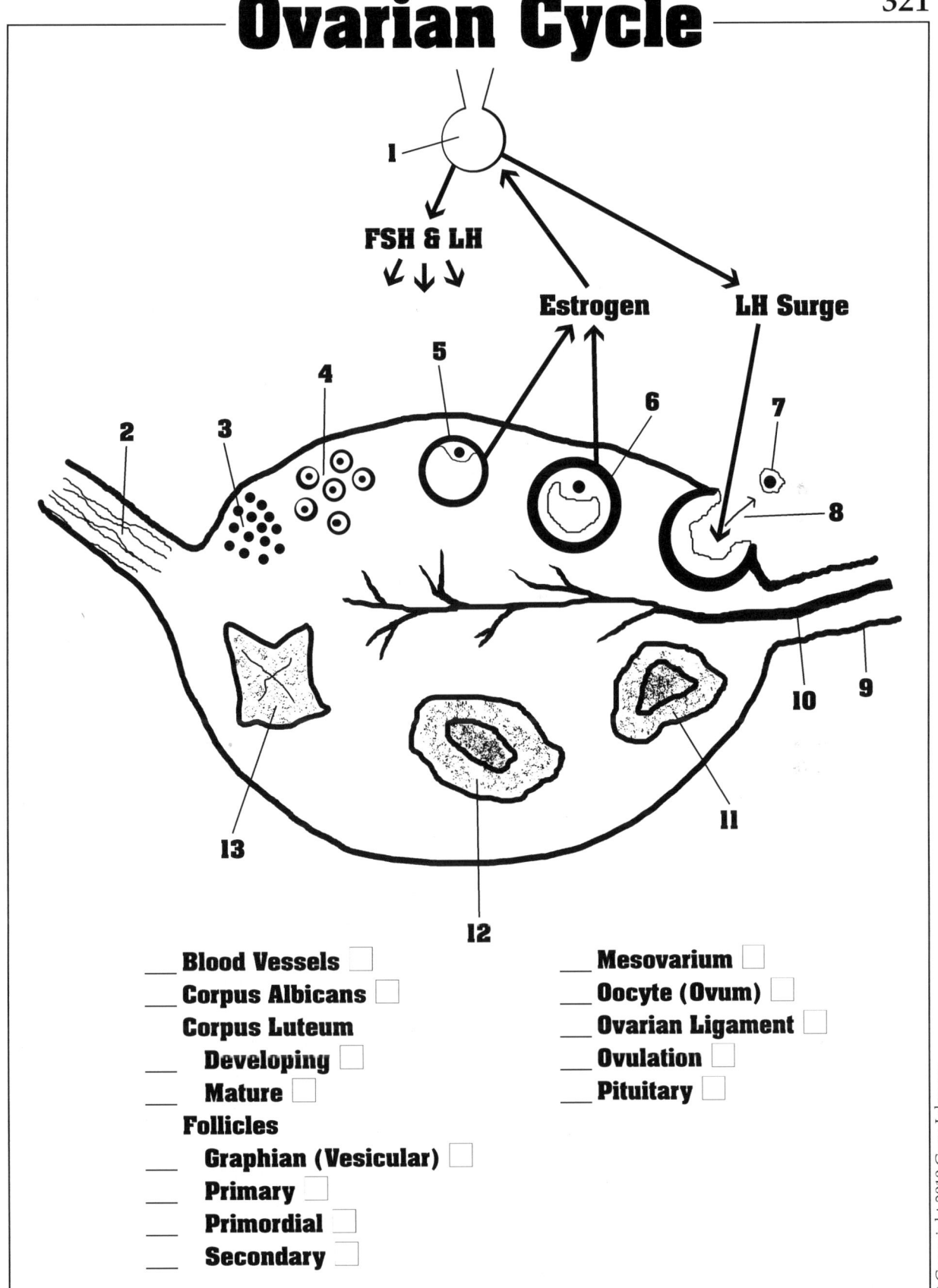

# 322   IDENTIFY THESE
## Famous Faces You Should Know

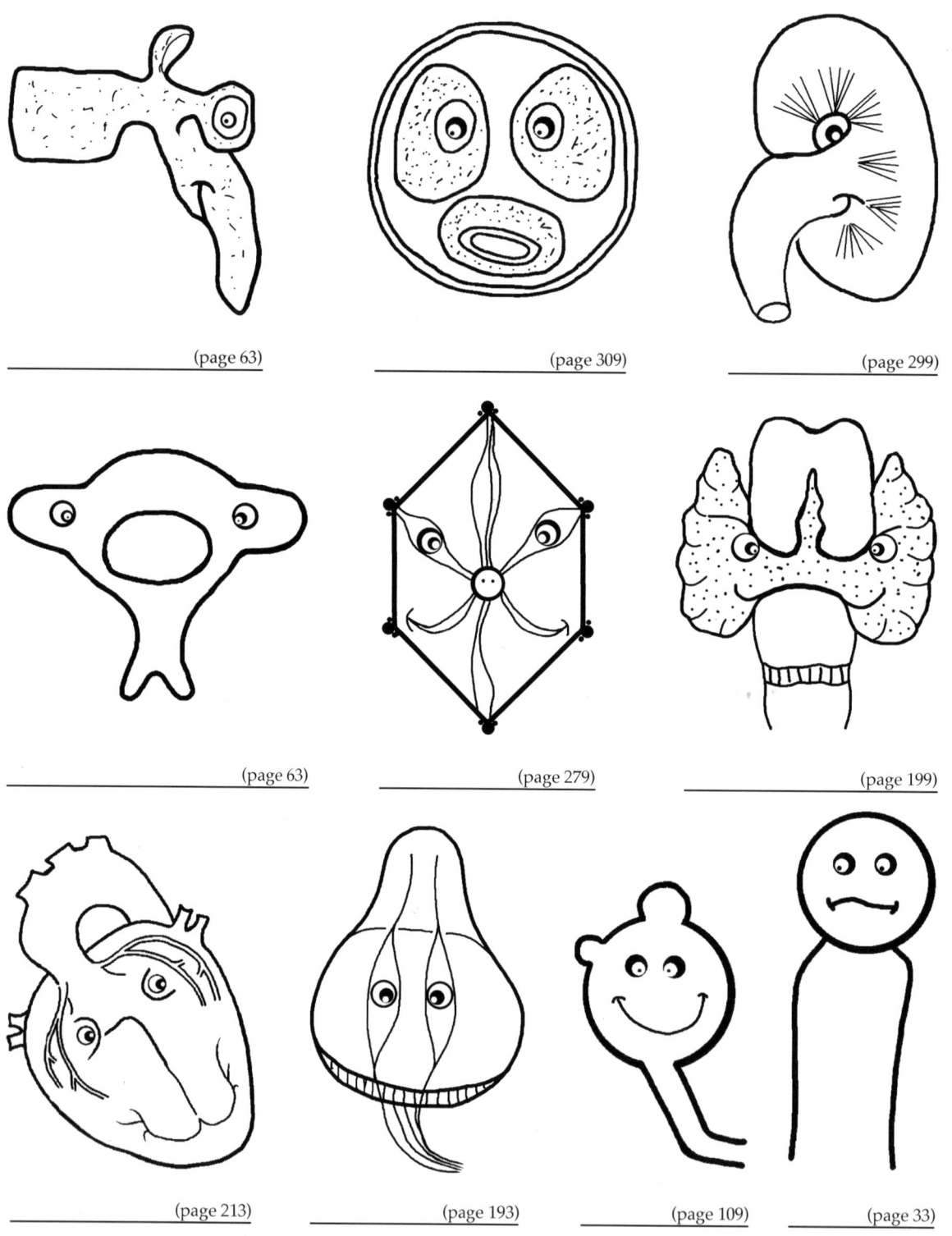

Calculate your score and submit it as your final grade.

Made in the USA
Columbia, SC
22 May 2018